Studies in Applied Philosophy, Epistemology and Rational Ethics

Volume 33

About this Series

Studies in Applied Philosophy, Epistemology and Rational Ethics (SAPERE) publishes new developments and advances in all the fields of philosophy, epistemology, and ethics, bringing them together with a cluster of scientific disciplines and technological outcomes: from computer science to life sciences, from economics, law, and education to engineering, logic, and mathematics, from medicine to physics, human sciences, and politics. It aims at covering all the challenging philosophical and ethical themes of contemporary society, making them appropriately applicable to contemporary theoretical, methodological, and practical problems, impasses, controversies, and conflicts. The series includes monographs, lecture notes, selected contributions from specialized conferences and workshops as well as selected Ph.D. theses.

More information about this series at http://www.springer.com/series/10087

James Trafford

Meaning in Dialogue

An Interactive Approach to Logic and Reasoning

Springer

James Trafford
London
UK

ISSN 2192-6255 ISSN 2192-6263 (electronic)
Studies in Applied Philosophy, Epistemology and Rational Ethics
ISBN 978-3-319-47204-1 ISBN 978-3-319-47205-8 (eBook)
DOI 10.1007/978-3-319-47205-8

Library of Congress Control Number: 2016954524

Printed on acid-free paper

This Springer imprint is published by Springer Nature
The registered company is Springer International Publishing AG
The registered company address is: Gewerbestrasse 11, 6330 Cham, Switzerland

To Lynda, in endless dialogue

Acknowledgements

The construction of a book is not something which simply stands upon the thoughts of others, but rather it is shot through with thinking with others from start to finish. As I suggest in this book, reasoning is always something that occurs in the context of the social: something that we do together. Whilst this is a book that is concerned primarily with philosophy of logic, I also suggest that by opposing the monological and universalist foundationalism of some approaches to logic and reasoning, we can foreground ways of reasoning that take difference and dialogue to heart. By shifting the important pragmatism of the inferentialist approach to logic from the monological to the dialogical, we are also led to a new way of thinking about the relationship between logic and reasoning. This, as I suggest towards the end of this book, has the (perhaps surprising) consequence that logic is inextricably bound up with the constitution of freedom. That is to say, far from seeing logic as floating free from the trappings of this world, I argue that it should be fully embedded within the mess of social structures, agents' bodies, and the power relationships between us. In this setting, logic and reasoning are not determined to be correct by dint of external authorities, whether social or metaphysical. Instead, their correctness is generated by the dynamic processes of our reasoning together.

In this sense, whilst a book like this one is a monograph, it is part of an ongoing dialogue, one that is deeply influenced and shaped by many thinkers whose bravery in the pursuit of counter-hegemonic ways of approaching their subject has radically changed the way that we think about ourselves and others. As such, I wish to thank all of those whose words and ideas have challenged and pushed me in the development of the approach discussed in this work. More specifically, I wish to thank those who have been supportive through years of conversation on some of the ideas discussed here: Alex Tillas; Fabien Schang; Alex Williams; Matt Hare; Luke Pendrell; Pete Wolfendale; Daniele Chiffi; Dave Ripley; Nicola Whittingham; Ray

Brassier; Luciana Parisi; Steve Goodman; Nick Srnicek; Benedict Singleton; Helen Hester; Tom Trevatt; Inigo Wilkins; Ben Woodard; Jan Ritsema; Mark Fisher; Reza Negarastani; Robin Mackay; Lendl Barcelos; Lorenzo Magnani; Ole Hjortland; Gianluigi Bellin; Guy Longworth; John Trafford; Ruth Trafford; Lizi Trafford; Lynda Fitzwater; Cosima Trafford; Zola Trafford; Magnus Trafford.

Contents

Chapter 1
Introduction: Reasoning in Time and Space

1 From Representationalism to Inferentialism

How do we use language to represent the world? It hardly needs stating that there is an intimate relationship between meaning, logic, and reason. Yet, increasingly, the standard approach to semantics has come under fire in the same moment as the nature of logic itself has been questioned.[1] The standard, representational, approach to meaning takes it that words are meaningful insofar as they represent something, have truth-conditions, and so on. This rests upon what Mark Wilson [2] calls the "classical" view of meaning, where a word-world connection (Wilson calls this semantic "gluing") determines how language users should correctly use linguistic expressions. So, the way in which we use these expressions is explained by a prior semantic relationship between these and the extra-linguistic world. In the domain of logic, this is captured by a model-theoretic approach that takes semantic consequence to be central. This view may be traced back to Frege's anti-psychologist account of logic (though it is certainly consolidated in the work of Tarski) where it is claimed that logic is based upon laws of truth, which map onto attributes of objective mathematical reality: 'If being true is thus independent of being acknowledged by someone or other, then the laws of truth are not psychological laws: they are boundary stones fixed in an eternal foundation, which our thinking can overflow, but never displace' [3, p. xvi]. As such, logic is supposed to concern propositional contents that are mind-independent, so it is unsurprising that, as a number of authors have pointed out (e.g. [4–7]), this brings with it a problematic epistemology, with little account of the justification of logical rules. Dummett [8], for example, argues that this brings with it an 'ontological mythology' (p. 25), in which the semantic relationship with these (supposedly) mind-independent entities is not really explained, nor is the ability of language users to cognitively access it.[2]

[1] For example, for an overview of non-classical logics, see [1].

[2] The issue of cognitive access to logical content on this view is discussed in [9, 10].

J. Trafford, *Meaning in Dialogue*, Studies in Applied Philosophy, Epistemology and Rational Ethics 33, DOI 10.1007/978-3-319-47205-8_1

This book will be concerned, rather, with the less developed yet increasingly well-known, inferentialist approach to logic and meaning, which has made attempts to surpass these issues (e.g. [11–17]). Inferentialism is a broad view which explains meaning by way of the inferential connections between linguistic expressions, rather than their denotation. As such, it is related to accounts of meaning-as-use (e.g. [18]), which resist a representational view of meaning in favour of a pragmatic approach based on know-how. In the latter tradition, in the 1950's, Wilfrid Sellars coined the phrase "the space of reasons", which has since become commonplace in philosophical discussion of reasoning and meaning. The space of reasons, according to Sellars, is a way of thinking both about the way in which language is used to reason and communicate, and also a domain that is characterised by the logical and the normative rather than the empirical and the descriptive. It is the former that is the appropriate language and explanatory system for the domain of human thinking and meaningfulness: '[t]he essential point is that in characterizing an episode or a state as that of knowing, we are not giving an empirical description of that episode or state; we are placing it in the logical space of reasons, of justifying and being able to justify what one says' [19, Sect. 36].[3] In this vein, contemporary versions of inferentialism, such as Brandom's (e.g. [5, 20, 21]), typically take the meaning of expressions to be explicable in terms of certain proprieties of inferential practices.

As such, it is unsurprising that inferentialism has gained traction as a way of explaining the meaning of logical expressions. Logical inferentialism is a specific branch of inferentialism, which gives explanatory priority to proof-theoretic tools over standard model-theoretic accounts of logical consequence. It also seeks to explain the meaning of logical expressions in terms of a specified set of inference rules. These are usually the introduction and elimination rules for classical (or intuitionistic) connectives, as set out in natural deduction frameworks, and where those rules are supposed to have a substantive connection with ordinary inferential practices. So, for example, the rules for conjunction may be given as follows:

$$\frac{\alpha \qquad \beta}{\alpha \wedge \beta}\ (\wedge\text{-I})$$

$$\frac{\alpha \wedge \beta}{\alpha}\ (\wedge\text{-E1})$$

$$\frac{\alpha \wedge \beta}{\beta}\ (\wedge\text{-E1})$$

These rules may be understood to determine the meaning of expressions involving conjunction, since they provide an account of how these expressions should function in order for arguments involving them to be valid. So, whereas the standard, representational approach suffers from epistemological issues regarding agents' contact

[3] In this sense, we may make an obvious allusion to the idea that, as mathematics is the descriptive and explanatory "language" of physical space, so with formal logic and the space of reasons. This would also cohere with Sellars' distinction between the "manifest" and "scientific" images, which, in brief, have to do with distinct languages of explanation, see (e.g. [19]).

with mind-independent entities, inferentialism looks set to alleviate this entirely, by shifting the explanation of logical validity to an account of the proprieties of inference.

2 A Puzzle Involving a Tortoise

What is important for logical inferentialism is not truth-preserving deductive consequence, but that there are certain normative rules that determine the meanings of logical expressions, and provide an account of the proprieties of inference practices. There are, however, a number of issues with this approach to logic. Arguably the most significant of these is an argument that providing a justification of basic logical rules such as *modus ponens* is inevitably circular, and so we require something like classical deductive consequence to adequately give us a foundation justifying those rules. The issue is that certain deductive rules must be construed to be both valid, and in some sense, normative for reasoning. Whilst an inferentialist account of logic provides a fairly clear way in which to think of how we come to use certain logical rules, it is difficult to hold together the idea that logical rules have normative force, with the ability to provide (and question) the justification of those rules.[4]

This is one interpretation of the story given by Lewis Carroll [22] in which Achilles and Tortoise are arguing about the first proposition of Euclid, with Tortoise offering the following fragment of argument:

> (A) Things that are equal to the same are equal to each other.
> (B) The two sides of this Triangle are things that are equal to the same.
> (Z) The two sides of this Triangle are equal to each other.
> "Readers of Euclid will grant, I suppose, that Z follows logically from A and B, so that any one who accepts A and B as true, must accept Z as true?" (p. 279)

In other words, the argument is supposed to lead us to the conclusion:

> (C) If A and B are true, Z must be true.

But, as Tortoise goes on to explain, it is possible for somebody to accept (A) and (B) whilst not accepting (C), and so not also being moved to accept (Z). Tortoise then challenges Achilles to 'force me, logically, to accept Z as true' (p. 279). By way of response, Achilles requests that Tortoise accept (C), which Tortoise does whilst also asking Achilles to write it down. Now, Tortoise restates the argument as follows:

> (A) Things that arc equal to the same are equal to each other.
> (B) The two sides of this Triangle are things that are equal to the same.
> (C) If A and B are true, Z must be true.
> (Z) The two sides of this Triangle are equal to each other. [...]
> "If you accept A and B and C, you must accept Z."
> "And why must I?"
> "Because it follows logically from them. If A and B and C are true, Z must be true. You

[4] A similar argument is due to Wittgenstein [18].

> don't dispute that, I imagine?"
> "If A and B and C are true, Z must he true," the Tortoise thoughtfully repeated. "That's another Hypothetical, isn't it? And, if I failed to see its truth, I might accept A and B and C, and still not accept Z. mightn't I?". (p. 280)

In other words, just as before, we now have another hypothetical statement that somebody may not accept:

> (D) If A and B and C are true, Z must be true.

Now, Achilles requests that Tortoise accepts (D), which he does once written down. Again, Achilles thinks that he has come to the end of the argument; '[n]ow that you accept A and B and C and D, of course you accept Z' (p. 280). As we may come to expect from the rather obtuse Tortoise, he asks, instead, what would happen should somebody accept A and B and C and D whilst refusing to accept Z: 'Then Logic would force you to do it! [...] Now that you've accepted A and B and C and D, you must accept Z' (p. 281). Tortoise is unpersuaded, and simply requests that Achilles write down another hypothetical statement as premise. As the story goes, the two go on like this for quite some time!

Let us schematise the argument a little, so that we can see the machinery underlying it. First, Achilles puts forward the inference of β from α and $\alpha \Rightarrow \beta$, whilst Tortoise accepts $(\alpha \wedge (\alpha \Rightarrow \beta)) \Rightarrow \beta)$ but refuses to infer β. In an attempt to persuade Tortoise, Achilles goes on to try to explain why Tortoise ought to infer β by suggesting $(\alpha \wedge (\alpha \Rightarrow \beta)) \Rightarrow \beta)$ and $\alpha \wedge (\alpha \Rightarrow \beta)$. But, again, Tortoise accepts $(\alpha \wedge (\alpha \Rightarrow \beta) \wedge (\alpha \wedge (\alpha \Rightarrow \beta)) \Rightarrow \beta)$, whilst refusing to infer β. And so on, *ad infinitum*. One lesson we may extract is that Achilles' attempts to explain the validity of the rule of modus ponens to Tortoise inevitably uses an explicit form of modus ponens itself. This has the significant consequence that, a simple inferentialist attempt to justify a basic logical rule looks to be circular.

One response to the regress identified by Carroll is to say that there is a difference between logical implication and rules of inference. Perhaps it is for this reason, that there is no normative force experienced by Tortoise to make the inference to β. This view is adopted by Priest [23], who argues for the separation of implication, which concerns logical relations between propositions, and the activity of making an inference. For example, the process of reasoning, making hypotheses, testing these against counter-examples, and so on is typically considered to be outside of the domain of logic because it is supposedly non-deductive. That is to say, if we take deductive consequence to concern abstract arguments that preserve truth from premises to conclusion, then clearly large swathes of reasoning and argumentation fall outside of this definition, since most of our inference practices do not carry on in that manner. This way of putting things may also be connected with the now widely accepted view suggested in [7], that there are obvious discrepancies between formal logic and ordinary reasoning, with the latter being far more flexible than deductive logic.[5] In distinction with the position argued for throughout this book, Harman goes

[5] As Harman suggests, there are many cases of ordinary reasoning where agents do not make an inference to β from α because, perhaps, α is unreasonable given the antecedent context.

on to argue that this should lead us to abandon the relationship between logic and reasoning altogether:

> There is a tendency to identify reasoning with proof or argument in accordance with rules of logic. Given that identification, logic obviously has a special role to play in reasoning. But the identification is mistaken. Reasoning is not argument or proof. It is a procedure for revising one's beliefs, for changing one's view [...] The question, then, is whether logic has a special role to play in this procedure of belief revision [7, p. 107].

The obvious issue with this response is that we have now simply shifted the problem to one which concerns the justification of rules by appeal to a metalinguistic account of the validity of certain implications. Perhaps, if we can mitigate problems facing regress by means of the semantic justification of logical implication, then we may be able to provide justification for certain logical rules, where logically correct rules are just those which preserve truth. Though, of course, this would be to give up on inferentialism, and logical implication would potentially be wrested away from playing any normative role over reasoning.[6] This would then cohere with the broader criticism of representational approaches to semantics alluded to above, in which representations end up playing roles that float free from the role that words play in the practices of reasoning and discourses. According to Tarski's analysis, there is nothing more to logical validity than truth-preservation over models, so the correct rules are just those which preserve truth in all models. In brief, this requires a semantical model, under which any interpretation in which the premisses of a rule are assigned the value "true", the conclusion is assigned "true" also. In this regard, Etchemendy [6, 25] argues that, if validity is merely truth-preservation over models, then deductive inference can not guide the formation of new beliefs. So, the reduction of validity to truth-preservation renders logic epistemically impotent for justifying a conclusion, given true premises:

> All we could ever conclude upon encountering an argument with true premises would be that either the conclusion is true or the argument is invalid. For if its conclusion turned out to be false, the associated class would have a non-truth-preserving instance, and so the argument would not be logically valid. Logical validity cannot guarantee the truth of a conclusion if validity itself depends on that self-same truth [25, pp. 265–6].

We can briefly flesh this out. If the logical validity of modus ponens consists only in the truth-preservation of all of its instances, then what can we say of any instance whose premises are true? Either that its conclusion is true or the argument form is invalid. What we can not do is deduce the truth of its conclusion from the truth of its premises, unless we also assume that modus ponens is valid (and so assume that all of its instances preserve truth). This, however, is to assume what we want to prove, which is the truth of the conclusion of this instance of modus ponens without assuming its validity, but this is not possible. If an argument form is logically valid, it is supposed to be truth-preserving, and this would give us a guarantee (independent

[6]This distinction between logic (proper) and reasoning is obviously not one that we will accept here, though see, for example, Beall [24] for a position that accepts the distinction with some interesting nuance, in the context of default classicality for paraconsistent logics.

Fig. 1 Matrices for conditional

α	β	$\alpha \Rightarrow \beta$
T	T	T
T	F	F
F	T	T
F	F	T

of the actual truth value of premises and conclusions) that all instances of that form preserve truth. This, of course, provides logical validity with its epistemic import, but, Etchemendy [25] argues: 'the reductive analysis just omits the guarantee, attempting to replace it with that which the guarantee is a guarantee of' (p. 8).

So, it is not at all clear how this could enable an escape from the regress identified above. We can see this quite clearly if we take the example of modus ponens in detail. Since modus ponens is deductively valid it is part of "logic", whereas other rules, such as induction (α, β (iterated) $\vdash \alpha \Rightarrow \beta$) and abduction ($\beta, \alpha \Rightarrow \beta \vdash \alpha$), are not. The reason for this demarcation is typically supposed to come down to issues regarding the justification of non-deductive rules. For example, if we take induction, then Hume's problem rears its head, i.e. an inductive principle cannot be proven deductively, since it is contingent, and only necessary truths may be proven deductively (according to tradition). But, it also can not be supported inductively, that is, by making an argument that it is typically reliable, since that would assume what we are attempting to prove. So, the demarcation seems to require that we have an adequate (and non-circular) way of accounting for the justification of logical (deductively valid) rules.

Say that we attempt to provide an argument that justifies the fact that modus ponens is deductively valid. Semantically, whilst a little vague, it is typical to say something like "if every way of making the premises true is also a way of making the conclusion true, then the rule is deductively valid".[7] So, then, we might start from the truth-table for conditionals, which is as follows (Fig. 1):

Then, we know, from the truth-table, that if α is true, then, if $\alpha \Rightarrow \beta$ is true, then β must be true also. So, if α is true, and $\alpha \Rightarrow \beta$ is true, then β must also be true, which is just the definition of deductive validity that we are working with. Let us lay out the argument schematically as follows:

(P1): If α is true, then, if $\alpha \Rightarrow \beta$ is true, then β must be true also (from the truth-table for $\Rightarrow$)
(P2): α is true (by assumption)
(C1): If $\alpha \Rightarrow \beta$ is true, then β is true (P1, P2, Modus Ponens)
(P3): $\alpha \Rightarrow \beta$ is true (assumption)
(C2): β is true (C1, P3, Modus Ponens)

[7] The following is folklore, but I particularly follow the account in [26].

As is clear, the argument attempting to justify the validity of modus ponens relies upon modus ponens itself at two stages of the argument. In other words, the defense of deduction looks circular in exactly the same way that Hume argued our inductive inferences are circular. So, we do not escape the Carrollian circularity by moving to the semantic justification of logical rules. It is possible, for example, to construct a semantics for a logical rule using a metalanguage that mirrors the structure of the rule just as we have done with modus ponens. But, again, this is both circular, and also open to the justification of non-deductive rules. As Celluci [27] points out, for example, the rule of abduction can equally be justified semantically using the same structure of argument as we gave above for modus ponens.[8]

3 Inferentialist Semantics and Dialogue

As I said above, an inferentialist approach to meaning and logic suffers from problems of providing justification for basic logical rules. In addition to the Carrollian regress, problems of paradoxical connectives such as Prior's [34] "tonk" (discussed in the following chapter) also provide reason to think that an inferentialist account of logical rules inevitably requires semantic justification to settle the issue. But, given that the latter fares little better, this is cold comfort.

Nonetheless, these issues highlight a widely held view that inferential rules are merely formal and abstract strings of syntactic objects that should be justified by means of soundness and completeness proofs. In this sense, it is thought, formal rules are purely syntactic entities, and, as such, can not be the sort of things that are meaningful in a true sense. This is an abstract version of the moral sometimes drawn from Chinese-room style arguments that 'syntax alone is not sufficient for semantics' [35]. These arguments work on the basis that an agent with access only to syntax (inference rules and linguistic expressions) can not deliver semantic meaning. In this respect, it is of central importance that inferentialism goes beyond formal syntax, so that rules of inference can be represented syntactically, but can also confer meaning

[8] An additional issue that is worth mentioning here is that any attempt to provide a reductive account of formal rules by means of model-theory brings with it a significant loss of information. A couple of examples suffice to highlight this. Classically, $\alpha \vee \beta$, and $\alpha \ddagger \beta$ (not both α and β), share the same semantic model. But, as Macfarlane [28] points out, they do not express the same kind of information: something will have been lost by their semantic identification. Similarly, a proof of an elementary mathematical conjunction and a proof of a conjunction involving a complex mathematical statement differ in terms of the kind of processes of verification involved, yet they are equivalent from the point of view of their being true or false. We shall also discuss in the following chapter the well-known "categoricity" problem [11, 14, 29–33], which has the result that the standard natural deduction framework for classical logic fails to rule out non-standard semantic models. In particular, that framework is easily shown to be sound and complete with respect to *both* the classical semantic model, and a model in which every formula is interpreted "true".

upon logical expressions. Such abstract arguments against inferentialism rest upon an incredibly impoverished view of syntax that fails to take into account that it is the dynamics of interaction with syntax, rather than the syntax itself, that gives rise to meaning. This coheres with Brandom's [5] argument that inferentialism is a species of pragmatism, and that pragmatics is essentially the interface between syntax and semantics:

> [...] it is pointless to attribute semantic structure or content that does no pragmatic explanatory work. It is only insofar as it is appealed to in explaining the circumstances under which judgments and inferences are properly made and the proper consequences of doing so that something associated by the theorist with interpreted states or expressions qualifies as a semantic interpretant, or deserves to be called a theoretical concept of content [5, p. 144].

As such, it is arguable that the issues facing inferentialism outlined above (and further discussed in the following chapter), occur due to the way in which it provides a formal structure for the inferential relations within the space of reasons. For example, standard approaches to inferentialism have emphasised the providing of a set of basic logical rules that can not be put into question (see e.g. [36] and the discussion in the following chapter). As such, these rules provide a foundation for reasoning that replaces the classical foundation in truth-conditional approaches to logic. This is perhaps most apparent in the tradition that privileges proofs as the central semantic notion such that a propositional content may be identified with a set of proofs (e.g. [13, 17]). There, proofs are not understood in terms of a constructive process of proving activity, rather they are taken to be objective and tenseless. Inference rules, are then something like "inference tickets", to use Prior's [34] term, and, as such, must be taken to be part of the foundational structure of the space of reasons. But, insofar as proofs are understood in objective terms, the act of proving is reduced to a kind of ratification, where proofs are unaffected by our interaction with them. The problem is that this leads us to problems of circularity and regress where the correct application of a rule is required for access to this realm of proof-objects.[9]

In distinction, I think that the above issues show us that we must reorientate ourselves with respect to the interface of syntax and semantics, and without giving in to this temptation of attempting to provide logical foundations. Instead, I shall suggest, the correct "foundation" for understanding inferentialism lies, not in monological deduction, but in interactive and dialogical reasoning activities. By reconsidering reasoning as a process of dynamic interaction (what is *done* with syntax) it becomes possible to provide a view according to which logic is not based upon a priori deductions that must be precise and true. Instead, this would take such deductions to be only the idealised result, and concretised remainder, of an active process of reasoning across multiple agents.[10] In this setting, meaning (and the justification of logical rules) is constituted by these more elementary forms of interaction.

---

[9] I discuss these issues in detail in Chap. 2.

[10] In this sense, there is also a clear connection, and also a distinction, with non-monotonic logics as developed to capture defeasible reasoning, artificial intelligence, and logic programming (e.g. [37]). The non-monotonicity of these logics allows for consequence relations that are not required (by monotonicity) to ensure that the addition of premises does not alter the validity of certain inferences, so that it is possible to retract consequences, for example. Whilst the approach to logic by means

This would be to reconsider the way in which we approach formal logic such that it coheres better with the broader account of inferentialism that we introduced in Sect. 1. There, Sellar's emphasis on the "space" of reasons may be taken as precursor to this shift, insofar as it emphasises the nature of reasoning as a process of giving and asking for reasons (to use Robert Brandom's [5] later interpretation), rather than as simply a series of deductive steps from premises to conclusions. This alternative view of reasoning places it primarily in the domain of the social, and also as an activity that is not just monological:

> [We] typically think about inference solely in terms of the relation between premise and conclusion, that is as a monological relation between propositional contents. Discursive practice, the giving and asking for reasons, however, involves both intercontent and interpersonal relations. The claim is that the representational aspect of the propositional contents that play the inferential roles of premise and conclusion should be understood in terms of the social and dialogical dimension of communicating reasons, of assessing the significance of reasons offered by others [20, p. 166].

To foreground this social dimension of reasoning is at odds with the logical tradition, which has emphasised the stringing together of propositional contents in deductions, over conversation, dialogue, and argument. Yet, as Mercier and Sperber [40] persuasively suggest, drawing upon considerable evidence from linguistics, cognitive, and behavioural sciences, the primary function of reasoning for human agents is in communicative argumentation by means of language:

> Reasoning contributes to the effectiveness and reliability of communication by allowing communicators to argue for their claim and by allowing addressees to assess these arguments. It thus increases both in quantity and in epistemic quality the information humans are able to share [...] To sum up, people can be skilled arguers, producing and evaluating arguments felicitously. This good performance stands in sharp contrast with the abysmal results found in other, non-argumentative, settings, a contrast made particularly clear by the comparison between individual and group performance [40, pp. 60–2].

This suggests that reasoning is inherently dialogical and argumentative, contrary to a view which takes these to piggy-back upon internal reasoning processes.[11] Moreover, it is surely only in the context of dialogue and social interaction that we can even begin to adequately examine thinking, since it is possible to use tools such as language to reflect upon thinking itself. Yet, if reasoning is inherently social and argumentative, this requires a rather different view of our "logical" practices than we find in the

---

(Footnote 10 continued)
of interaction and dialogue requires a form of "non-monotonicity", rather than writing this in to the semantics as is often the case there (with the so-called closed- or open-world assumptions, for example), the approach given here makes aspects of non-monotonicity part-and-parcel with the dynamics of syntax, with the result that semantics that does not require standard completeness results. Resultantly, it is possible to provide a syntactical basis accounting for closed- and open-world assumptions without writing them into the semantics, whilst also dealing with issues arising with default negation, for example (as discussed in (e.g. [38])). I will discuss this in detail in Chap. 5, where I show how the formal framework for interactions given in Chap. 4 can be developed into a framework for interactive proof and refutation search (a similar approach though developing a different set of formal tools is taken in [39]).

[11] I marshal more evidence to this effect in Chap. 3.

tradition, especially that of classical, truth-conditional, approaches. It is arguable that this distinction can be traced back to one made by Aristotle between the dynamics, or dialectics, of arguments between agents, and the abstract and static structure of an argument from a set of premises to a conclusion. Whilst, for Aristotle, both of these belong to the realm of "logic", it is the latter that was subsequently privileged as properly logical, and later came to be the model of logical validity.[12]

In this respect, in his excellent discussion of the epistemology of mathematics, Paul Ernest [42] argues that if we accept that conversation is central to mathematical reasoning, then this also brings with it an account of the history of mathematical inquiry, whilst also requiring that we understand mathematics to be a social activity. There, it is argued that, in addition to the "maverick" literature, which argues against the view of mathematics as *a priori*, analytic, and so on, we should also include argument against: 'the claim that mathematics is monological as opposed to dialogical, consisting of monological claims and proofs, based on a unique, firm foundation, and that no conversation, dialogue or dialectic is needed' [42, p. 35].[13] We might also consider the psychological work of Lev Vygostky and John Shotter, whose work, as Ernest points out, makes central the notion of thought as internalised conversation. Perhaps, then, reasoning itself is constructed along the lines of roles of proponent, and critic: 'the role of proponent, in which a line of thinking or a thought experiment (Peirce, Rotman) is followed through sympathetically, for understanding; and the role of critic, in which an argument is examined for weaknesses and flaws' [42, p. 37]. Even if we do not accept that this is the case for all forms of thought, what is important is that conversation and dialogue are central to the activity of reasoning. These involve the making of assertions, which are subject to defense and the asking of reasons, allowing that an assertion may be tested through other reasons, counterexamples, and so on.[14]

[12]See, for example, the discussion in [41].

[13]In the tradition to which Ernest refers, the work of Imre Lakatos (e.g. [43]) is, perhaps, most exemplary, and will be discussed in detail in Chap. 3, where I also discuss how this requires a distinct approach to more standard formal approaches to dialogue and games. For example, Lorenzen-style dialogical logic [44], is effectively a deterministic formula-checking approach that requires each dialogical move subsequent to the first to be determined by a set of rules about which inference rule must be applied. Hintikka's [45] approach to game-theoretical semantics, on the other hand, has more to do with determining relationships between syntax and objects in semantic models, and, as such, do not really reflect the dialogical approach that we are interested in here. Moreover, just as Lorenzen's dialogue games are deterministic evaluations of formulae, so too are Hintikka's winning strategies fixed as soon as the moves are fixed by a set of game rules (see e.g. [46] for an excellent overview of the two approaches).

[14]I also take the overarching picture of reasoning presented by Laden [47] to be allied to the view discussed here, particularly since that work "[...] describes reasoning as the responsive engagement with others as we attune ourselves to one another and the world around us" (p. 8).

4 From the Foundation of Rules to the Action of Reasons

The account suggested here, particularly with its emphasis on the underlying dynamics of reasoning is deeply at odds with the Fregean picture of logical meaning. It is the latter which upholds the problematic foundationalism of both static contents (truth-conditional approaches) and static proofs (monological inferentialist approaches). On Frege's view, logical formulas should be understood as propositional contents, whilst the turnstile may be interpreted as signifying judgement in the form of assertion. Judgments are something that may be done to contents (whilst not affecting those contents), so, for example, $\vdash \alpha$ indicates that α may be asserted unrestrictedly since no assumptions or premises are required for its assertion. It is notable also that there is only one type of judgment for Frege, assertion. There exists no dual, negative, judgment such as denial or rejection, so any attempt to construe logic as inherently dialogical can not even get off the ground. In contrast, then, we would be interested much more (or at least as much) in the *acts* of judging than the contents of judgment, where these judgments may be more than just assertion in order to deal with the different forms of activity that dialogue requires.

The Fregean viewpoint underlies the standard inferentialist theory of meaning, which is monological and unilateralist. In other words, it provides an account of meaning in terms of the validity of certain inferences, which is itself explained in terms of the conditions under which certain propositions may be asserted. More recently, a number of authors have argued for an approach to inferentialism based on a bilateralism (e.g. [48–51]). Bilateralism is the view that meaning is to be construed in terms of the conditions on both assertion and denial. So, the speech act of denial is taken to be on a par with, and equally foundational as, assertion. In particular, this view holds that the grasp that we have on speech acts of assertion and denial is prior to (in the order of explanation) our understanding of negation.[15]

This view helps us to consider a logical framework that is more adequate to the view of reasoning discussed briefly in Sect. 1, where, as Ernest [42] puts it; 'the form or underlying logic [...] might even be termed dialectical, because of the way assertions are met with counter assertions or antitheses, leading to new assertions' (p. 31). Unfortunately, this proposal, on its own, does not adequately solve the problems of justification, nor does it provide a framework for accounting for the dynamics of reasoning.[16] For example, assertion and denial are not related to any sort of context of reasoning on these proposals, they are free-floating. Furthermore, this does nothing to provide an account of the dynamics of reasoning since judgments are still taken to operate on contents that are, presumably, pre-existing, so we do nothing to assuage the concerns about the foundation of rules. An additional problem for these views is that they fail to adequately provide any account of what it would take to provide a proof, or refutation. This is because the balance between assertion and denial in

[15]So, rather than derive the denial of α from the assertion of $\neg\alpha$ as Frege has it, rather $\neg\alpha$ is explained in terms of the fundamental role of denial.

[16]I will discuss this further in Chap. 2.

bilateralism offers only an account of indirect proof (or refutation), and it typically deals with non-decidable logical structures such as classical logic.[17]

It is worth considering further the role that assertions and denials play in a dialogical scenario. For example, it is typically the case that we think of assertions in terms of prescriptive norms such as "aiming at the truth", or "on condition of a proof", and so on (presumably a symmetric kind of story is to be offered by bilateralists for denial). However, whilst these fit fairly well with the Fregean view of logic as concerned with propositional contents, this does not adequately capture the dynamics of reasoning that we require. It is more plausible to consider norms on assertion and denial in terms of challenge, testing, and so on. This is aptly summarized by Marion [52] as follows:

> [...] an act of asserting a statement brings with it a commitment to defend the assertion, if challenged, so to make an assertion is to make a move in a game, in which one is asked for and must provide grounds or reasons justifying the making of that assertion. In other words, the "game of asking for and giving reasons" is embedded in the very nature of assertions. (p. 20)

The emphasis here is not on *pre*scribing certain assertions, but rather on a kind of *pro*scriptive yet dynamic relationship of making an assertion, which immediately brings with it a commitment to its defence. In this context, the making of a denial involves a commitment to challenging the initial assertion. So, for example, an agent making an assertion that "the swatch is red", does not merely submit the statement as a propositional content to an evaluation that it is "true", or "false", rather, as Lecomte [53] rightly notes, it is much more like: 'playing it as a token in a game, knowing that other players can ask for reasons for saying it, either by challenging the choice of the name swatch or by contesting that it is red. It is only after the game has come to an end that the assertion can be evaluated' (p. 26). It is this interaction that I will argue is constitutive of meaning, and, in this way, we can reconsider the way in which certain logical rules are justified by these more basic constraints on interaction.[18] Propositional content, rather than being given by objective proofs or representational fiat, is something that is built-up through this process of interaction between agents. In addition, rather than thinking of logical rules as in place prior to rational thought being possible, we should rather think of the justification of logical rules and the construction of a proof (or refutation) subsequent to an interaction that has *terminated*. In other words, the view of logic that is presented here is one that does not operate upon pre-established propositional contents, but one in which that content is constructed over the course of our activities of reasoning. Furthermore, this is something than may be evaluated only when that activity has come to an end (in agreement or disagreement, in providing a verification or falsification, and so on).

---

[17] I discuss this further in Chap. 2.

[18] This shift in perspective will obviously also bring with it a change in the way in which we think about soundness and completeness proofs, since semantic models can no longer play the role of keeping check on where our reasoning goes astray. As such, we shall discuss internal constraints on reasoning that give rise to an account of validity and local completeness results in Chaps. 4 and 6.

It is whatever is left-over from this process that we typically call "deductive" logic, and which may be written down as a proof.[19]

There are a number of ways in which we might search for foundations, to provide a sure footing for deductions, rationality, and semantics. Returning, briefly, to the metaphor of the "space" of reasons. In much the same way Hilbert attempted for mathematical logic, Euclidean geometry attempted to provide foundations for geometry in the form of a set of intuitive axioms and definitions from which the geometric theorems may be deduced. The standard inferentialist picture attempts to displace this kind of foundationalism, but since it simply replaces deductive axioms with logical rules, it effectively fails to do so. There is yet, a different kind of search for foundations, that is in keeping with the story we tell here.

Whilst the "space of reasons" is primarily treated as metaphorical, this alternative approach to foundations can be illuminated by analogy with shifts in the mathematics of space.[20] There, the idea that a (Euclidean) geometry is constructed on a foundation of points was, historically, taken to be fundamental. Nonetheless, the development of non-Euclidean and noncommutative geometry in the work of Alain Connes and Alexander Grothendieck, for example, show us that it is incredibly fruitful to consider pointfree geometries by means of an algebra of functions that generalises the functions of traditional geometry.[21] Of course, this does not bring with it the elimination of geometric points, nor of Euclidean approaches to geometry. Rather, it augments our traction upon space through a generalisation of functions, whilst also highlighting the way in which Euclidean geometry may be thought of as a kind of static distillation of a much wider domain. As such, developments in Riemannian and algebraic geometry provide a new foundation for geometry. Rather than furnish a small set of axioms from which theorems may be deduced regarding points, lines and planes, this approach is an abstraction from Euclidean space that allows us to consider space as a manifold of points that are defined without recourse to embedding in Euclidean space (Riemann). This means that Euclidean geometry is like a local model which is encapsulated by this more universal theory of space. This is perhaps most clear in the work of Grothendieck, where the topological properties of space are explored using the tools of algebraic topology and category theory. These are shown to encompass other kinds of geometry as a kind of arché structure.

A similar shift is argued for here, where the axioms and rules that are ordinarily taken to be constitutive of deductive logic are taken to be secondary to a more abstract structure of interaction. In this setting, it is not that the space of reasons

---

[19] This is not dissimilar to the view presented in Lakatos [43] on the nature of mathematical inquiry, but also the view of Per Martin-Löf [54] that a 'proof is, not an object, but an act [...], and the act is primarily the act as it is being performed, only secondarily, and irrevocably, does it become the act that has been performed. (p. 231)' There are also connections with the project initiated by Jean-Yves Girard (e.g. [55]), especially the relationship between proof-search and termination. I will situate the project outlined here in relation to all of these in Chaps. 2 and 4.

[20] I do not intend or suggest that we use these tools to attempt to uncover the geometry of the space of reasons, though, in Chap. 5 I will suggest that the mathematical structure of the formal logic given in Chap. 4 can be mapped on to categorical and topos-theoretic mathematical universes.

[21] See [56] for an overview of the field.

is structured by a set of inferential rules that are pre-given, which then allows for rational reasoning to take place. Rather, it is the interactive (dynamic and shared) nature of reasoning through which this stable framework is constituted, and so may be formally accounted for.

In this regard, this book argues for a view in which processes of dialogue and interaction are taken to be foundational to reasoning, logic, and mathematics. This is both a continuation, and a substantial modification, of the inferentialist approach to logic and meaning. As such, this volume will articulate the foundations of the inferentialist approach to meaning, situating it within a variety of contexts, including philosophy of language, philosophy of mathematics, and epistemology of science. In addition, it will draw out a number of reasons for reassessing foundational approaches to logic, reasoning, and computation in terms of dialogue and interaction. As a consequence of this shift in approach, I will show that inferentialism is greatly strengthened, such that it can answer the most stringent criticisms of the view. These criticisms include, for example, that logic and reasoning are disconnected [7], that the justification of logical rules is circular [22], that an inferentialist epistemology is too demanding [57], and that the inferentialist approach to logic fails to fully determine the meaning of logical connectives [29, 34, 58]. More positively, this leads to a view of logic that emphasises the dynamics of reasoning, provides a novel account of the justification and normativity of logical rules, and leads to a new and attractive approach to foundations of logic that helps to clarify the issues surrounding the relationship between logic and reasoning.

5 Structure

In the hope that it may help to orientate the reader, what follows is an overview of the argument that will be advanced in the rest of the book.

I begin, in chapter one, by providing a more detailed overview of the contemporary literature on logical inferentialism, identifying a number of objections frequently made against it. These are grouped into three main categories. (1) The justification of logical rules. The general issue here is that the rules of logic can not be justified without making appeal to the rules themselves. This kind of circularity that uses a rule in providing an explanation of its validity is problematic since no justification for the rule itself has been provided. Furthermore, it paves the way for appeals to the justification of clearly non-deductive rules such as the rule of abduction. (2) The determination of meaning by rules. In model-theoretic semantics for propositional logics, compositionality of meaning is unproblematic whenever the semantics is truth-functional. This is not the case for an inferentialist approach to semantics, where failures of compositionality are well known for the semantics determined by monological entailment structures for classical and intuitionistic logic. This is obviously problematic if we want to construct an approach to the meaning of logical constants as being determined by their rules. A slightly different, though equally well-trodden, problem comes in the form of the paradoxical connective, "tonk" [34]. Tonk is a binary connective defined as follows:

$$\frac{\alpha}{\alpha Tonk\beta}\ (Tonk\text{-I}) \qquad\qquad \frac{\alpha Tonk\beta}{\beta}\ (Tonk\text{-E})$$

Tonk appears to have been provided with a definition, but successive application of the rules allows us to infer any conclusion from any premise:

$$\dfrac{\dfrac{\alpha}{\alpha Tonk\beta}\ (Tonk\text{-I})}{\beta}\ (Tonk\text{-E})$$

Something has gone awry with the definition. The attempt to figure out exactly what has become a philosophical industry involving the various constraints one might place on a definition to avoid the problem. (3) A failure to provide analogues to soundness and completeness theorems. This problem is deeply connected with the previous two, and concerns how we think of the full determination of meaning. Semantical approaches to consequence usually take precedence over proof-theoretic approaches, with soundness and completeness theorems connecting the two. Here, we are after an approach to completeness that neither requires semantical models, nor an idealization beyond ordinary reasoning (as is typically the case).

Chapter two begins by arguing that the issues raised in chapter one are exacerbated by taking a monological approach to logic, which fails to capture the dynamics of reasoning. For example, the problem of categoricity disappears if we move to a symmetrical (multiple-premises and multiple-conclusions) logical structure. This has been argued for perhaps most forcefully in [59], where denials are treated symmetrically to assertions so that a valid sequent $\Gamma \vdash\!\dashv \Delta$ can be interpreted as saying "it is incoherent to simultaneously assert all of Γ, and deny all of Δ". This is tricky, however, because (a) it is unclear exactly what role assertions and denials are supposed to play without reference to a conversational context; (b) it is unclear how one of the central inferentialist notions, "proof", works, since we do not have that, for example, (classically) $(\alpha \vee \neg\alpha)$ to be provable, rather it is merely incoherent to deny it. The rest of the chapter takes up this second issue, leaving the former for Chapter three. I go on to provide a detailed analysis of the notion of proof, which is supposed to be a stand-in capable of playing a similar role to truth in an inferentialist approach to semantics. In particular, I discuss different approaches to proof from the constructivist tradition, including the platonistic conception of proof and the notion of proofs as acts. I go on to argue that for proof to play a central semantic role in a theory of meaning, the latter conception of proof is required. Moreover, if this semantic role is going to reflect the dynamics of reasoning, then it must be possible to draw a distinction between a valid proof and a proof-attempt. I then position this discussion in the literature on the norms of asserting, with particular relation to the notion of the "game of giving and asking for reasons". I suggest that the making of an assertion brings with it a commitment, not to its truth, but to defend its truth. That is to say, assertoric norms do not restrict what an agent ought to assert, rather they constrain how agents respond to challenge and dialogue. On this view, making an assertion is a matter of bringing that assertion into "play", at which point, it is subject to norms involving a commitment to its defense, to providing reasons for it, and allowing it to be "tested" through interaction with other reasons, counterexamples and so on. In this setting, I finish by arguing that, if inferential rules are going to be the sort of thing that can determine meanings, then we will need to consider refutations alongside, and symmetric to, proofs.

Chapter three provides an overview and argument to the effect that reasoning, including scientific and mathematical inquiry, is inherently both dialogical and dynamic. That is to say, reasoning is social and interactive, and requires an account of history that is not captured by an absolutist and monological account of proofs. I begin by considering the role of dialogue in the history of logic, before moving on to discuss processes of scientific and mathematical inquiry. In particular, I argue that assertion and denial, dialogical response and testing, are central to the dynamics of reasoning and the creation of new knowledge. This is placed squarely in the tradition of Lakatos' [43] approach to inquiry, which is understood to operate as a kind of dialogue between "prover" and "refuter", where a conjecture is considered by means of proof attempts and tests in the form of refutation attempts, disproofs and counter-examples. Such "proofs that do not prove" are part and parcel with the development of inquiry, until a conclusive proof or refutation of the conjecture is reached. Lakatos considered this process to be one of dialectics, as distinct from logic. Here, I will suggest, on the contrary, that this process may be construed to be logical, so long as logic itself is not also understood to be an absolute and monological structure. I finish by setting out the constraints that this would place on the development of a logical structure by drawing out the distinctions between this approach and more traditional dialogical and game-theoretic approaches to logic. I go on to provide a discussion of the broader approach to dialogue, with particular reference to the interactional work of Ginzburg (e.g. [60]). There, I discuss a number of constraints over dialogues taking interaction, rather than rules, to be *a priori*, such that cooperative interactions may be understood to form the basis of the kinds of approaches to inquiry previously discussed. Briefly, I turn to considering how this approach fares in relation to providing a way of accounting for "reasons", which does not fall into the trap of important criticisms in feminist and post-colonialist theory.

In Chapter four, I return to some of the foundational issues discussed in the introduction and first chapter, particularly regarding the justification of logical rules. It is argued that standard accounts from both proof-theoretical and semantical points of view do not suffice to account for the justification of basic logical rules. By way of response, an analysis of logical inference as acts taking place in dialogical situations is provided. This is used to underpin a novel account of the proof-theoretic notion of harmony as a way of balancing dialogical concepts of proofs and tests. I show that this leads us to a constructive logic over both proofs and refutations, and where there exists an internal justification for logical rules at the termination of dialogue. This means that, rather than start with logical rules as a priori, the rules themselves are derivative of dialogical balance. This is the chapter where I introduce and discuss a possible formal structure, which provides a framework for this dialogical approach to logic. I discuss its relationship to bi-intuitionistic, linear, and non-monotonic logics, and provide a kind of proof-theory that replaces rules with schematic moves.

Chapter five considers different approaches to providing mathematical structures accounting for the abstract processes of interactive approaches to proof and refutation. I first consider mode standard constructive semantic approaches to the formal propositional logics for proofs and refutations. Whilst these illuminate aspects of the structure of interactive logic, they also highlight the shortcomings of standard semantics, even in dualised form. Nonetheless, the way in which these structures "objectify" propositions is useful for considering propositions as stable types *subsequent to their construction in interaction*. These ideas are made more precise through the notion of bisimulation equivalence, and closure under bisimulation. To finish, I provide a number of examples, using these to highlight a slightly more liberal conception of the coherence of interactions.

Chapter six suggests that the approach argued for up to this point provides a novel approach to understanding logical paradoxes. In particular, it is argued that what is typically thought to be captured by paraconsistency and paracompleteness is best understood as forms of epistemological instability that can be dealt with in the context of the dialogical dynamics of reasoning. This is used to motivate an approach called paracoherence, which can deal with strengthened versions of paradox, and is compatible with a bivalent approach to "local" truth and falsity. This is clarified by first considering logic and paradox in relation to standard arguments regarding logical revision. Arguments for and against logical revisionism share assumptions regarding logical rules that have been argued against in previous chapters. As such, I suggest that the interactive approach provides not a revision of logic, but an attempt to rethink logic from the ground up.

Chapter seven, the final chapter, suggests a number of ways in which the view of logic suggested here can be more broadly considered in relation to other theories of reasoning. To do so, I situate the interactive approach discussed in this book in relation to the project of the naturalisation of logic argued for, primarily, by John Woods and Lorenzo Magnani. On their view, reasoning is typically non-deductive, and there may be no appeal to "reason" as a normative domain (typically supposed to be governed by logic) as holding above the way in which human agents *actually* reason. I argue that, whilst the interactive view shares some of the critical gesture of their work, it does, in fact, offer a means by which normative authority can, and indeed *should* hold such that reason and reasoning can not be collapsed into one another so easily. In discussing these issues, I garner support from evidence from enactive and interactive theories of sub-personal activities, as well as the ethical and political dimensions of dialogue and social reasoning.

What follows will be of interest to those wishing to deepen their engagement with philosophy of language, philosophical and mathematical logic, theories of reasoning. It will also interest those who actively engage in current debates involving, for example, logical pluralism and revision, and the relationship between logic and reasoning, from advanced undergraduates, to professional philosophers, mathematicians, and linguists.

References

1. Graham Priest. *An Introduction to Non-Classical Logic: From If to Is*. Cambridge University Press, 2008.
2. Mark Wilson. *Wandering Significance: An Essay on Conceptual Behavior.* Oxford: Clarendon Press, 2006.
3. Gottlob Frege. *The basic laws of arithmetic: Exposition of the system*, volume 137. Univ of California Press, 1964.
4. Paul Boghossian. Knowledge of logic. *New essays on the a priori*, pages 229–254, 2000.
5. Robert Brandom. *Making It Explicit: Reasoning, Representing, and Discursive Commitment*. Harvard University Press, 1994.
6. John Etchemendy. *The Concept of Logical Consequence*. Harvard University Press, 1990.
7. Gilbert Harman. *Change in View*. MIT Press, 1986.
8. Michael A. E. Dummett. *Origins of Analytical Philosophy*. Harvard University Press, 1993.
9. Mark Eli Kalderon. Reasoning and representing. *Philosophical Studies*, 105(2):129–160, 2001.
10. James Trafford. Inference and rational commitment. *Prolegomena*, 12(1):5–20, 2013.
11. Nuel D. Belnap and Gerald J. Massey. Semantic holism. *Studia Logica*, 49(1):67–82, 1990.
12. Paul Boghossian. Epistemic analyticity: A defense. *Grazer Philosophische Studien*, 66(1):15–35, 2003.
13. Michael A. E. Dummett. *Elements of Intuitionism*. Oxford University Press, 2000.
14. James W. Garson. Expressive power and incompleteness of propositional logics. *Journal of Philosophical Logic*, 39(2):159–171, 2010.
15. Christopher Peacocke. *Thoughts: An Essay on Content*. Blackwell, 1986.
16. Jaroslav Peregrin. Meaning as an inferential role. *Erkenntnis*, 64(1):1–35, 2006.
17. Dag Prawitz. Meaning approached via proofs. *Synthese*, 148(3):507–524, 2006.
18. Ludwig Wittgenstein. *Philosophical Investigations, 4th Edition (Trans. Hacker and Schulte)*. Wiley-Blackwell, 2009.
19. Wilfrid Sellars et al. Empiricism and the philosophy of mind. *Minnesota studies in the philosophy of science*, 1(19):253–329, 1956.
20. Robert Brandom. *Articulating Reasons: An Introduction to Inferentialism*. Harvard University Press, 2000.
21. Robert Brandom. *Between Saying and Doing: Towards an Analytic Pragmatism*. Oxford University Press, 2008.
22. Lewis Carroll. What the tortoise said to achilles. *Mind*, 4(14):278–280, 1895.
23. Graham Priest. Two dogmas of quineanism. *Philosophical Quarterly*, 29(117):289–301, 1979.
24. J. C. Beall. Free of detachment: Logic, rationality, and gluts. *Noûs*, 49(1), 2013.
25. John Etchemendy. Reflections on consequence. *New essays on Tarski and philosophy*, pages 263–299, 2008.
26. Susan Haack. The justification of deduction. *Mind*, 85(337):112–119, 1976.
27. Carlo Cellucci. The question Hume didn't ask: why should we accept deductive inferences. In C. Cellucci and P. Pecere (eds.), editors, *Demonstrative and Non-Demonstrative Reasoning*, pages 207–235. Edizioni dell'Università, Cassino, 2006.
28. John MacFarlane. Logical constants, 2009.
29. Rudolf Carnap. *Formalization of Logic*. Cambridge, Mass., Harvard University Press, 1943.
30. J Michael Dunn and Gary Hardegree. *Algebraic methods in philosophical logic*. OUP Oxford, 2001.
31. Gary M. Hardegree. Completeness and super-valuations. *Journal of Philosophical Logic*, 34(1):81–95, 2005.
32. Lloyd Humberstone. *The Connectives*. MIT Press, 2011.
33. David J Shoesmith and Timothy John Smiley. *Multiple Conclusion Logic*. CUP Archive, 1978.
34. A. N. Prior. The runabout inference ticket. *Analysis*, 21:38–39, 1960.
35. John R. Searle. *Minds, Brains and Science*. Harvard University Press, 1984.
36. Paul Boghossian. Analyticity reconsidered. *Noûs*, 30(3):360–391, 1996.

37. Dov M. Gabbay and Christopher John Hogger and J. A. Robinson. *Handbook of Logic in Artificial Intelligence and Logic Programming*. Monograph Collection (Matt - Pseudo), 1993.
38. José Júlio Alferes, Luís Moniz Pereira, and Teodor C Przymusinski. 'classical'negation in non-monotonic reasoning and logic programming. *Journal of Automated Reasoning*, 20(1-2):107–142, 1998.
39. Alexis Saurin. Towards ludics programming: Interactive proof search. In *Logic Programming*, pages 253–268. Springer, 2008.
40. Hugo Mercier and Dan Sperber. Why do humans reason? arguments for an argumentative theory. *Behavioral and brain sciences*, 34(02):57–74, 2011.
41. Lorenzo Magnani. The eco-cognitive model of abduction. *Journal of Applied Logic*, 13(3):285–315, 2015.
42. Paul Ernest. The dialogical nature of mathematics. *Mathematics, education and philosophy: An international perspective*, pages 33–48, 1994.
43. Imre Lakatos. *Proofs and Refutations: The Logic of Mathematical Discovery*. Cambridge University Press, 1976.
44. Paul Lorenzen and Kuno Lorenz. *Dialogische logik*. Wissenschaftliche Buchgesellschaft, 1978.
45. Jaakko Hintikka. *Logic, Language-Games and Information: Kantian Themes in the Philosophy of Logic*. Oxford, Clarendon Press, 1973.
46. Tero Tulenheimo. On some logic games in their philosophical context. In *Ludics, dialogue and interaction*, pages 88–113. Springer, 2011.
47. Anthony Simon Laden. *Reasoning: a social picture*. Oxford University Press, 2012.
48. Huw Price. 'Not' again. unpublished.
49. Greg Restall. Multiple conclusions. In *Logic, methodology and philosophy of science: Proceedings of the twelfth international congress*, pages 189–205. Kings College Publications, 2005.
50. Greg Restall. Assertion, denial and non-classical theories. In Francesco Berto, Edwin Mares, Koji Tanaka, and Francesco Paoli, editors, *Paraconsistency: Logic and Applications*, pages 81–99. Springer, 2013.
51. Ian Rumfitt. Yes and no. *Mind*, 109(436):781–823, 2000.
52. Mathieu Marion. Why play logical games? In Ondrej Majer, Ahti-Veikko Pietarinen, and Tero Tulenheimo, editors, *Games: Unifying Logic, Language, and Philosophy*, pages 3–26. Springer Verlag, 2009.
53. Alain Lecomte. Ludics, dialogue and inferentialism. *Baltic International Yearbook of Cognition, Logic and Communication*, 8(1), 2013.
54. Per Martin-Löf. On the meanings of the logical constants and the justifications of the logical laws. *Nordic Journal of Philosophical Logic*, 1(1):11–60, 1996.
55. Jean-Yves Girard. On the meaning of logical rules i: syntax versus semantics. In *Computational logic*, pages 215–272. Springer, 1999.
56. Alain Connes and Matilde Marcolli. *Noncommutative geometry, quantum fields and motives*, volume 55. American Mathematical Soc., 2008.
57. Timothy Williamson. *The Philosophy of Philosophy*. Blackwell Pub., 2007.
58. Nuel Belnap. Tonk, plonk and plink. *Analysis*, 22(6):130–134, 1962.
59. G. Restall. Multiple conclusions. In Valdes-Villanueva L. Hajek P. and Westerstpahl D., editors, *Logic, Methodology, and Philosophy of Science: Proceedings of the Twelfth International Congress*, pages 189–205. Kings College Publications, 2005.
60. Jonathan Ginzburg. *The Interactive Stance: Meaning for Conversation*. Oxford University Press, 2015.

Chapter 2
Inferentialism and Its Discontents

A perennial question in consideration of logic concerns where the rules of logic come from? This question is overflowing with sub-questions regarding, for example, what is meant by *the* rules of logic; whether or not we are concerned with the justification of rules, or the meaningfulness of certain rules, or both (these are often run together); how do we understand the relationship between logical rules and truth; what is the normative role of logical rules? In this chapter, I provide a brief overview of inferentialism in Sect. 2, before going on to identify a number of concerns frequently made against it. Section 3 develops the discussion of the circularity of justifying logical rules that I outlined in the introduction. These problems are compounded by the issues raised by paradoxical connectives such as "tonk", discussed in Sect. 4. One suggestion made in the literature is to combine the inferentialist and referentialist approaches in a "modest" form of inferentialism, which we discuss in Sect. 5, pointing out that this approach runs afoul of issues facing the full determination of meaning in terms of categoricity and compositionality. I finish, in Sect. 6 by discussing a more promising response that does not require semantic justification or rules, but looks to internal constraints on rules, namely that they are "harmonious" However, this, it is suggested, is still problematic from the point of view of circularity, and, moreover, it also does not circumvent the issues of categoricity and compositionality.

1 Inferentialism (A Brief Overview)

Inferentialism is an approach to the meaning of expressions that takes the agents' inferential activity to be the primary semantic notion. Rather than take reference, or denotations of expressions, to establish meaning, it is certain rules of inference (and inferences themselves) that do this job. This belongs to a wider tradition in philosophy of language, perhaps most clearly articulated in the work of Robert Brandom (e.g. [1]), who argues that our language use ought to be understood in terms of a "game of giving and asking for reasons" (GOGAR) [2]. Broadly speaking, then, inferentialism is a kind of conceptual role semantics, in which the meaning of a con-

J. Trafford, *Meaning in Dialogue*, Studies in Applied Philosophy,
Epistemology and Rational Ethics 33, DOI 10.1007/978-3-319-47205-8_2

cept is determined by its conceptual role.[1] This, following the work of Wilfrid Sellars (e.g. [4]), and also the later Wittgenstein (e.g. [5]) may be seen as constructing an alternative position that is distinct from the much more standard truth-conditional semantics, or, more broadly, representationalism. It is not, however, the case that the inferentialist picture of meaning must be at odds with more traditional approaches to semantics, as we shall see later (Sect. 3). Rather, it is perfectly plausible that talk of "truth" and "reference" can be maintained. What is important is that these should not be considered to be *epistemically* or *explanatorily* prior. As such, as I show later in this chapter, one might maintain a "modest" form of inferentialism in which rules of inference may be said to determine (in some way) a referential semantics. More standard approaches to inferentialism have rather suggested that inferentialism brings with it a revision of standard semantic theories such that truth is displaced by other more epistemically tractable notions such as "proof" or "assertibility". [2] It is, on this view, central to an approach to meaning that it is "graspable" in agents' ordinary understanding, and for many authors, this has led to consideration of constructive approaches to logic (discussed further below in Sect. 4).[3]

What coalesces inferentialist approaches is their opposition to representationalist approaches to meaning that are based upon word-world relationships that are supposed to be established directly. For example, Jerry Fodor's [10–12] externalist approach to meaning takes it that the meaning of a concept is constituted, in the main, by a nomological relation between mind and world, where concept acquisition is a matter of 'getting nomologically locked to the property that the concept expresses' [12, p. 125]. So, for example, the content of a concept such as CUP is explained by its being "locked" to the property of "being a cup" Understanding the meaning of a concept has to do not with accepting a specific range of inferences (or even associated beliefs), rather it is simply a matter of having the ability to make accurate tokenings of one's concepts. It is plausible, then, that agents can be locked to the same properties whilst entertaining vastly different inferential patterns regarding the linguistic expression that tokens that concept. Without going into too much detail, consider that this view is fairly intuitively too coarse-grained. It seems, for example, that the expressions "water is clear" and "H_2O is clear" are distinct, whilst, on the locking view, the expressions have the same content. Relatedly, Georges Rey [13] has argued against this view with what he calls the "fortuitous locking problem" The kernel of his worry is this: if we pay no attention whatsoever to the inferential relations that agents associate with a given expression, then the nomic relationships in virtue of which they have those specific concepts (as well as any related propo-

[1] Block [3] provides an excellent overview of the collection of positions that fall under this view.

[2] Michael Dummett is perhaps the most well-known proponent of this account (e.g. [6–8]).

[3] Perhaps most famously, Dummett's "manifestation thesis" is used to argue for an anti-realist approach to logic in which the manifestation of our ability to verify a sentence is constitutive of agents' knowledge of that sentences meaning. This requirement, that an agent have some justification (such as a proof, or verification) of a statement, fairly quickly leads (according to Dummett) to a rejection of classically valid logical laws such as law of excluded middle (which is supposed to be valid even in the absence of verification of either disjunct). Further discussion of these issues is in [9].

Fig. 1 Matrices for conjunction

α	β	$\alpha \wedge \beta$
T	T	T
T	F	F
F	T	F
F	F	F

sitional attitudes and so on) will be largely opaque and subjectively inaccessible to those agents. In that case, it is possible that we might stand in a nomic relationship with a property that is, intuitively, not the subject matter of our conversations. As he puts it, '[i]insofar as one is moved by standard externalist claims about the reference of terms, one supposes that they refer to whatever real phenomenon the people in one's community are getting at in their uses of the terms' (p. 317); but, Rey goes on to say, the lockings must be fortuitous, otherwise, they would "not give rise to any meaning intuitions, they would not enter into cognitive deliberations, and, most importantly, they would not seem to figure in any cognitive psychological laws" (p. 318). The concern is that purely covariational relations would not seem to safeguard against agents being linked up with properties that are not those to which they think we refer. The account threatens to make the reference-fixing facts both opaque and implausibly separate from rational inquiry and agents' understanding.

Of course, these issues may well be adequately dealt with, but even so it is widely accepted (even by Fodor at times) that an inferentialist story is better suited for logical expressions. Say, for example, that we want to take a representationalist approach to the meaning of logical expressions. Then we might say that the meaning of a logical expression is the contribution that the expression makes to the truth-conditions of propositions in which it occurs, and where those truth-conditions are constraints on the way that the world must be if that proposition is true.[4] Then, the occurrence of $\wedge$ in the proposition expressed by the sentence "$\alpha \wedge \beta$" contributes to the truth-conditions of $(\alpha \wedge \beta)$ is its Boolean function as captured in the standard truth-table (Fig. 1).

We can then say that $\wedge$ has the semantic function of conjunction because its meaning is such that it requires the sentence to conform to the logical properties of conjunction. Even if it is possible to deal with obvious issues facing abstract entities, this view presents overwhelming problems when it comes to providing an epistemology of logical meanings. How, for example, could a token mental state come to have these specific logical properties as its content, and thus play a role in intentional psychology? Alone, they seem *prima facie* inadequate to type token mental states. The problem is one of determining how it is possible for mental transitions to come into contact with those (presumably abstract) truth-conditional contents. This is somewhat difficult since it does not seem possible, even in relatively simple cases, to determine every instance of the truth-preserving schema for conjunction, say, in advance of being able to know whether or not it is valid. Recall that an inference of

[4] Peacocke [14] makes a similar suggestion, though his position is more nuanced as discussed below.

the form: "$\alpha \wedge \beta \vdash \beta$" is supposed to be valid whenever β is a logical consequence of $\alpha \wedge \beta$, so that, when $\alpha \wedge \beta$ is true, β is true also. The inference, however, is typically thought to hold between speech-acts, judgements, assumptions, and so forth, whereas its validity is, on this story, grounded on propositional contents. Of course, on this view, in order for an agent to entertain a proposition involving conjunction, the agent must first grasp the truth-conditional content of conjunction. Then, famously, grasping conjunction will involve conjunction itself: otherwise, how could the agent antecedently grasp that truth-conditional content? If what we are after here is an account of the epistemological justification of conjunction, we are unlikely to be satisfied by any account which (a) utilises the concept in its explanation (b) presupposes antecedent knowledge of the proposition expressed by a sentence involving the logical term. As Mark Kalderon puts it;

> [One] cannot coherently claim that a speaker could come to entertain the content of [conjunction] by knowing the standard explanation since the standard explanation presupposes the antecedent intelligibility of [conjunction]. [15][5]

There is also a difficulty in explaining how agents such as children can make inferences involving conjunction without an ability to know its truth-conditional structure, since they would be disqualified from grasping the relevant logical content. Any such account would appear to be too conceptually demanding, since many people cannot state the truth-tables or other truth-functional explanation of the logical constants, whilst, (at least seeming to be) making logical inferences.[6]

The inferentialist approach to meaning, on the other hand, seems to be especially well-suited to logical expressions, suggesting that their meaning is constituted by a set of inferential rules, and where these rules have a substantial connection with ordinary reasoning practices with those expressions.

2 Dispositions and Circular Justifications

Unfortunately, however, there are also several problems facing attempts to justify the validity of inferential rules, which, together, may push us back towards the representationalist account. As we discussed in the introduction, whilst the inferentialist approach looks to offer a clear way in which to think of how we come to use certain logical rules, it does not yet provide an account of why *these* rules are justified. But, even asking questions about the latter is tricky, for a number of well-worn reasons that I briefly rehearse here. Following in Carroll's footsteps, Quine [17], for example, argues that if we think that rules originate in some form of convention, then a regress argument takes hold:

[5]Conjunction replaces existential quantification in the original.

[6]For extended discussion, see [16]. This issue is also raised by Dummett's (e.g. [6], p. 216–7) argument against grounding logical content in truth-conditional semantics due to concerns regarding the acquisition of the relevant concept.

> [...] Derivation of the truth of any specific statement from the general convention thus requires a logical inference, and this involves us in an infinite regress [...] logic is needed for inferring logic from the conventions. [17, p. 351]

A more generalized form of the problem that Quine identified is that it is difficult to hold together the idea that logical rules have normative force, with the ability to question the justification of those rules. Providing justification for basic logical rules brings with it the issues of circularity that we discussed in the introduction. One option, that we shall discuss in this section, is to say that this is an acceptable form of circularity since it is just a *pragmatic*, rather than a *justificatory* circularity. Dummett [7], for example, argues that the circularity is unproblematic since it is not a gross circularity as the validity of a rule like *modus ponens* does not need to be used as a premise in an explanation of its own validity:

> We have some argument that purports to arrive at the conclusion that such-and-such a logical law is valid; and the charge [...] is that at least one of the inferential steps in the argument must be taken in accordance with that law. We may call this a "pragmatic circularity" [...]. [7, p. 202]

As such, pragmatic circularity is judged, by Dummett, to be safe from the point of view of explaining the validity of modus ponens. But, as he goes on to say, this will do nothing to assuage concerns for somebody who genuinely doubts its validity[7]:

> If the justification is addressed to someone who genuinely doubts whether the law is valid, and is intended to persuade him that it is, it will fail of its purpose, since he will not accept the argument. [7, p. 202]

But, this misses the force of the problem, by serving only to highlight the way in which logical rules can not be put in question. Rather, pragmatic circularity clarifies the rather autocratic nature of logical rules insofar as the attempt to dispute those rules brings with it their infinite re-establishment.[8] That an argument involving pragmatic circularity uses a rule in providing an explanation of its validity is problematic since no justification for the rule itself has been provided. Moreover, as pointed out by Celluci [19], this only paves the way for appeals to the justification of clearly non-deductive rules such as the rule of abduction.

One common suggestion that is found in the literature that is supposed to uphold the "basicness" of rules such as modus ponens is that pragmatic circularity is not question begging because the correct use of a logical rule is grounded in agents' dispositions to use that rule (or otherwise some kind of implicit grasp of that rule) (e.g. [14, 15, 20–24]). There are a variety of different positions that orbit around a central idea: a circularity of justification for a specific set of rules is not problematic for a view on which even to think with a concept requires one to be disposed to infer according to those rules. For example, Boghossian [21, p. 230] thinks that an agent's

---

[7] These points hold for any logical rule, but it is worth noting that modus ponens is not unrestrictedly valid in logics such as paraconsistent LP, see [18].

[8] It is also worth noting that pragmatic circularity quickly leads us to a form of epistemic circularity once we bring in talk of validity and truth.

dispositions to infer according to the classical natural deduction rules determines what that agent means by logical expressions such as "and", and "if, then":

> [...] the logical constants mean what they do by virtue of figuring in certain inferences and / or sentences involving them and not in others. If some expressions mean what they do by virtue of figuring in certain inferences and sentences, then some inferences and sentences are constitutive of an expression's meaning what it does, and others aren't. [20, p. 353]

With these in place, Boghossian takes it that such an agent is justified in making inferences according to those rules, since those rules are understood to be stipulated to be valid in this way. Then, "and", for example comes to have the meaning required for the validity of the inference rules defining conjunction.[9] Furthermore, according to Boghossian, it is necessary for a thinker to have those inferential dispositions to even have the relevant logical concept (and grasp the content of the expression); without those dispositions, a thinker could not even entertain beliefs involving that content.

On Christopher Peacocke's [14, 23, 24] view, a thinker must find certain inferences "primitively compelling" [10] An inference is primitively compelling if a thinker finds it compelling, it is underived from other principles, and its correctness is not answerable to anything else for possession of a certain concept [25, p. 6]. Thus, in order for a thinker to have a logical concept (or to grasp a content in Peacocke's terms), there are a substantial set of constraints, called possession conditions, that a thinker must meet, and these constraints also determine the meaning of that concept. These, again, are certain inferential practices to which that thinker must be disposed. Roughly speaking, Peacocke endorses a supervenience thesis, on which, A and B grasp the same content N iff they have the same set of fundamental inferential dispositions I (involving N). According to Peacocke, these inferential dispositions are also supposed to individuate content, constituting what is required of a thinker to grasp a given content, so the corollary also holds: A and B have a specific set of inferential dispositions I iff they grasp the same content N. Despite their differences, the general position can be distilled as follows:

(Supervenience): Where two thinkers A and B have attitudes involving logical expressions $N\alpha$ and $N\beta$, and they are disposed to accept the validity of the same set of fundamental inferences I involving $N\alpha$ and $N\beta$, then A and B grasp the same logical content.[11]

So, for example, on Peacocke's view, the possession condition for "and" is that the thinker finds "primitively compelling" the standard introduction and elimination rules for conjunction. That is to say, for a thinker to grasp the content of conjunction, they must have a disposition to infer in accordance with the introduction and elimination rules as a matter of semantic competence. A thinker would not be ascribed beliefs or judgments involving the content of conjunction if they did not have those specific dispositions to infer.

---

[9] Whilst similar, Mark Kalderon [15] emphasises agents' dispositions to accept the validity of certain inferences.

[10] As with Boghossian, these inferences conform to the classical natural deduction rules.

[11] Kalderon [15] has a similar account.

Let us briefly review a number of criticisms of this view.[12] Timothy Williamson [29, Chaps. 4, 5] has perhaps been most vocal in providing reasons to think that any such account will fail to provide justification for the validity of inferential rules. Williamson argues convincingly, that there is no straightforward connection between understanding the meaning of a logical expression and assent to the validity of arguments in which it figures. The argument is two-pronged. First, Williamson argues that there are many cases where speakers of a language fail to be disposed to assent to instances of the rules that are taken to define logical connectives whilst yet being judged as linguistically competent. Call this the argument from dispositional discrepancy. Let us briefly flesh out this issue. Any approach that rests upon a principle, such as (Supervenience), provides conditions that a token cognitive state must meet, by metaphysical necessity, in order to type that state as a specific propositional attitude. In relation to belief-formation, these dispositions are manifested in fixing propositional attitudes relevant to a thinker's judgments. However, it does not seem acceptable to say that in order to grasp a content, every inference of introduction or elimination needs to be manifested in judgments. If that were so, then thinkers grasping a content would not be capable of making errors of judgments, or basic inferential mistakes; if they did, they could not be ascribed a judgment with that content. This seems too inflexible a requirement to deal with the practice of actual thinkers. One way that the problem might be dealt with is by allowing for performance errors. It may be possible, still, to introduce a fairly standard performance/competence distinction, in which a thinker's actual inferential practices do not tell against their competence, because of various performance errors, interference factors, memory limitations, distractions, and so on.[13] One might appeal, for example, to the idea that there are cases in which thinkers make mistakes regarding inferences of introduction rules, because those dispositions are masked. Though, for this to have traction, one has to be careful, since, it is required that whatever is constitutive of content ought to be manifested in thinker's actual dispositions in order to ascribe the correct content to that thinker.[14] Now, say, for example, that it is allowed that a thinker can be ascribed an attitude with a specific content in terms of a particular competence that thinker possesses, but that this competence is not performed in the thinker's judgments. Then, it becomes tricky to see how a thinker can be attributed with that content rather than some other content that does seem to be performed in their judgments.[15] In this regard, even in the simple case of conjunction, the inferentialist account seems to be on shaky ground. For example, there are exemplary failures of closure under conjunction. In cases such as the preface paradox (as well as complicated tautologies), routine failures to infer according to conjunction introduction without restriction occur. Regarding the

[12] For discussion and argument against these views, see [26–29].

[13] See also the discussion in [26] on these issues.

[14] Otherwise, it wouldn't seem possible to say that *by those dispositions* a specific content determined by a specified set of rules is being manifested, rather than some other content altogether. Whilst, for conjunction, this does not look to be a difference that makes a difference, since most accounts of conjunction agree, this is not the case for other logical constants such as negation, for example.

[15] In essence, this objection to dispositionalism is in Kripke [27, p. 30].

elimination rules, consider cases of the conjunction fallacy as in the infamous Linda problem [30]. In such cases, participants routinely and systematically fail to infer in accordance with the conjunction elimination rule; and rather, suggest that the probability of a conjunction is greater than one of its conjuncts. Whilst many philosophers have discussed the relevance of heuristics and biases literature to performance ability, the relevance to competency is not often recognized, for if the fallacy corresponded to performance error alone, there should be random failures with little correlation, but those failures are systematic and consistent:

> If each departure from normative responding represents a momentary processing lapse due to distraction, carelessness, or temporary confusion, then there is no reason to expect covariance among biases across tasks (or covariance among items within tasks, for that matter) because error variances should be uncorrelated. [31, p. 646]

The issue may be understood by means of the distinction between whether or not the dispositions in (Supervenience) are supposed to be read *descriptively*, or *normatively*.[16] On the latter view, the normativity would then be cashed out as inhering in the meanings of the logical constants, so that having a concept of a logical constant requires having a disposition to make certain inferences as normatively correct.[17] The problems raised above can be put like this: say that the specified set of inferences are supposed to be descriptive of actual use (e.g. [25]), then this does not seem to adequately map onto agents' ordinary reasoning; if they are supposed to be *normative* over use (e.g. [22]) then the flexibility of use of expressions would make it hard to see why we should think this offers a way of grounding the justification of rules (i.e. why *this* rule, rather than *that* rule?). It seems, for example, that Williamson is correct to say that thinkers routinely understand even simple logical contents such as conjunction, whilst nonetheless failing to instantiate the relevant inferential competency. But, conceiving of such inferential transitions as required to have thoughts involving a content (and, so also attribution of content to a thinker) would effectively entail routine content-failure. This seems both implausible and unwarranted, particularly given the significant literature regarding the flexibility regarding mental states that is nonetheless consistent with thinkers having the capacity for attitudes with the relevant content. Routine failures to infer according to a set of natural deduction rules does not provide good evidence for content-failure, in which case, the semantic content of logical expressions is not settled by the inferential rules thinkers are disposed to make.[18]

---

[16] Simplifying a little, Peacocke takes the former view, whilst Boghossian takes the latter.

[17] We may presume, then, that for Boghossian, elements of the problem posed by Carroll are immunised since, it can not be possible for Tortoise to have the concept of logical conditional without feeling some sort of normative "pull" to make the correct inference. In Chap. 3 I shall discuss alternative approaches to the normativity of rules.

[18] These problems are not assuaged by an appeal to implicit conceptions as in Peacocke [32] (but note that implicit conceptions are not supposed to be necessary for the simple cases discussed here). Briefly, an implicit conception is taken as a possession condition for a concept that influences and explains our judgments, but is less stringent than possession conditions. Nonetheless, for implicit conceptions to do any serious explanatory work, they must be guaranteed to be truth-preserving

Second, Williamson argues that there are cases where competent speakers of a language may be disposed to assent to the validity of deviant sets of inference rules. This may be the case, for example with material modus ponens, $\{\alpha, \alpha \supset \beta\} \therefore \beta$, which, whilst classically valid is paraconsistently invalid.[19] Williamson argues that such deviance should neither be taken as evidence that a paraconsistent logician fails to understand the meaning of the expression "if, then", nor as evidence that they are using the expression with a different meaning. Williamson points to Vann McGee's supposed counterexample to modus ponens, with the suggestion that modus ponens is not, therefore, a valid rule of inference. Williamson goes on to say that, supposing McGee is incorrect, it would still not be the case that we should say that McGee does not grasp the meaning of the expression "modus ponens" Generalising further, the suggestion is this: given that there are philosophical experts that deny that basic inference forms are valid, it cannot be the case that for an inference form to be meaning-determining, any expert speaker competent with the relevant expressions must be disposed to accept it. Call this the argument from deviance.[20]

Both arguments are taken by Williamson to give grounds to argue against the idea that there are such things as meaning-determining rules, and, as such, that inferentialism fails to provide a decent account of logical meaning. I think this is rather too quick, but nonetheless, it does seem to tell against any position that rests upon a principle that looks like (Supervenience).[21] The important moral, for our purposes, is that these accounts do little to dispel the concerns over circularity of justification as identified in our discussion of Carroll's short story. That is, even if we could find some way of adequately spelling out inferential rules (by means of thinkers' dispositions) in terms of introduction and elimination rules in natural deduction systems for classical logic, this will not provide an account of their justification on pain of circularity.[22] One response that the inferentialist might take here is to say that it is simply not possible to challenge the classical meaning of logical expressions such as "and" or "not" in the sense that they would fail to be genuinely meaningful. Boghossian [22] makes an argument of this form, suggesting that all that is required to immunise the inferentialist against objections from circularity is that "and" means what we all take it to mean. However, Glüer [36] has pressed an objection against this view, to the effect that this entails that we are required to know that the classical rules are

(Footnote 18 continued)
(according to Peacocke). The problem is that Peacocke stipulates as an additional premise that the implicit conception of "or" is: (α or β is true iff either α is true or β is true) [32, p. 46], so the regress is reinstated. Similar problems occur for Peacocke's [33] qualification that it is a fundamental "rule of reference" that determines meaning, since, there the constraints on grasping meaning are even more substantive than those discussed here.

[19]Loosely, a paraconsistent logic is just one in which the inference $\alpha, \neg\alpha \vdash \beta$ (where β is an arbitrary formula) is invalid.

[20]Note that, if, as Priest [34] suggests, a difference in truth-conditions is taken to be sufficient for a difference in meaning, then a representationalist account (such as Williamson's) will equally face difficulties regarding similarity of meaning for proponents of classical and paraconsistent logics.

[21]That is, I accept that the arguments put pressure upon the attempt to justify certain inferential rules in terms of the dispositions of ordinary agents with expressions of natural languages.

[22]See also the discussion in [35].

valid on the meaning of "and" prior to knowing that "and" means what we take it to mean, that is, if "and" is to be understood as the content that makes those rules valid.

Without getting ahead of ourselves, we might also think that this calls into question a prescriptive and monological approach to inferentialism. For our present purposes, the issue is that a view such as Boghossian's commits an agent who understands the meaning of a logical expression to being disposed to make an inference in the correct linguistic context, as a kind of directive such as: "on assuming α, infer β!". Whereas, as suggested above, there are many cases of ordinary reasoning where agents do not infer β from α. Given this, we may well anticipate that any attempt to straightforwardly justify some specific formal inferential rules by appeal to ordinary reasoning will be tricky, to say the least.

3 Paradoxical Connectives

A slightly different, though equally well-trodden, problem that also pushes inferentialism towards semantic justification comes in the form of the paradoxical connective, "tonk" [37]. Tonk is a binary connective defined as follows:

$$\frac{\alpha}{\alpha Tonk\beta}\ (Tonk\text{-I}) \qquad\qquad \frac{\alpha Tonk\beta}{\beta}\ (Tonk\text{-E})$$

Tonk appears to have been provided with a definition, but successive application of the rules (as long as the logical structure is transitive) allows us to infer any conclusion from any premise:

$$\dfrac{\dfrac{\alpha}{\alpha Tonk\beta}\ (Tonk\text{-I})}{\beta}\ (Tonk\text{-E})$$

Something has gone awry with the definition. Tonk trivializes inferential practice since it allows a thinker to infer arbitrary propositions from any premise, and so it does not seem to express a meaning. It is, therefore, necessary to provide a way of discerning which rules of inference confer genuine meanings on a logical connective. The attempt to figure out exactly what has become a philosophical industry involving the various constraints one might place on a definition to avoid the problem.[23] A number of possibilities for ruling out such paradoxical connectives are available in the literature. For example, one suggestion [40] is that the rules must give a conservative extension whereby a logical constant introduced into the vocabulary should not allow for an inference not involving the new constant to be deducible that was not deducible before the constant was added (e.g. [6, p. 454]). Tonk clearly fails on this account, but, given however that conservative extension is relative to the underlying logical system, this does little to settle the problem of whether or not it expresses a genuine meaning. As Enoch and Schecter [28] put it:

[23] See, for example, [24, 38, 39].

> [...] considerations of conservativeness only apply to a concept given some inferential background; whether a rule is a conservative extension can only be evaluated with respect to some presupposed derivability relation.

For example, as Priest [34] points out, adding the inferential rules for Boolean negation to classical logic is conservative, but it is not for intuitionist logic. There are also well-known occasions where conservative extension fails, such as for second-order logic, which are nonetheless taken to be valid.[24] The issue with such paradoxical rules is that we are pushed towards thinking that proof-theoretic validity may well serve to give us an epistemological grip on certain inferences according to a proof-theory, but these may well bring with them unwanted consequences, potentially trivializing the derivability relation over a language. The lesson, according to Prior [37], is that a semantic justification of logical implications must be given in advance of their inferential characterization, since it is entirely plausible that a set of rules will fail to be valid. If true, this would not, necessarily, mean that we must leave the ambit of inferentialist approaches behind, since, as we now discuss, there are "modest" forms of inferentialism which are perfectly compatible with certain kinds of representationalism.

4 Modest Inferentialism and Its Problems

The kind of thought described above leads us back to considerations regarding the semantical characterisation of consequence with which we began. However, it is possible, as suggested above, to maintain an inferentialist approach whilst combining this with a truth-conditional semantics. For example, there is a suggestion, perhaps most strongly argued for by Christopher Peacocke [25, 32, 42, 43] (taking up a response originating in [38, 39]), which suggests that certain rules may fail because they are not sound with respect to truth-conditional semantics. Whilst the details of these earlier responses have an air of too-swiftly leaving aside motivations for the proof-theoretic approach, Peacocke attempts to resolve this issue by ensuring that no assumptions are made regarding the underlying semantics other than their interaction with the relevant inference rules. What results is a *modest* form of inferentialism where a specified set of inference rules determine the truth-conditional content of logical constants. Let us momentarily set aside our worries regarding dispositions and primitive compulsions, so that we may discuss this as a possible response to problems of circularity and paradoxical connectives.[25] This is to combine the benefits of inferentialist and representationalist approaches so that, roughly, a specified

---

[24] For example, [41] discusses the way in which second-order quantifiers may lead to systems that are not conservative extensions of e.g. Peano arithmetic.

[25] It is plausible, for example, to hold a form of modest inferentialism in which a set of formal rules determine truth-conditional semantics without taking a stand upon how those rules are related to our reasoning practices, e.g. [44–46]. Note, though, that, whilst this may be thought to deal with paradoxical connectives, it would not dispel our concerns with the circularity of rule justification.

set of natural deduction rules are taken to determine the truth-conditional content of logical constants. On Peacocke's [24, 25] view, this is put in terms of "reading-off" valuational semantics from inference rules that a thinker finds "primitively compelling". The innovation here is that only inferences that are truth-preserving can be genuinely content-determining. This constraint, as applied to tonk, shows that there is no binary function on truth-values that validates both its introduction and elimination rules, and, hence, there are no coherent semantics for tonk. The introduction rule requires that "α tonk β" is true when α is true and β is false, but the elimination rule requires that when α is true and β is false, "α tonk β" is false. Hence, there is no coherent semantic assignment of truth-values for tonk since there is no truth-function that makes tonk-inferences truth-preserving. Then, as Boghossian [47] has it; there can be no determinate way the world has to be, if "α tonk β" is to come out true. Resultantly, this not a "pure" inferentialism, but rather a modest form of inferentialism, in which truth-conditional semantics still plays a role in ensuring that our meanings do not misfire. According to Peacocke, this is practically forced on us by considerations regarding tonk-like connectives:

> [w]hat I have argued equally implies that the roles that determine genuine warrant cannot be picked out unless we rely on reference and truth. Even if mere warranted assertibility were the aim of judgement, we would still need to say which sets of rules determine genuine meanings and which do not. Unless this theorist has some new resource for ruling out spurious meanings, he will need to rely on considerations having to do with reference, semantic value and truth; and then his conceptual role theory is no longer pure. [48, p. 389]

It is not quite clear in Peacocke's work, for example, how exactly a set of inferential rules determine truth-conditional content. He does offer, however, a "determination theory" invoked in [23], with the general requirement that:

> The given rules of inference, together with an account of how the contribution to truth-conditions made by a logical constant is determined from those rules of inference, fixes the correct contribution to the truth-conditions of sentences containing the constant. [49, p. 172]

The trouble with Peacocke's account, however, is that we are given little more than a means of checking the credibility of a definition (in his preferred classical natural deduction setting) to see if there is a corresponding classical truth-function. Peacocke takes the existence of a meaningful definition to depend upon the existence of an associable Boolean truth-function. For example, to check that, the classical inference rules define conjunction, we need only look around for a binary-function making those rules truth-preserving under all assignments [25, p. 18ff]. Take, for example, the classical sequent rules for conjunction, where, following Gentzen's suggestion, the commas on the left of $\vdash$ are interpreted as "and", and those on the right as "or":

$$\frac{\Gamma, \alpha, \beta \vdash \Delta}{\Gamma, \alpha \wedge \beta \vdash \Delta}\ \wedge\text{-L} \qquad\qquad \frac{\Gamma \vdash \Delta, \alpha \qquad \Gamma \vdash \Delta, \beta}{\Gamma \vdash \Delta, \alpha \wedge \beta}\ \wedge\text{-R}$$

We can read these as telling us the conditions under which it is legitimate to derive a conjunction from its conjuncts (and a conjunct from a conjunction). But, we can equally read the rules as placing constraints upon when a conjunction may be judged to be true or false, given the truth or falsity of its conjuncts. On the latter, we can extract, from the rules, the classical truth-function for $\wedge$:

$$f^{\wedge}(x, y) = \begin{cases} 1 & if\ x = 1\ and\ y = 1 \\ 0 & otherwise \end{cases}$$

Peacocke fails to offer any formal account of how the determination can be made precise or generalisable (see for example, [25]). As such, let us consider a formal framework that does this, if only in order to point out a number of problems with the position.

First, let us provide a generalised and suitably abstract definition of a "logic" which we will think of a structure of entailment, rather than a consequence operation, since we are primarily interested in an inferentialist account.[26]

Definition 1 Let S be a set of well-formed formulae in some propositional language $\mathcal{L}$, and $\vdash$ is a binary relation $P(S) \times P(S)$ (where $P(S)$ is the set of all finite subsets of S). Then, let a "logic" L, (as entailment structure) be just an ordered pair, $(S, \vdash_L)$. We say that L is normal whenever $\vdash_L$ is reflexive, transitive, monotonic and finitary over each element of S.

Entailment structures can be restricted in different ways, which we shall think of in terms of sequents in a logic.[27]

Definition 2 A sequent is an ordered pair, $\langle \Gamma, \Delta \rangle$ where Γ, Δ are finite (possibly empty) sequences of formulas of S. Say that a right-asymmetric sequent (Γ, α) is restricted to at most a single formula on the right; a left-asymmetric sequent (α, Γ) is restricted to at most a single formula on the left, and a symmetric sequent (Γ, Δ) has no such restrictions. A sequent rule $\mathcal{R}$ in any logic L is an ordered pair consisting of a finite sequence of sequent premises and a sequent conclusion $\mathcal{R} = \langle \{SEQ^P\}, SEQ^C \rangle$, and, in case the list of premises is empty, the instance of a rule is called an axiom.

In this way, we may think of a specific logic to be determined by a proof structure (set of axioms and sequent rules), where any collection of sequents $\mathcal{S}$ that is closed under standard structural rules determines a finitary, normal, logic. For example, it is the case that $\Gamma \vdash \alpha$ iff for some finite $\Gamma_0 \subseteq \Gamma$, we have $(\Gamma_0 \vdash \alpha) \in \mathcal{S}$.[28]

There are a number of ways in which we can think of truth-conditional semantics as being determined by a set of rules, but, to my mind at least, the most obvious from an inferentialist point of view is to construct it from chains of entailments in a logic. In this setting, we are only interested in standard, right-asymmetric proof-structures, so we can give a simple definition in terms of closed and consistent theories of a logic.

---

[26]For similar approaches to logic as abstract structures, see [50–52].

[27]On the relationship between sequent calculus and natural deduction frameworks, see [53, 54].

[28]For symmetric sequents, this is $\Gamma \vdash\dashv \Delta$ iff for some finite$\Gamma_0 \subseteq \Gamma$, $\Delta_0 \subseteq \Delta$ we have $(\Gamma_0 \vdash\dashv \Delta_0) \in \mathcal{S}$. See also [50, p. 113]. Note that I use $\vdash\dashv$ rather than typical $\vdash$ for symmetric sequents to highlight that they can be read in both directions as discussed in the following chapter.

Definition 3 (Closure) Say that a theory is some subset, $\Sigma \subseteq S$ which is $\vdash_L$-closed, for some L, when, for all $\Sigma \vdash_L \alpha$, $\alpha \in \Sigma$.

Then, let us define completeness and consistency for a logic L.

Definition 4 A logic L is consistent when there is no formula α, such that $\emptyset \vdash_L \alpha$ and $\alpha \vdash_L \emptyset$.

We also require a general definition of a maximal theory for a logic. The obvious thing to do would be to follow a standard Lindenbaum-Asser construction, but we do not necessarily want to prejudice the construction towards any specific set of proof-theoretic constraints on L, so we do not want to employ classical negation. To deal with this, we utilise what Béziau (originally developed in [55]) terms a relatively maximal theory.[29]

Definition 5 For a theory Γ, a formula α, and a relation $\vdash_L$, say that Γ is relatively maximal with respect to α in L when $\Gamma \nvdash_L \alpha$, and, for all $\beta \notin \Gamma$, $\Gamma, \beta \vdash_L \alpha$.

The first clause ensures that Γ avoids the formula α, and the second ensures maximality. Since, when Γ is closed, we have the equivalence $\Gamma \vdash_L \alpha$ iff $\alpha \in \Gamma$, any closed set Γ obeying the above clauses is a relatively maximal theory. Then, any proper superset of Γ will be trivial, such that, for any $\alpha \notin \Gamma$, $\Gamma \cup \{\alpha\}$ is trivial. This allows us to state the following theorem.

Theorem 6 *For any finite normal logic L, given a formula β, and a theory Γ (where $\Gamma, \beta \in S$) such that $\Gamma \nvdash \beta$, Γ can be extended to Γ' (where $\Gamma \subseteq \Gamma'$ and Γ' is relatively maximal with respect to β (in L) so that for no proper superset Γ'' of Γ' do we have $\Gamma'' \nvdash \beta$.*

Here is a sketch of a proof.

Proof Take the enumerable formulas of S, $\{\alpha_1, \alpha_2 ... \alpha_i, \alpha_{i+1}\}$, and a theory $\langle\Gamma\rangle$ (which is closed under $\vdash$, and $\Gamma \subseteq S$), where $\Gamma \nvdash \beta$. Then, for some formula $\alpha_i \in S$, we know that either $\Gamma \cup \{\alpha_i\} \nvdash \beta$, or $\Gamma \nvdash \beta \cup \{\alpha_i\}$. When the former is the case, we can extend Γ to $\Gamma \cup \{\alpha_i\}$. In other words, by induction on the enumerable formulas of S, we can build up the relatively maximal theory Γ', which avoids each $\alpha_i \in S \backslash \Gamma'$ through the construction of a chain:

i. $\langle\Gamma_n\rangle = \langle\Gamma\rangle$

ii. $\langle\Gamma_{n+1}\rangle = \begin{cases} \langle\Gamma_n \cup \{\alpha_{i+1}\}\rangle \text{ if } (\Gamma_n \cup \{\alpha_{i+1}\} \nvdash \beta) \\ \langle\Gamma_{n+1} = \Gamma_n\rangle \text{ otherwise.} \end{cases}$

The limit of this construction is:

$$\langle\Gamma'\rangle = \bigcup_{n \in I} \Gamma_n$$

It is simple to see that Γ' is relatively maximal *w.r.t* β in L, since, on the above chain, we have $\Gamma_n \nvdash \beta$, and, so $\Gamma' \nvdash \beta$. If not, there must be some finite subset $\Gamma \subseteq \Gamma'$,

[29]Thanks to an anonymous referee for pointing me towards the original source.

for which $\Gamma \models \beta$ (by finiteness). This means that, for some $\Gamma_{n+i} \supseteq \Gamma$, $\Gamma_{n+i} \vdash \beta$ (by$\mathbb{M}$), which contradicts the definition of Γ_{n+i}, (where $\Gamma_{n+i} \nvdash \beta$). Then, take a formula α_i, where $\alpha_i \notin \Gamma'$. By definition, $\Gamma_{n+1} \subseteq \Gamma'$, so $\alpha_i \notin \Gamma_{n+1}$. Then, we know that $\Gamma_{n+1} = \Gamma_n$, and $\Gamma_n, \alpha_i \vdash \beta$. By monotonicity, and the fact that $\Gamma_n \subseteq \Gamma'$, we have $\Gamma', \alpha_i \vdash \beta$. □

In [56], it is proven that a standard semantics for a logic, L, may be constructed in this way by simply taking the characteristic function of relatively maximal theories for L as follows: given a sequent $\alpha_1, \ldots, \alpha_n \models \beta$ in a logic L, and a relatively maximal theory Γ' of L, we say that Γ' satisfies this sequent iff, whenever $\Gamma' \models \alpha$, for each $\alpha_1, \ldots, \alpha_n, \Gamma' \models \beta$.

With this in place, we have a formal construction for modest inferentialism, with which we can highlight two problems that occur for it. The first issue is the well-known "categoricity" problem [50, 57–62], which has the result that the standard inferential rules for classical logic fail to rule out non-standard interpretations, so they do not suffice to determine the meaning of logical constants. In particular, that framework is easily shown to be sound and complete with respect to *both* the classical semantic model, and a model in which every formula is interpreted "true". It is also the case, as pointed out in [58], that a valuation can be defined such that $\alpha \vee \neg\alpha$ comes out as true, whilst each disjunct is false, whilst respecting the standard rules of classical propositional logic. This is also problematic from the point of view of compositionality, which is an issue levelled at modest inferentialism, perhaps most clearly in [11, 12], that it fails to conform to certain intuitively correct constraints upon the compositionality of language. In brief, the claim there is that compositionality requires that the content of complex concepts is fully derivable from the content of their constituent concepts. But, it is simple to show that the rules defining disjunction, for example, do not ensure that the truth-values of the sub-formulas for a formula $\alpha \vee \beta$ always determine the truth-value of that formula.[30]

Take the rules ordinarily used to define classical negation (in a right-asymmetric entailment structure):

$$\frac{\Gamma, \neg\alpha \models \beta \qquad \Gamma, \neg\alpha \models \neg\beta}{\Gamma \models \alpha}\,(Reductio) \qquad \frac{\Gamma \models \alpha \qquad \Gamma \models \neg\alpha}{\Gamma \models \beta}\,(EFQ)$$

The issue is that there are possible relatively maximal theories where both α and $\neg\alpha$ are in Γ', and which are not ruled out by the rules used above (essentially, this occurs when every formula is in Γ'). Now, take $\vee$ as example. Sticking with the natural deduction form for a moment, we schematise $R^{\vee}$ as follows:

$$\frac{\alpha}{\alpha \vee \beta}\,\vee I^{\alpha} \qquad\qquad \frac{\beta}{\alpha \vee \beta}\,\vee I^{\beta}$$

$$\frac{\alpha \vee \beta \qquad \begin{matrix}[A^u] \\ \sigma\end{matrix} \qquad \begin{matrix}[\beta^v] \\ \sigma\end{matrix}}{\sigma}\,\vee E^{u,v}$$

[30] See [46, 63].

The issue arises for cases in which α and β are not in Γ' for the rule $E^{u,v}$. In this case, we can not ensure that $\alpha \vee \beta$ is not in Γ' since we are able only to conditionally infer σ from $\alpha \vee \beta$, given independent sub-derivations to σ (which will not figure in the immediate sub-formulas of complex formulas involving $\vee$). In other words, what we do not have in schematic form is all of the relevant information encoded within the immediate sub-formulas involved in the derivation. What $E^{u,v}$ tells us is just that, if there are proofs available from α to σ and β to σ, then we have proofs of $(\alpha \supset \sigma)$, and $(\beta \supset \sigma)$. With these, and the disjunction elimination rule we can then show only $(\alpha \vee \beta), (\alpha \supset \sigma), (\beta \supset \sigma) \vdash \sigma$. With this in mind, we may rewrite the rules in sequent form (again, call this $R^{\vee}$):

$$\frac{\Gamma \vdash \alpha \vee \beta \qquad \Gamma, \alpha \vdash \sigma \qquad \Gamma, \beta \vdash \sigma}{\Gamma \vdash \sigma}\ \vee\text{-E} \qquad \frac{\Gamma \vdash \alpha}{\Gamma \vdash \alpha \vee \beta}\ \vee\text{-I}\alpha$$

$$\frac{\Gamma \vdash \beta}{\Gamma \vdash \alpha \vee \beta}\ \vee\text{-I}\beta$$

However, this won't fix matters, since, again, we have a situation in which there are relatively maximal theories agreeing on the immediate sub-formulas, α, β, but not on the formula $\alpha \vee \beta$ itself. Again, the issues arise with the elimination rule, which only gives us something along the lines of: if we can infer σ from α, and we can infer σ from β, and $\alpha \vee \beta$ is in Γ' then σ is in Γ'. If we consider the derivations of σ from α, β, however, we require for $\Gamma, \alpha \vdash \sigma$ only that $\sigma \notin \Gamma'$ when $\alpha \in \Gamma'$. Then, for $\Gamma, \alpha \vdash \sigma$ (and equally, $\Gamma, \beta \vdash \sigma$), we need either that $\alpha \notin \Gamma'$, or $\sigma \in \Gamma'$. This provides a counterexample to compositionality for $R^{\vee}$ since we need only find an pair of relatively maximal theories Γ^1, Γ^2, that are identical for α, β, whilst $(\alpha \vee \beta) \in \Gamma^1$ and $(\alpha \vee \beta) \notin \Gamma^2$. This is possible when $\alpha, \beta \notin \Gamma^1$ and $\alpha, \beta \notin \Gamma^2$, whilst $\sigma \notin \Gamma^1$ and $\sigma \in \Gamma^2$.[31] For example, if $\alpha = \sigma$, then both of the conditional premises are satisfied by Γ^1, and, in order for Γ^1 to satisfy the rule it must be the case that the major premise is satisfied, and so $(\alpha \vee \beta) \in \Gamma^1$. Hence, there are relatively maximal theories that are equivalent on the sub-formulas but that do not agree on the formula itself, and so $R^{\vee}$ is not compositional. Things becomes particularly interesting when we bring both issues together by considering formulas involving negation such as $\alpha \vee \neg\alpha$, since this may be satisfied by some relatively maximal theory, Γ' whilst neither α nor $\neg\alpha$ are.

We began this section by considering a possible way of dealing with tonk-like issues by appealing to some form of truth-conditional semantics to settle the issue of which rules determine "genuine" meanings, and which do not. Problems of categoricity and compositionality suggest that this is unlikely to provide an adequate response.[32] Resultantly, whilst issues of circularity and paradoxical connectives suggest that an inferentialist account of logical rules inevitably requires semantic justi-

[31] This example is discussed in detail in [44, 46].

[32] In the following chapter I discuss alternative accounts for which these issues do not arise, though others do.

fication, a modest form of inferentialism provides little hope since it fails to provide an account of how rules completely determine meaning, and even where it does, these meanings do not obey intuitive constraints upon compositionality. We may be tempted, then, to jettison inferentialism altogether, but, given the problems (raised above and in the introduction) with the representationalist approach, this looks to provide meagre consolation. In the following section, we turn to an alternative, stringently inferentialist, suggestion, in which it is argued that internal features of rules suffice to determine genuine logical meanings, and so may be thought of as "self-justifying".

5 An Alternative in Harmony?

Rather than look to soundness *w.r.t* a semantic model as a deciding factor for the consideration of logical rules, a fairly common alternative response is to consider internal, syntactic, constraints such as conservativeness (e.g. [40]); inversion principles (e.g. [64]); harmoniousness (e.g. [7, 65]). Whilst these approaches have significant differences, they share the view that introduction and elimination rules in natural deduction calculi (or left and right rules in sequent calculi) are in balance with each other. Gentzen [66], for example, puts this as follows:

> The introductions represent, as it were, the "definitions" of the symbols concerned, and the eliminations are no more, in the final analysis, than the consequences of these definitions. This fact may be expressed as follows: In eliminating a symbol, we may use the formula with whose terminal symbol we are dealing only in the sense afforded it by the introduction of that symbol. (p. 80)

This is developed by the idea of an inversion principle, in which a rule is considered invertible if (broadly speaking) its conclusion implies its premises. Importantly, this constraint requires that anything "extraneous" to a proof can be eliminated through normalization (or cut-elimination) such that Gentzen's *Hauptsatz* can be proved for the logical system.[33] Dummett (e.g. [7]) uses the term "harmony" to describe this constraint upon genuine logical rules, suggesting that, for rules to be meaningful requires that the grounds for making an assertion should be in harmony with the consequences that we are entitled to draw from it. For example, in [7], Dummett argues that it is possible to think of certain basic logical rules as self-justifying by means of introduction rules for a logical connective providing what is called the *canonical* way of introducing an expression involving a logical connective. For example, a canonical argument for $\alpha \wedge \beta$ must end with an application of $\wedge$-introduction (i.e. an inference from α, β to $\alpha \wedge \beta$). Then, say we want to know if the inference from $\alpha \wedge \beta$ to β is valid, all we need do is look for a canonical argument for β, which we already have to hand since we have a canonical argument for the premise, $\alpha \wedge \beta$.

---

[33] As developed by the so-called Curry-Howard correspondence between proofs and programs (e.g. [67]) the inversion principle has computational significance, which we discuss in more detail in Chap. 6.

Thus, according to Dummett, the inference from $\alpha \wedge \beta$ to β must be valid because it is possible to take the argument for the premise $\alpha \wedge \beta$, and transform it into a canonical argument for β. This balance is given a fairly intuitive spin by Dummett [7], and Tennant [68], the latter of which puts it as follows:

> In the natural-deduction setting the constitutive balance can be described as follows. The introduction rule states the conditions under which a conclusion with that operator dominant can be inferred. It describes, if you like, the obligations that have to be met by the speaker in order to be justified in asserting the conclusion in question. The corresponding elimination rule states what the listener is entitled to infer from the speaker's assertion. (p. 628)

So, bluntly, the balance internal to logical rules is also supposed to track central features of agents' use of logical expressions. Moreover, this restriction to harmonious rules is supposed logical meanings with a form of "self-justification" as Tennant [68] has it:

> [...] a logical word's being governed by harmonious rules is what confers upon it its precise logical sense. The rules have to be in place for the word to mean what it does. There is no independent access to the meaning of the word, with respect to which one could then raise the question whether the rules governing it 'respect' that sense, or properly transmit truth to and from sentences involving it. (p. 629)

So, perhaps, in this way, the justification of logical rules may be achieved by imposing restrictions on the admissibility of certain rules. Whilst tonk falls by the wayside since it is clearly non-harmonious, this is tricky, since, analogous to the justification of non-deductive rules raised above, so too there seem to be harmonious, yet non-deductive rules. Then, harmony, by itself, does not seem adequate to demarcate those rules which are deductive from those that are not. Perhaps most famous, are the concepts "flurg" and "aqua", which have introduction and elimination rules as follows:

$$\frac{\text{x is water}}{\text{x is aqua}}\ \text{Aqua-I} \qquad\qquad \frac{\text{x is aqua}}{\text{x is } H_2O}\ \text{Aqua-E}$$

and;

$$\frac{\text{x is an elliptical equation}}{\text{x is flurg}}\ \text{Flurg-I}$$

$$\frac{\text{x is flurg}}{\text{x can be correlated with a modular form}}\ \text{Flurg-E}$$

As Enoch and Schecter [28] put it;

> Simply introducing the terms "aqua" and "flurg" with their associated rules of inference is insufficient for being justified in employing those rules. Had it been, science and mathematics would have been much easier than they actually are.[34]

Steven Read [70] has also defined a harmonious, yet inconsistent pair of introduction of elimination rules for the operator •. In brief, he defines the introduction rule for • as:

---

[34] Though see (e.g. [69]) for discussion of possible responses.

$$\begin{array}{c}[\bullet\alpha]\\ \dfrac{\alpha}{\bullet\alpha}\ \bullet\text{-I}\end{array}$$

which, yields the clearly harmonious elimination rule:

$$\dfrac{[\bullet\alpha] \qquad [\bullet\alpha]}{\alpha}\ \bullet\text{-E}$$

But, as Read points out, the pair quickly lead to triviality (as long as standard structural rules exist in the system), since we can prove α for any α:

$$\dfrac{\dfrac{\dfrac{\bullet\alpha \qquad \bullet\alpha}{\alpha}\,\bullet\text{-E}}{\bullet\alpha}\,\bullet\text{-I} \qquad \dfrac{\dfrac{\bullet\alpha \qquad \bullet\alpha}{\alpha}\,\bullet\text{-E}}{\bullet\alpha}\,\bullet\text{-I}}{\alpha}\,\bullet\text{-E}$$

As such, Read argues that harmony is insufficient to ensure non-triviality. Furthermore, as with conservative extension, harmoniousness is arguably dependent upon the underlying structure of the proof system employed. So, as Priest [34] points out, harmony does little to justify the structure itself. Therefore, it looks like, whilst harmony may be a decent requirement on a rule, it is insufficient for ensuring the validity of that rule.

Even so, let us momentarily accept that some form of harmony does indeed establish a condition on the admissibility of logical rules. Even then, it is difficult to see how this approach fares much better when it comes to avoiding the circularity identified above. For example, even given harmony, we are surely still warranted in asking why, for example, the set of introduction rules that were considered to define a meaningful assertion are correct to begin with. Otherwise, we would also require a way of accounting for the way in which agents correctly grasp a logical rule, and these will lead us back to the circularity arguments with which we began. Unless this can be shown to bottom out somewhere, regress follows from the fact that the justification of a logical rule relies on correct application of an initial rule (presumably an introduction rule).[35]

In the context of discussions of balance and harmony, it is often suggested that these (amongst other factors) should cause us to consider constructive, rather than classical, logics. In the terminology introduced above, this is usually to take the sequents in right-asymmetric form to define intuitionistic, rather than classical, logic. We will examine the account in detail in the following chapter, but, in brief, in this setting, the validity of a proof is not dependent upon an interpretation (even one developed in terms of relatively maximal theories), but rather on local constraints such as canonicity and harmony (e.g. [71]). This provides additional reasons for thinking that cut-elimination procedures (and harmony) are important in considerations of validity, since, according to Dummett's (e.g. [7]) “fundamental assumption”’if we have a valid argument for a complex statement, we can construct a valid argument for it which finishes with an application of one of the introduction rules governing its principal operator' (p. 254). This way of thinking about valid proofs is not appropriate

[35]We will return to a discussion regarding the reasons for considering harmony beyond the typical attempt to show that certain rules are self-justifying in the following chapter, by considering the bifurcation of proofs as “objects” and proofs as “acts”.

to classical logic because canonicity requires the disjunction property to hold: a proof of a disjunction must be given in (or reduced to) canonical form, which requires that it is also possible to provide a proof of one of the disjuncts. A lack of disjunction property is problematic from the *p.o.v* of constructing a proof-theoretic semantics, since we require an ability to determine, in a fine-grained manner, the manner in which a proof is "valid" Moreover, if we accept that harmony is a necessary condition on logical rules then standard rules for classical negation may fall by the wayside with tonk (e.g [7, 65]). For example, the classical rule for negation elimination;

$$\frac{\neg\neg\alpha}{\alpha}\ \neg\text{-E}$$

is not in harmony with the rule of negation introduction[36]:

$$\frac{\begin{matrix}[\alpha]\\ 0\end{matrix}}{\neg\alpha}\ \neg\text{-I}$$

whilst the intuitionistic rule for negation elimination is in harmony with this introduction rule[37]:

$$\frac{\alpha \qquad \neg\alpha}{0}\ \neg\text{-E}$$

In [72] it is argued that the problem of categoricity is avoided by this manoeuvre. In this context, the problematic situations would be a scenario in which every statement is provable is compatible with the inferential rules, or it is possible for $\alpha \vee \neg\alpha$ to be provable, whilst neither α nor $\neg\alpha$ are.[38] The reason that these situations can not arise, according to [72], is that there can be no canonical proof of 0, since there exists only the null introduction rule for 0 and the elimination rule is just the absurdity rule (e.g. [74]):

$$\frac{0}{\alpha}\ 0\text{-E}$$

As such, this is rules out a scenario in which there exists a proof of both α and $\neg\alpha$ (where $\neg\alpha$ is equivalent to $\alpha \Rightarrow 0$). But, as pointed out in [75], the self-justification of this rule makes use of the rule for 0-elimination, unlike other rules for the connectives. The sticking point is that:'one also needs to regard, controversially, the rule of 0-elimination as justified on the basis of a non-existent rule of 0-introduction, taken as

[36]Note that I use 0 in place of the usual $\bot$ throughout.

[37]The issue of intuitionistic negation, and also of proof-systems that are classical for which the negation rules are harmonious will be discussed in the following chapter.

[38]We might, instead, take an approach analogous to the "modest" inferentialism discussed for the classical logician above. Since Kripke-style semantics are typically taken to be standard for intuitionistic logic, this would take the idea that the sequent rules determine valuations in a Kripke-model. But, in this context, as discussed in detail in [46], problems of categoricity (for negation) and compositionality (for disjunction) are not assuaged. The latter is perhaps most surprising given the canonicity requirement (that a proof of "$\alpha \vee \beta$" requires a canonical proof of "$\alpha \vee \beta$" from either α or β). However, as [72] point out, this may be put down to an insistence, by Dummett, for example, that Kripke semantics do not adequately specify the semantics of connectives; they'are not to be thought of as giving the full picture of the way in which the intuitionistic logical constants are given meaning: that can only be done directly in terms of the notion of a construction and of a construction's being recognized as a proof of a statement'. [73, p. 287].

saying that there is no canonical proof of 0' [75]. So, it is difficult to maintain the requirement that the rules defining 0 are harmonious.[39] An alternative introduction rule, suggested by Dummett [7], and discussed in [75], is:

$$\frac{\alpha, \beta, \sigma \ldots}{0}\ \text{0-I}$$

The idea being that the premises of the rule include every atomic sentence of $\mathcal{L}$, and, since 0-E effectively allows us to infer any atomic sentence, the rules are in harmony. This is not so simple, however, since this alternative rule does, in fact, allow for situations in which proofs for α and $\neg\alpha$ may exist (specifically, where there exists a proof for each atomic formula of the language). This is similar to the argument made in [77], where it is pointed out that, whilst the 0-elimination rule only tells us that anything may be inferred from 0, this does not ensure that 0 has the meaning of false.[40] For example, it is possible to consider a language in which all atoms are true, and in which case 0 will be true rather than false, and in which case the 0-elimination rule does not determine the (intended) meaning of 0. Of course, this means that $\neg\alpha$, defined as $\alpha \Rightarrow 0$, must also be true by vacuous discharge, and so, in this language we would have both α and $\neg\alpha$ true. Of course, we could just stipulate that 0 "means" false, so that these situations can not arise, but this would not be in keeping with the idea that we are defining the behaviour of 0 by means of a set of rules for negation. So, again, the rules that are supposed to determine the meaning of negation, intuitionistically, do not do so.

6 Soundness and Completeness

For all of the aspirations of inferentialist approaches to meaning, the above, albeit fairly brief, survey, suggests that matters are not so simple. Indeed, we may well be attracted to attempts to mitigate problems facing justification of rules by means of the semantic justification of logical implication, by which we then provide justification for certain rules, i.e. those which preserve truth in all semantic models. But, as we argued in the introduction, and also in §2, this is no salve whatsoever. Nevertheless, the feeling that this is the correct way of construing logic is deeply held. For example, Read [79] forcefully argues this:

> What is good about the notion of proof-theoretic validity is that it recognises that what rules one adopts determines the meaning of the logical terms involved and commits one to accepting certain inferences as valid. What is bad is to infer from this that those inferences really are valid. Proof-theoretic validity serves an epistemological function to reveal how those inferences result from the meaning-determining rules alone. But it cannot serve the metaphysical function of actually making those inferences valid. Validity is truth-preservation, and proof must respect that fact.

[39] A detailed discussion can be found in [76].

[40] The argument is rehearsed in [78].

Typically, model-theoretic consequence is thought to delineate the correctness of inference because it is (in some sense) reducible to the categorical notion of truth-preservation. In rough:

β is derivable from $\alpha_1, \alpha_2 \ldots \alpha_n$ whenever β is a logical consequence of $\alpha_1, \alpha_2 \ldots \alpha_n$..

In the Bolzano-Tarski tradition, this is typically thought of as follows. If Γ is a theory of $\mathcal{L}$, and V is a semantic structure (typically thought of as a model) for $\mathcal{L}$, then V is a model of Γ: $V \models \Gamma$ if, for every $\alpha \in \Gamma$, $V \models \alpha$. And, α is a logical consequence of Γ ($\Gamma \models \alpha$) if, for every model V of Γ, $V \models \alpha$. Inferential derivability, on the other hand, is usually understood as derivability in a formal system (sequent calculus; natural deduction; Hilbert system), where $\Gamma \vdash_L B$ is valid in a formal system L, whenever B can be derived from Γ by means of the axioms and inference rules of L.

As is instilled in any introductory logic course, soundness and completeness ensure that the proof-theory is both correct (according to the semantics), and that it is also strong enough to prove any valid entailment. So, the story goes, the entailment structure gives us epistemic access to validity in the form of proofs, derivable sequents, and so on. But, we require the semantic structure to tell us where the counter-models are, lest we attempt an invalid proof. So, typically, when considering the relationship between an entailment structure and a semantic model, counter-models play the role of ruling out invalid formulae. For example, take Γ^+ as a set of sentential theorems, V as some model, with $\vdash$ a derivability relation, and $\models$ a model-satisfaction relation. Then, we ordinarily require that, for every formula α: Either $\exists\Gamma^+(\Gamma^+ \vdash \alpha)$ or, $\exists V(V \not\models \alpha)$.[41] That is to appeal to the idea that a sequent $\Gamma \vdash \alpha$ is invalid iff some model V makes all the formulas of Γ true, whilst making α false. It is precisely this appeal to counter-models that is not in keeping with a proof-theoretic approach to semantics, but it is also what it looks as though is required given the inability of the right-asymmetric entailment structures to rule out inadmissible cases. So-called paradoxical connectives such as "tonk" [37] provide grist to the mill for this order of priority. Given this, there is a residual feeling that we are not able to provide an account of "real" validity limited to entailment structures alone.[42] Nonetheless, as we have seen, neither a standard proof-theoretic, nor semantic, account of logical rules gives us an adequate solution to the problem of their justification. Furthermore, attempts to bring the two together in so called "modest" versions of inferentialism expose significant fissures between entailment structures and semantics that are easily missed by focusing upon soundness and completeness results alone. Problems facing proof-theoretic validity of rules pushes us toward a semantic justification in terms of logical consequence; in turn, this pushes us back towards the problems originally identified with inferentialist approaches to the validity of logical rules. We are in a tight circle indeed.

[41] This is just a form of Gödel's completeness theorem.

[42] As pointed out in the introduction, more broadly, these are linked to the issues raised by Chinese-room style arguments.

References

1. Robert Brandom. *Making It Explicit: Reasoning, Representing, and Discursive Commitment.* Harvard University Press, 1994.
2. Robert Brandom. Asserting. *Noûs*, 17(4):637–650, 1983.
3. Ned Block. Conceptual role semantics. In Edward Craig, editor, *Routledge Encyclopedia of Philosophy*, pages 242–256. Routledge, 1998.
4. Wilfrid Sellars et al. Empiricism and the philosophy of mind. *Minnesota studies in the philosophy of science*, 1(19):253–329, 1956.
5. Ludwig Wittgenstein. *Philosophical Investigations, 4th Edition (Trans. Hacker and Schulte).* Wiley-Blackwell, 2009.
6. Michael Dummett. *Frege: Philosophy of Language*. Duckworth, 1973.
7. Michael A. E. Dummett. *The Logical Basis of Metaphysics*. Harvard University Press, 1991.
8. Michael A. E. Dummett. *Origins of Analytical Philosophy*. Harvard University Press, 1993.
9. Shahid Rahman, Giuseppe Primiero, and Mathieu Marion. *The realism-antirealism debate in the age of alternative logics*, volume 23. Springer Science & Business Media, 2011.
10. Jerry A Fodor. *The language of thought*, volume 5. Harvard University Press, 1975.
11. Jerry Fodor and Ernest Lepore. Why meaning (probably) isn't conceptual role. *Mind & Language*, 6(4):328–343, 1991.
12. Jerry A Fodor. *Concepts: Where cognitive science went wrong*. Clarendon Press/Oxford University Press, 1998.
13. Georges Rey. Semantic externalism and conceptual competence. In *Proceedings of the Aristotelian Society*, pages 315–333. JSTOR, 1992.
14. Christopher Peacocke. *Thoughts: An Essay on Content*. Blackwell, 1986.
15. Mark Eli Kalderon. Reasoning and representing. *Philosophical Studies*, 105(2):129–160, 2001.
16. Gilbert Harman. *Change in View*. MIT Press, 1986.
17. W. v. O. Quine. Truth by convention. In P. Benacerraf and H. Putnam, editors, *Philosophy of Mathematics*, pages 329–354. Cambridge University Press, 1983.
18. Graham Priest. *An Introduction to Non-Classical Logic: From If to Is*. Cambridge University Press, 2008.
19. Carlo Cellucci. The question Hume didn't ask: why should we accept deductive inferences. In C. Cellucci and P. Pecere (eds.), editors, *Demonstrative and Non-Demonstrative Reasoning*, pages 207–235. Edizioni dell'Università, Cassino, 2006.
20. Paul Boghossian. Analyticity reconsidered. *Noûs*, 30(3):360–391, 1996.
21. Paul Boghossian. Knowledge of logic. *New essays on the a priori*, pages 229–254, 2000.
22. Paul Boghossian. Epistemic analyticity: A defense. *Grazer Philosophische Studien*, 66(1):15–35, 2003.
23. Christopher Peacocke. What determines truth conditions? In J. McDowell and P. Pettit (eds.), editors, *Subject, Thought, and Context*, pages 181–207. Oxford, Clarendon Press, 1986.
24. Christopher Peacocke. Understanding logical constants: A realist's account. In T. J. Smiley and Thomas Baldwin, editors, *Studies in the Philosophy of Logic and Knowledge*, page 163. Published for the British Academy by Oxford University Press, 2004.
25. Christopher Peacocke. *A Study of Concepts*. MIT Press, 1992.
26. Corine Besson. Propositions, dispositions and logical knowledge. In M. Bonelli and A. Longo, editors, *Quid Est Veritas? Essays in Honour of Jonathan Barnes*. Bibliopolis, 2010.
27. Saul A. Kripke. *Wittgenstein on Rules and Private Language*. Harvard University Press, 1982.
28. Joshua Schechter and David Enoch. Meaning and justification: The case of modus ponens. *Noûs*, 40(4):687–715, 2006.
29. Timothy Williamson. *The Philosophy of Philosophy*. Blackwell Pub., 2007.
30. Amos Tversky and Daniel Kahneman. Judgments of and by representativeness. Technical report, DTIC Document, 1981.
31. Keith E. Stanovich and Richard F. West. Advancing the rationality debate. *Behavioral and Brain Sciences*, 23(5):701–717, 2000.

32. Christopher Peacocke. Implicit conceptions, understanding and rationality. *Philosophical Issues*, 9:43–88, 1998.
33. Christopher Peacocke. Interrelations: Concepts, knowledge, reference and structure. *Mind & Language*, 19(1):85–98, 2004.
34. Graham Priest. *Doubt Truth to Be a Liar*. Oxford University Press, 2006.
35. C. S. Jenkins. Boghossian and epistemic analyticity. *Croatian Journal of Philosophy*, 8(1):113–127, 2008.
36. Kathrin Glüer. Analyticity and implicit definition. *Grazer Philosophische Studien*, 66(1):37–60, 2003.
37. A. N. Prior. The runabout inference ticket. *Analysis*, 21:38–39, 1960.
38. J. T. Stevenson. Roundabout the runabout inference-ticket. *Analysis*, 21(6):124–128, 1961.
39. Steven Wagner. Tonk. *Notre Dame Journal of Formal Logic*, 22(4):289–300, 1981.
40. Nuel Belnap. Tonk, plonk and plink. *Analysis*, 22(6):130–134, 1962.
41. Dag Prawitz. Book reviews. *Mind*, 103(411):373–376, 1994.
42. Christopher Peacocke. *Being Known*. Clarendon, Oxford, 1999.
43. Christopher Peacocke. *Truly Understood*. Oxford University Press, 2008.
44. James W Garson. *What logics mean: from proof theory to model-theoretic semantics*. Cambridge University Press, 2013.
45. Ole Thomassen Hjortland. Speech acts, categoricity, and the meanings of logical connectives. *Notre Dame Journal of Formal Logic*, 55(4):445–467, 2014.
46. Jack Woods. Failures of categoricity and compositionality for intuitionistic disjunction. *Thought: A Journal of Philosophy*, 1(4):281–291, 2012.
47. Paul Boghossian. How are objective epistemic reasons possible? *Philosophical Studies*, 106(1-2):340–380, 2001.
48. Christopher Peacocke. Three principles of rationalism. *European Journal of Philosophy*, 10(3):375–397, 2002.
49. Christopher. Peacocke. Proof and truth. In Haldane and Wright (eds.), editors, *Reality, Representation and Projection*. New York: Oxford University Press., 1993.
50. Lloyd Humberstone. *The Connectives*. MIT Press, 2011.
51. Arnold Koslow. *A Structuralist Theory of Logic*. Cambridge University Press, 1992.
52. Jaroslav Peregrin. Inferentializing semantics. *Journal of Philosophical Logic*, 39(3):255–274, 2010.
53. Sara Negri and Jan von Plato. *Structural Proof Theory*. Cambridge University Press, 2001.
54. Sara Negri and Jan von Plato. Sequent calculus in natural deduction style. *The Journal of Symbolic Logic*, 66(04):1803–1816, 2001.
55. R. Wojcicki. *Theory of Logical Calculi. Basic Theory of Consequence Operations*. Kluwer, 1988.
56. Jean-Yves Béziau. Sequents and bivaluations. *Logique Et Analyse*, 44(176):373–394, 2001.
57. Nuel D. Belnap and Gerald J. Massey. Semantic holism. *Studia Logica*, 49(1):67–82, 1990.
58. Rudolf Carnap. *Formalization of Logic*. Cambridge, Mass., Harvard University Press, 1943.
59. J Michael Dunn and Gary Hardegree. *Algebraic methods in philosophical logic*. OUP Oxford, 2001.
60. James W. Garson. Expressive power and incompleteness of propositional logics. *Journal of Philosophical Logic*, 39(2):159–171, 2010.
61. Gary M. Hardegree. Completeness and super-valuations. *Journal of Philosophical Logic*, 34(1):81–95, 2005.
62. David J Shoesmith and Timothy John Smiley. *Multiple Conclusion Logic*. CUP Archive, 1978.
63. James Trafford. Compositionality and modest inferentialism. *Teorema: Revista internacional de filosofía*, 33(1):39–56, 2014.
64. Dag Prawitz. *Natural Deduction: A Proof-Theoretical Study*. Dover Publications, 1965.
65. Neil Tennant. *The Taming of the True*. Oxford University Press, 1997.
66. Gerhard Gentzen. *The Collected Papers of Gerhard Gentzen*. Amsterdam, North-Holland Pub. Co., 1970.

67. Morten Heine Sørensen and Pawel Urzyczyn. *Lectures on the Curry-Howard isomorphism*, volume 149. Elsevier, 2006.
68. Neil Tennant. Rule-circularity and the justification of deduction. *Philosophical Quarterly*, 55(221):625–648, 2005.
69. Paul Boghossian. Blind reasoning. *Aristotelian Society Supplementary Volume*, 77(1):225–248, 2003.
70. Stephen Read. Harmony and autonomy in classical logic. *Journal of Philosophical Logic*, 29(2):123–154, 2000.
71. Dag Prawitz. Remarks on Some Approaches to the Concept of Logical Consequence. *Synthese*, 62(2):153–171, 1985.
72. Julien Murzi and Ole Thomassen Hjortland. Inferentialism and the categoricity problem: Reply to Raatikainen. *Analysis*, 69(3):480–488, 2009.
73. Michael A. E. Dummett. *Elements of Intuitionism*. Oxford University Press, 2000.
74. Dag Prawitz. On the Idea of a General Proof Theory. *Synthese*, 27(1-2):63–77, 1974.
75. Luca Incurvati and Peter Smith. Rejection and valuations. *Analysis*, 70(1):3–10, 2010.
76. Neil Tennant. Negation, absurdity and contrariety. In *What is Negation?*, pages 199–222. Springer, 1999.
77. Michael Hand. Antirealism and falsity. In *What is Negation?*, pages 185–198. Springer, 1999.
78. Nils Kürbis. Proof-Theoretic Semantics, a Problem with Negation and Prospects for Modality. *Journal of Philosophical Logic*, pages 1–15, forthcoming.
79. Stephen Read. Proof-theoretic validity. In C. Caret and O. Hjortland, editors, *Foundations of Logical Consequence*. Oxford University Press, 2014.

Chapter 3
Proof and Assertion

In this chapter, I shed light on the issues raised in the above chapter by providing an overview of the concepts of proof and assertion, both of which are central to inferentialist approaches to semantics. In so doing, I shall argue that, in contrast to the prescriptive and monological approach to logic that underlies the problems discussed in the previous chapter, we should think of both proof and assertion in a much more liberal way. In brief, the suggestion is that, to take seriously the dictum that "meaning is determined by use" requires us to provide an account of the processes of our proving activities, which take place in the context of assertion games.

I begin by showing, in Sect. 1, that the problem of categoricity disappears if we move to a symmetrical (multiple-premises and multiple-conclusions) logical structure. Such structures are tricky to interpret, though Greg Restall provides one interpretation in terms of the relationship between assertions and denials, where denial is a symmetrical speech-act to assertion. I argue that this is problematic since: (a) there is no account of the role of assertions and denials provided there; (b) there is no way of giving an account of *proofs* within this framework. This discussion makes central the notion of "proof", and its relationship with inferentialist semantics, so in Sect. 2, I provide an overview of "proofs" and "proving", in terms of the standard bifurcation in that tradition between proofs as objects, and proofs as acts. I argue that for proof to play a central semantic role in a theory of meaning, the latter conception of proof is required. Since this runs against the grain of some standard accounts of assertion, in Sect. 3 I show how the approach to proofs finds a home in social, commitment based, approaches to assertion. On this view, making an assertion is a matter of bringing that assertion into "play", at which point, it is subject to norms involving a commitment to its defense, to providing reasons for it, and allowing it to be "tested" through interaction with other reasons, counterexamples and so on. With this in hand, in Sect. 4, I provide a way of thinking about proofs as acts by means of a distinction between hypothetical and actual proofs, which I use to provide a more substantial basis for a theory of meaning. In this setting, I finish in Sect. 5, by arguing that, refutations ought to be taken to be on a par with proofs, so, to some extent upholding Restall's account, whilst also pointing towards a response to the issues raised against it in Sect. 1.

J. Trafford, *Meaning in Dialogue*, Studies in Applied Philosophy, Epistemology and Rational Ethics 33, DOI 10.1007/978-3-319-47205-8_3

1 From the Monological to the Dialogical

Arguably, as the above discussion clarifies, attempts to self-justify logical rules run into problems due to the requirement that these rules provide us with "inference tickets" that are taken to prescribe the correctness of assertions in the context of open assumptions, or justified assertions.[1] This is part and parcel of the standard approach to the inferentialist theory of meaning as monological and unilateralist. That is, it provides an account of meaning in terms of the validity of certain inferences, which are explained in terms of the conditions under which certain propositions may be asserted (e.g. [3, 4]).

More recently, a number of authors have argued for an approach to inferentialism based on a bilateralist approach (e.g. [5–8]). Bilateralism is the view that meaning is to be construed in terms of the conditions on both assertion and denial, rather than assertion alone. So, more broadly, bilateralism is a form of rejectivism, which takes rejection to be on a par with, and equally foundational as, assertion (e.g. [5, 9, 10]). In particular, this view takes it that the grasp that we have on speech acts of assertion and denial is prior to (in the order of explanation) our understanding of negation. So, rather than derive the rejection of α from the assertion of $\neg\alpha$, in fact $\neg\alpha$ is to be explained in terms of the fundamental role of denial. According to one prominent bilateralist proponent, Restall [6], whilst natural deduction is so-called because of its supposed intimate connection with ordinary reasoning, it exacerbates the problems raised above by emphasising a prescriptive approach to inference.[2] For example, a valid entailment, $\Gamma \vdash \alpha$ is typically read as saying that, under the assumption that each $\beta \in \Gamma$ holds, α holds also, where this is taken to provide a normative prescription upon what thinker's ought to infer, (bringing with it the issues discussed in the previous chapter). Jaroslav Peregrin, [11, 12] argues against this approach: '$\Gamma \vdash \alpha$ is much more reasonably construed as a constraint: the exclusion of the possibility to deny α when one has asserted Γ' [11, p. 3], rather:

> [I]nferential rules do not prescribe what ought to be done, but what is allowed and what is not allowed to do when one asserts or judges that p. Inferential rules do not primarily consist in commands, obligations or incentives for speakers or believers; they rather constrain our linguistic practices by delimiting what, on an inferential point of view, we may and may not do by entertaining conceptual contents [11, p. 3].

Whilst Peregrin's view of inference has certain obvious advantages over the typical account of the justification of inference discussed in the previous chapter, the

---

[1]The concept of "inference tickets" is central to the discussion of tonk [1], though it possibly originates in the work of Gilbert Ryle [2]: 'Knowing if p, then q is, then, rather like being in possession of a railway ticket. It is having a license or warrant to make a journey from London to Oxford' (p. 308).

[2]In other rejectivist approaches, for example that of Rumfitt [8], there is an emphasis on the distinction between force and content. There, for any content, it is the case that it can be marked with the force of assertion, or the force of denial. This is formalised with " signed" sentences $+\alpha$ and $-\alpha$, signifying the force-markers on sentential metavariables. This is taken to formally capture the English language form of a question-forming sentence α?, with the answer " yes" or "no".

primacy awarded to rules, and the reliance on natural deduction formalism runs into similar problems. In distinction, Restall considers a view which awards primacy to a dialogical, rather than monological view of logic. This is far more consistent with the broader inferentialist approach discussed in Chap. 1, in the tradition of Wilfrid Sellars, and more recently, Robert Brandom, which emphasises the social and interpersonal in its approach to reasoning. However, whilst Brandom, for example, also sticks to a natural deduction framework to formalize this view, it is important that Restall [6] argues that the symmetry inherent in sequent calculi provides a better means by which we can understand logic as constraining the legitimate combinations of speech attitudes.

According to Restall, we are to read a symmetric sequent of the form $\Gamma \vdash\!\dashv \Delta$ as saying that "it is incoherent to simultaneously assert all of Γ, and deny all of Δ".[3] Furthermore, according to Restall, the symmetry of the turnstile is more basic than logical rules. So, for example, Restall argues that, contrary to the received view, "[d]enial is not the be analyzed as the assertion of a negation", going on to consider assertion and denial in terms of the geometrical duality of the turnstile.[4] So, incoherence is supposed to play a normative constraint over agents' commitments, whilst also defining the dynamic of the structure of sequents. To put this in context a little, Restall's [6] suggestion is to think of logic as governing positions involving asserted and denied sets of formulas.

Definition 7 (*Position*) A position $[\Gamma : \Delta]$ is a pair of sets of formulae where Γ is the set of asserted formulas, and Δ the set of denied formulas.

A position expressed in a language may be used to represent an agent's rational "state" in terms of the coherence between assertions and denials. So, where $[\Gamma : \Delta]$ is a position, we say that $[\Gamma, \alpha : \Delta, \beta]$ is the state formed by adding the formula α to the left set Γ, and β to Δ. Think of the above coherence constraints over rational commitment as saying that, a position $[\Gamma : \Delta]$ is incoherent if it contains some formula in both the left set and the right set, so that $\Gamma \cap \Delta \neq \emptyset$ since this indicates that the formula is both asserted and denied. This account of incoherence over positions allows us to characterise sequent provability:

Definition 8 (*Sequent provability*) If $[\alpha : \beta]$ is incoherent, then $\alpha \vdash\!\dashv \beta$.

This is because, if a position consisting of asserting α and denying B is incoherent, then $\alpha \vdash\!\dashv \beta$, since an agent who asserts α and denies β, as said above, has made a mistake. It is obvious, from this construction, that the notion of a position requires the generalisation of sequent provability whilst maintaining symmetry over the turnstile. That is, we need to shift from a right-asymmetric sequent structure to a symmetric structure (more commonly known as "multiple-conclusion" sequents).

[3] Restall uses the standard turnstile, but I use $\vdash\!\dashv$ to bring the duality to the surface.

[4] So, importantly, the sequent calculus is to be understood not as a metacalculus for single-conclusion sequents, but as directly marshalling the relation between formulas of a language.

Definition 9 (*Sequent provability generalised*) If $[\Gamma : \Delta]$ is incoherent, then $\Gamma \vdash\!\dashv \Delta$.

Now, if a position, $[\Gamma : \Delta]$, consisting of asserting Γ and denying Δ is incoherent, then $\Gamma \vdash\!\dashv \Delta$, since an agent is mistaken to assert all of $\alpha \in \Gamma$ and deny all $\beta \in \Delta$. The general idea is to construct a sequent calculus out of these constraints over assertion and denial. For example, since both asserting and denying the same formula is (assumed to be) incoherent, from $[\Gamma, \alpha : \Delta, \alpha]$ we can give the typical identity axiom over atomic formulas. We also have weakening, since, if a position is incoherent, then the addition of asserted or denied formulas will not bring it back to a coherent position. Contra-positively, if $[\Gamma : \Delta]$ is coherent, and $\Gamma' \subseteq \Gamma$ and $\Delta' \subseteq \Delta$, then $[\Gamma' : \Delta']$ will be coherent. Restall also suggests that we construct cut by thinking of extensibility constraints on asserted and denied formulas. For a position $[\Gamma : \Delta]$ is coherent, if the positions $[\Gamma, \alpha : \Delta]$ or $[\Gamma : \Delta, \alpha]$ are incoherent, then the original position $[\Gamma : \Delta]$ must already be incoherent. In other words, if a position is coherent, it should be extensible by a formula α to a coherent position where α is either asserted or α is denied. So, where $[\Gamma : \Delta]$ is coherent, either $[\Gamma, \alpha : \Delta]$ or $[\Gamma : \Delta, \alpha]$ is coherent.

More importantly, rules for the connectives can also be constructed out of positions. For example, if the position $[\Gamma : \Delta, \alpha \wedge \beta]$ is coherent, then $[\Gamma : \Delta, \alpha]$, $[\Gamma : \Delta, \beta]$, or both, are coherent. Contra-positively, if $[\Gamma : \Delta, \alpha]$ and $[\Gamma : \Delta, \beta]$ are incoherent, then so to is $[\Gamma : \Delta, \alpha \wedge \beta]$. In this case, we know that $\Gamma \vdash\!\dashv \Delta, \alpha$, and $\Gamma \vdash\!\dashv \Delta, \beta$, so that $\Gamma \vdash\!\dashv \Delta, \alpha \wedge \beta$. This gives us:

$$\frac{\Gamma, \alpha, \beta \vdash\!\dashv \Delta}{\Gamma, \alpha \wedge \beta \vdash\!\dashv \Delta}\ \wedge\text{-L} \qquad \frac{\Gamma \vdash\!\dashv \Delta, \alpha \qquad \Gamma \vdash\!\dashv \Delta, \beta}{\Gamma \vdash\!\dashv \Delta, \alpha \wedge \beta}\ \wedge\text{-R}$$

We can also construct the rules for classical negation by taking a negation $\neg\alpha$ to be assertible when α is deniable, and vice-versa. So, if $[\Gamma : \Delta, \alpha]$ is incoherent, then so too is $[\Gamma, \alpha : \Delta]$. This gives us Gentzen's classical negation rules:

$$\frac{\Gamma \vdash\!\dashv \alpha, \Delta}{\Gamma, \neg\alpha \vdash\!\dashv \Delta}\ (\neg\text{-L}) \qquad \frac{\Gamma, \alpha \vdash\!\dashv \Delta}{\Gamma \vdash \neg\alpha, \Delta}\ (\neg\text{-R})$$

Analogous accounts can be provided for all of the classical sequent rules [6], giving us a way of constructing the classical sequent rules in multiple-conclusion form, which is built out of a simple and plausible account of positions involving sets of assertions and denials.

It is simple to see that this offers a (partial) response to Carroll's argument, as Restall [13] has it:

> The force of modus ponens is not merely to recommend to us a statement $((\alpha \rightarrow \beta) \wedge \alpha \rightarrow \beta)$ for our acceptance, but to render the combination - accept $\alpha \rightarrow \beta$, accept α, reject β - as out of bounds (3 fn.2).

As is clear, the reason that we have something of a response to the problem of justification is that we have moved from the standard inferentialist view that consists of *pre*scribing certain assertions, to one that is *pro*scriptive over the combination of assertions and denials. However, the response is (at best) partial because we would also require a far clearer account of what being "out of bounds" consists in, and also more precisely what the relationship of coherence and incoherence is operating upon. For example, whilst we have moved away from the prescriptive monological account of inference rules, we are offered only an account of a set of assertions and denials for individual agents. However, to consider assertion and denial as unrelated to any sort of context of reasoning on these proposals makes them effectively free-floating. Since assertions (and denials) are supposed to be part and parcel of the social activities of reasoning, discussing, and interacting, this is a significant flaw. For example, this account presupposes that the joint assertion and denial of some α is incoherent. Now, whilst it may seem plausible to say that it is irrational (leaving aside how to make precise the notion of rationality at work), to do so, it seems more than plausible to think of a conversational context in which it is the case that some agent asserts α, whilst another denies α. Then, if we were to attempt to formalise the above by taking a position to involve multiple-agents, we would not end up with the sort of classical logical structure given above (for example, "cut" will not hold).[5]

It is, nonetheless, worth noting that there is a significant advantage to adopting a symmetric sequent structure to define a logic. Recall in the previous chapter the issues raised regarding the categoricity of logics defined by right-asymmetric sequent structures. In this setting, these issues no longer arise. For example, the extension theorem in this context is slightly different since we can say that, for any finite normal logic L, and $\Gamma \not\vdash\!\dashv \Delta$, Γ, Δ can be extended to $\Gamma \subseteq \Gamma'$, $\Delta \subseteq \Delta'$, with $\Gamma' \not\vdash\!\dashv \Delta'$ (and no proper superset $\Gamma'' \supseteq \Gamma'$, $\Delta'' \supseteq \Delta'$ do we have $\Gamma'' \not\vdash\!\dashv \Delta''$). Let us call such maximal Γ', Δ' quasi-partitions.[6] The semantics defined this way also differs: for a symmetric sequent, $\alpha_1, \ldots, \alpha_n \vdash\!\dashv \beta_1, \ldots, \beta_m$, in a logic L, and a quasi-partition Γ', Δ' of L, we say that the quasi-partition satisfies this sequent iff it is not the case that $\Gamma' \vdash\!\dashv \alpha$, for each $\alpha_1, \ldots, \alpha_n$ whilst $\Delta' \vdash\!\dashv \beta$, for each $\beta_1, \ldots, \beta_m$. Now, we have a situation in which the non-normal interpretations are straightforwardly ruled out, primarily because we have symmetry over the turnstile. So, for example, the scenario in which α, $\neg\alpha$ end up in Γ' is ruled out simply by ensuring that we have $\alpha, \neg\alpha \vdash\!\dashv \emptyset$ for all formulas of S. Similarly, the trouble with disjunction is now assuaged since we have $\Gamma, \alpha \vee \beta \vdash\!\dashv \alpha, \beta, \Delta$. The symmetry manifest itself most clearly in the fact that we now have a way of refuting propositions that is symmetrical with the way in which they are proved, and collecting these refuted propositions together in Δ'. So, for example, we have the following:

- $\alpha \vee \beta \in \Gamma'$ iff $\alpha \in \Gamma'$ or $\beta \in \Gamma'$.
- $\alpha \vee \beta \in \Delta'$ iff $\alpha \in \Delta'$ and $\beta \in \Delta'$.
- $\alpha \wedge \beta \in \Gamma'$ iff $\alpha \in \Gamma'$ and $\beta \in \Gamma'$.
- $\alpha \wedge \beta \in \Delta'$ iff $\alpha \in \Delta'$ or $\beta \in \Delta'$.

---

[5]See [14] for discussion of a classical logic without cut, and Chap. 5 for further discussion of these issues in relation to paradoxes.

[6]This is following [15], and these are "quasi"-partitions since we allow that either of Γ', Δ' may be empty.

- $\neg\alpha \in \Gamma'$ iff $\alpha \in \Delta'$.
- $\neg\alpha \in \Delta'$ iff $\alpha \in \Gamma'$.

It is by the fact that the above hold in the symmetrical logic that the issues regarding the determination of meaning and compositionality can be dealt with. However, the trouble with this "fix" is that I do not think it gives us any sort of proof-theoretic semantics at all. Say that we attempt to replace the role of counter-models in model-theoretic semantics with the notion of quasi-partition described above. For instance, we have that $\alpha \wedge \beta$ is refuted whenever one of the conjuncts is refuted. Then, one, fairly obvious, issue is that we do not have a "conjunction" property for refutations, just as we do not have a disjunction property for proofs. That is, we can refute $\alpha \wedge \beta$ without having a refutation of either of the conjuncts since the "conjunction" property for refutation fails in classical logic. This is just as problematic from the *p.o.v* of constructing proof-theoretic semantics, since we require an ability to determine, in a fine-grained manner, the manner in which a proof or refutation is "valid".[7]

In fact, this belies a much greater issue with this approach: it does not offer us any account of a proof (or refutation) *at all*. For example, the above response to Carroll also illuminates its own shortcomings, since it does not yet tell us why Tortoise may seem to be *required* to assert α, only that the simultaneous combination of incoherent assertions and denials is somehow proscribed by logic. That is to say, the price that we pay for an account that can deal with Carroll's problem is a loss of a direct account of proof (or refutation). Loosely, we can think of standard proof-theoretic approaches to logic as providing an account of the assertibility conditions for a set of formulae, with the general condition that a formula α is assertible whenever there exists a proof of α (we will flesh this out in detail below). Treating denial symmetrically suggests the need for an account of the deniability conditions for a set of formulae, perhaps with the general condition that a formula α is deniable whenever there exists a refutation of α. This would then be in keeping with the general inferentialist approach that the meaning of a logical constant is determined by a set of logical rules, with the proviso that we have, in addition to a set of rules defining its assertibility conditions, a set of rules defining its deniability conditions. But, pursuing this kind of project requires a rather different approach to Restall's, since his notion of the incoherence of assertions and denials would then be derivative of these explanatory structures. Steinberger [16] makes a similar point:

> Rather than offering us an explanation of the assertion conditions and the denial conditions for each of the connectives, Restall offers neither. Instead, Restall's account purports to elucidate the meanings of the constants in terms of their role in the interplay between assertions and denials. This is an indirect approach to consequence by means of "incoherence" i.e. indirect proof/refutation [...] Restall's characterization of a connective's meaning in terms of how they constrain our assertions and denials does not succeed in fully characterizing the meanings of the logical constants.

[7]The other obvious issue with the symmetric classical solution, for the inferentialist, is that whilst relatively maximal theories (and quasi-partitions) can stand in as syntactic counterparts to truth and falsity, this is only possible by idealization over the entire language.

Additionally, it is arguably the case that the basic structure of inference is one that is typically construed in argument form from (possibly) multiple premises, and a single-conclusion. There are several well-known issues with such multiple-conclusion structures, which are well documented (e.g. [16, 17]).[8] The most prominent of these is that they fail to adequately capture the structure of ordinary argument, which is a key desiderata for the inferentialist. As Rumfitt [17] has it:

> [...] the rarity, to the point of extinction, of naturally occurring multiple-conclusion arguments has always been the reason why mainstream logicians have dismissed multiple-conclusion logic as little more than a curiosity (p. 79).

As discussed in the previous chapter, we should beware of any over-simplistic identification between logical formalism and actual linguistic practices. Nonetheless, the basic structure of inference is one that is typically construed in argument form from (possibly) multiple premises, and a single-conclusion, which underwrites the notion of a "proof". Perhaps if we had some other, independent, account of assertion and denial, then these issues might be mitigated against, but as it stands, there are intuitive problems with this view. For example, for a classical logician, we can expect $(\alpha \vee \neg\alpha)$ to be always correctly assertible, but Restall's account gives us only the indirect "it is incoherent to deny $(\alpha \vee \neg\alpha)$".[9] The issue is that the role of logic is reduced to a structure, not of entailment and inference, but of coherence management.

Relatedly, Restall argues that coherence and incoherence are metalinguistic features of the turnstile that can deal with a number of alternative logics employing different rules of inference. This is tricky, however, should we want to employ this more basic feature as a way of explaining those rules. For example, if the symmetry of the turnstile is to provide us with an *explanation* for logical rules, then only the rules defining classical logic may be admissible.[10] Nonetheless, even if this can be adequately dealt with, let us consider what may be lost by moving to a symmetric sequent calculus for consideration of proof-theoretic semantics. Take a constructive logic such as intuitionistic logic. There, we have asymmetry of proof and refutation which is reflected by the restriction on intuitionistic sequents to at most a single formula on the *r.h.s* of the turnstile; the indirect interpretation of refutation as reduction to absurdity; the fact that a proof of α is inequivalent to $\neg\neg\alpha$. Nonetheless, it is arguable that it is precisely these restrictions that ensure constructivity, for example, ensuring by the disjunction property, that a (cut-free) proof of $\alpha \vee \beta$ (where both α and β have single, cut-free proofs), must be a proof of one of the disjuncts. In logics without this restriction, such as classical logic, this fails by contraction, so there may exist a disjunction $\alpha \vee \beta$ without any means of deciding which formula is proved, which is problematic if we pursue a semantics of proof. Lafont [18, Appendix B.1] goes as far as arguing that this renders classical logic inconsistent; 'not from a logical viewpoint (falsum is not provable), but from an algorithmic one', since proofs

[8]If we try to read the turnstile in terms of consequence, then we quickly run into familiar issues facing multiple conclusion consequence.

[9]This is Steinberger's [16] example.

[10]That is, unless structural rules are altered as in (e.g. [14]).

can not be considered algorithmically, and so 'classical logic has no denotational semantics, except the trivial one which identifies all the proofs of the same type'. Lafont considers the following example:

$$\dfrac{\dfrac{\dfrac{\begin{matrix}[\pi_1]\\ \vdash \alpha\end{matrix}}{\vdash \alpha, \beta}\,(Weak_R) \qquad \dfrac{\begin{matrix}[\pi_2]\\ \vdash \alpha\end{matrix}}{\beta \vdash \alpha}\,(Weak_L)}{\vdash \alpha, \alpha}\,(Cut)}{\vdash \alpha}\,(Contr_R)$$

which can be reduced to either;

$$\begin{matrix}[\pi_1]\\ \vdash \alpha\end{matrix}$$

or;

$$\begin{matrix}[\pi_2]\\ \vdash \alpha\end{matrix}$$

which are identical, classically. We should note that this result holds for any logic in symmetric form. For example, we may construct a paraconsistent or paracomplete logic in which the left or right negation sequent rule is inadmissible, yet is symmetrical in terms of the structure of sequent derivation.[11] But, since Lafont's example does not involve negation, there is no reason why such logics should be capable of dealing with this form of "inconsistency".[12]

2 The Nature of "Proofs" and "Proving"

Let us examine more precisely the nature of proofs, and their role in an inferentialist semantics in the constructivist, and intuitionistic, traditions that we introduced in the previous chapter. For example, in [19], Kahle and Schroeder-Heister say that:

> Proof-theoretic semantics [assigns] proofs or deductions an autonomous semantic role from the very onset, rather than explaining this role in terms of truth transmission. In proof-theoretic semantics, proofs are not merely treated as syntactic objects [...], but as entities in terms of which meaning and logical consequence can be explained.

This view is fairly typical to accounts of meaning of the kind developed in the proof-theoretic approach to meaning that we discussed above. So, for some α, we do not talk about the meaning of α in terms of the denotation of α, rather we talk about *establishing* the meaning of α by providing a set of proofs of α. That is, we can say that the meaning of some proposition is equivalent with a set of proofs for that

---

[11] See [6, 7] for further details.

[12] By way of response to this issue, Lafont suggests that we might restrict sequents to asymmetric form, as in intuitionistic logic, in which case $(Weak_R)$ will not hold. Lafont also considers linear logic, but I will not pursue that here, returning to this issue in Chap. 5.

proposition, so the meaning of α is something that has been determined, and, which should be communicable to others, and so we should be capable of carrying out time and again. There is, however, at least a bifurcation in the way in which we understand the nature of proofs that alters their semantic role. We can clarify this distinction in terms of the way in which the inferentialist account differs from the standard truth-conditional approach to logical validity.

The truth-conditional approach may be characterised as follows:

1. β is a logical consequence of α

That is, we can think of the truth-conditional account of the relationship between α and β just as an ordered pair that is interpreted by a model-theory, where it is the case that, whenever α is satisfied in a model, β is satisfied also (for every α, β). A proof, on the other hand, is typically understood as a function that *assigns* ordered pairs of formulae. This is perhaps clearest in right-asymmetric sequent calculus, since there we have a sequent which is just an ordered pair of formulae, and we may think of a valid proof as a function mapping the conjunction of premises into a conclusion. So, the proof-theoretic approach may be characterise as follows:

1. f is a function mapping α into β in some entailment structure

The key distinction is that consequence is just an abstract relation between propositional contents, which holds whether or not an agent has any knowledge of it. A proof, on the other hand, concerns a function that is supposed to represent something like an inference, which is carried out by an agent. So, the latter concerns a relation, not between propositional contents, but between *judgments*, or *speech-acts*. This is all a little imprecise, but we can put some bones on it by means of Prawitz and Dummett's definition of valid proofs.[13]

2.1 *Proofs as Reducible to Formal Derivation*

According to the Prawitz-Dummett view, validity is definitional of what a proof is, and its validity is relative to a formal entailment structure. So, for Prawitz, to ask whether or not a proof is valid is nonsensical. Instead, Prawitz uses a distinction between "closed" and "open" arguments to provide a definition of validity. Open arguments are just those which involve undischarged assumptions, or unbound variables. In contrast, a closed argument has no assumptions, and is valid just in case it is *either* canonical, such that it ends with an instance of an introduction rule, *or* it can be reduced to a canonical argument for the conclusion. It is reductions that ensure that a derivation can be normalized, since successive reduction procedures ensure that any "roundabouts" in the derivation can be eliminated.[14] This also clarifies how

[13] The following is based upon the accounts in [4, 20, 21].

[14] Dummett, in [20, p. 254], calls the fact that every closed derivation in an intuitionistic entailment structure can be reduced to a canonical derivation, the "fundamental assumption".

fundamental harmony is to the account. Whenever the introduction and elimination are in "harmony", this ensures that if the conclusion of an introduction rule is also the major premise of an elimination rule (at some point in a derivation), then it is possible to reduce that derivation to one with the same premises and conclusion, without the "detour" through those steps. Then, according to Prawitz, it is also possible to say that a closed argument is valid iff it can be identified with a proof. The validity of an open argument, rather, is dependent upon its reduction to a closed canonical argument.[15] As is clear, the validity of an argument (and, therefore, its equivalence with a proof) places the notion of *reduction* at its centre, so:

Definition 10 (*Prawitz-Dummett definition of validity*): An argument A is valid whenever:

- A is closed and canonical;
- or A is closed and reduces to a canonical argument;
- or A is open and reduces to a closed canonical argument;

The emphasis here is on closed and canonical arguments in an entailment structure, which allows for the reduction of non-canonical to canonical arguments, and also ensures that for a canonical argument to be valid requires its immediate subproofs to be valid.[16] The justification of open arguments relies upon this prior notion, by taking the open argument and replacing all open assumptions with closed proofs (or open variables with closed terms). So, in general, a proof of β under the assumption α is valid whenever it is possible to replace the assumption α with a (valid) closed proof of α.[17] I do not wish to repeat the arguments made in the previous chapter, but the central role of reduction in this account also clarifies the degree to which this approach to proof-theoretic semantics falls foul of the issues regarding the self-justification of logical rules discussed there.

In addition, the emphasis upon reduction to closed canonical proofs also indicates how proofs are understood on this account.[18] Whilst it looks as though Prawitz intends to provide an account of the validity of arbitrary inferences, or arguments, his approach relies on the identification of arguments with formally valid proofs. As such, there is little reason to think that this definition of validity renders epistemically tractable the notion of valid proofs as a way of representing arbitrary arguments and

---

[15] This account also relies upon the subset of atomic formulas being taken to be "valid" in the sense that evidence is available that they hold. Then, a set of logical rules extends atomic validity to an inference structure as defined in the previous chapter.

[16] I have already mentioned the relationship between this and issues of compositionality in the previous chapter. As I said there, I think that the problems facing the classical inferentialist are not assuaged by shifting to intuitionistic versions.

[17] As [22] puts it, according to Prawitz, '[a]n argument is valid if either it reduces to a non-logical justification of an atomic sentence, or it reduces to an argument whose last inference is an introduction inference and whose immediate subarguments are valid', (p. 7).

[18] We should note that this emphasis is clearly linked to the formalisation of proofs in a natural deduction framework, which, as Schroeder-Heister [23] has explained at length, allow open assumptions only to be placeholders for closed derivations. This is not the case in sequent-calculi, which is one reasons why later the proof-theoretic framework will be developed primarily in that framework.

inferences. So, it is not clear that we have adequately provided a way of thinking about proofs as more than just syntactic entities. What the account leaves aside is any role for the agents making actual arguments, or constructing proofs: we just have that proofs are valid by definition as formal derivations (with the provisos given above). Perhaps, for example, we should think of the function in our simplified definition as underpinned by a *construction* showing that each α can be mapped into each β; or an inference, whose validity is now to be explained by its coinciding with the definition given above. However, these notions would be reduced by the above approach to the objective and formal account of a proof. That is, the actual processes of inference, or construction, would be jettisoned insofar as any account of the *actual* validity of the proof is concerned, and, it is precisely for this reason that Carroll's regress looms large.

2.2 *Proofs as Objects, and as Acts*

Let us spend a little time unpacking this since there are a number of ways in which to construe the notion of "proof". Above, I have mentioned that typically constructive requirements such as disjunction property may be technically necessary in providing an account of proof-theoretic semantics. But, nonetheless, these do not seem to capture what is central to an account of proof-theoretic semantics. Rather, these properties are typical features of intuitionistic logic, whose constructive nature lies in an ability to understand proofs epistemically by requiring the existence of inferences, or constructions, and also takes a proposition to be determined by its proofs, rather than by an interpretation in a model. Nonetheless, pinning down exactly what this requires is notoriously tricky, resulting in a bifurcation of the understanding of what is meant by proof in the literature [24–26] into:

- Actualism (there actually exists a proof of α)
- Potentialism (α is potentially provable).[19]

Actualism is a tensed, or time-bound, notion of a proof that is given in a specific moment, and is typically thought to have been held by early constructivists such as Brouwer and Heyting (e.g. [27]).[20] In general, actualism emphasises the role of agents' *activities*, as Dummett [3] describes the view:

---

[19] I will not enter into a historical exegesis regarding the two positions (on this, the references above are more than adequate).

[20] As such, this approach is usually thought to cohere with the standard Brouwer-Heyting-Kolmogorov (BHK) interpretation of logical connectives as providing intuitive constraints on the complex composition of proofs involving logical constants. But, this is clearer in Kolmogorov's [28] interpretation of a constructive approach to logic than even Brouwer's, which argues that constructive logic has to do 'not with theoretical propositions but, on the contrary, problems':

> In addition to theoretical logic, which systematizes a proof schemata for theoretical truths, one can systematize a proof schemata for solutions to problems [...] In the second section, assuming the basic intuitionistic principles, intuitionistic logic is subjected to a critical study;

> [...] an understanding of a mathematical statement consists in the capacity to recognize a proof of it when presented with one; and the truth of such a statement can consist only in the existence of such a proof [...] They exist only in virtue of our mathematical activity, which consists in mental operations, and have only those properties which they can be recognized by us as having. (pp. 4–5)

So, on this view, for there to be a valid proof from α to β requires some agent to be capable of presenting that proof. Nonetheless, this is out of keeping with the, now, fairly dominant understanding of proofs as objective. This is to take the potentialist approach to proof, a view found in Prawitz [4] and Dummett (at certain points in any case [e.g. [3]]). The central distinction is usually taken to rest upon whether or not for α to count as constructively true, actual possession of a proof of α is required of an agent, or it is required only that there exist an effectively decidable possible proof. To take the latter view, (as pointed out in [e.g. [24], pp. 84–85]) seems to require commitment to an objective realm of propositions that ultimately faces similar objections to the truth-conditional approach, leading to an "inert platonism of proofs" [30]. This is because potentialism leads to a problematic account of truth regarding propositions that is both tenseless and independent of subjective understanding:

> [This] pays too much tribute to a platonician conception of mathematical inference: according to this conception, far from being deduced or extracted by ourselves, the consequences of an hypothesis follow from it by themselves, or rather in virtue of the existence of certain objects that it is none of our responsibility to conceive, or to make up, but only to discern; these objects may be likened to documents namely of documents already written, perhaps never read by anybody yet, but in which we could not fail to recognize, were we to become acquainted with them [...] By identifying proofs with sequences of formulas or, more generally, with objects that are independent from us, one almost unavoidably reduces the activity of justification to a scanning and control process that requires no cognitive or physical particular resource, and that consequently could not be affected by the limits of the cognitive capacities of the agents [31].

Potentialism, then, would not seem to adequately capture what is "constructive" about constructivism, which, traditionally at least, awards a key role to the actions of agents involved in the process of constructing proofs. This has significant consequences on the arguments that the view is more epistemically tractable than denotational approaches, reducing the role of agents to one of mere confirmation of proofs that are objectively valid. This has the result that, on fairly basic assumptions, a form of law of excluded middle for potential proofs may be valid:

> [...] Potential Excluded Middle: (PEM) α is potentially true or α is not potentially true becomes intelligible, and valid, in its classical reading. For, on this reading, it simply means that all propositions, as they are conceived by the potential intuitionist, are atemporally determinate, and this is clearly true: if it were indeterminate whether α is provable or not, the provability of α would be for ever prevented since, according to the conception at issue, a proposition cannot become provable. Therefore, such a hypothetical state of indeterminateness of α could be nothing but a state of well-determined unprovability of α. Whether

---

(Footnote 20 continued)
it is thus shown that it must be replaced by the calculus of problems, since its objects are in reality problems, rather than theoretical propositions [28, p. 58] (Translated in [29]).

> α is provable or not is a fact concerning the immutable world of propositions, where there is no room for any indeterminateness [24].

The issue here is that this renders agents proving activities essentially impotent *w.r.t* propositional content, and, as a result, we have no way of accounting for propositions that have not yet been proved, but may become proved at some later time. This issue is also related to the built-in asymmetry between proof (and truth) and refutation (and falsity) in constructive logic. We shall return to this below in Sect. 7, but, in brief, whilst truth is directly attainable (by constructing a proof), accounting for falsity, as Dummett [32] is aware, presents difficulties for the potentialist that mirror those identified above:

> There is a well-known difficulty about thinking of mathematical proofs [...] as existing independently of our hitting on them, which insisting that they are proofs we are capable of grasping or of giving fails to resolve. Namely, it is hard to see how the equation of the falsity of a statement (the truth of its negation) with the non-existence of a proof or verification can be resisted: but, then, it is equally hard to see how, on this conception of the existence of proofs, we can resist supposing that a proof of a given statement either exists or fails to exist. We shall then have driven ourselves into a realist position, with a justification of bivalence. If we refuse to identify falsity with the non-existence of a proof, we shall be little better off, because we shall find it hard to resist concluding that there are statements which are determinately neither true nor false [...] [32, p. 285].

That is, insofar as proofs are understood in objective terms, the *act* of proving is reduced to a kind of ratification, where a proof itself is unaffected by our interaction with it. The immediate problem with this, as identified in the previous chapter and above, is that this leads us to problems of circularity and regress where the correct application of a rule is required for access to this realm of proof-objects.

Instead, the actualist view of proof, according to Dubucs and Marion [30] gives us a way of accounting for proofs as "acts", rather than "objects":

> We propose that one distinguishes between two different notions of proof, namely those of proof as 'object' and as 'act'. According to the first conception, a proof is something like an assemblage of strings of symbols satisfying such and such property. From the second, more dynamic, conception, a proof is a process whose result may be represented or described by means of linguistic symbols.

On this view, we retain the constructive nature of a proof-theoretic semantics such that a proof is understood as an *act* that is undertaken by agents. So, for example, we can allow that *something occurs* whenever an agent comes to have an effective decision procedure for a proposition α, for example, that law of excluded middle holds for that α (e.g. [33]). Perhaps then, what is key to considering proof-theoretic semantics, is that formulae such as α may be reasoned with, considered, and so on, without the presumption of the existence of such a procedure, *and*, moreover, without even taking α to be a proposition, at least insofar as a proposition is supposed to be semantically evaluable, or equated with an equivalence class of proofs. If this is correct, then at the heart of constructivism, it allows for the existence of undecidability in the structure of syntax itself, and that our actions can render what is undecidable now to become decidable at some future moment. That is, we may take the act of proving seriously insofar as logic is taken not just to deal with the propositional and objective, but with the *actions* of reasons themselves.

2.3 Liberalising Proofs

On the proofs as *acts* view, a proof is a kind of activity that is undertaken by agents, allowing that there are some sort of dynamics to proofs insofar as they are carried out in time. This is suggestive of a much more liberal approach to proofs than is given by the Dummett-Prawitz account discussed above. Consider, for example, the role of the conditional, by which it is possible to transform an open proof into a closed proof on their account. Take a standard derivation introducing a conditional:

$$\frac{\begin{matrix}[\alpha]\\ \beta\end{matrix}}{\alpha \to \beta}\ (\to\text{-I})$$

Here, we have a proof of $\alpha \to \beta$, which, given a proof of β, no longer depends upon α as assumption. That is to say, it is closed and canonical, so it is a valid proof, by definition, that introduces the conditional statement whether or not α holds. In contrast, on *actual* proofs interpretation, a proof of a conditional, $\alpha \to \beta$, is just a function that maps *actual* proofs of the α into *actual* proofs of β.[21] Such a function can not map onto assumptions, according to this view, since then, the function could not be a map at all: 'as long as no proof of α is known, [the function] f has nothing to map. So we can still define f as the constant function which, once a proof π of α is known, maps every proof of α into the proof of β' [24, p. 91].[22]

Whilst at first sight this might be at odds with the idea that proofs are acts, rather than objects, in fact it provides us with a way of thinking about proofs as *both* acts and objects. In this regard, of particular note is the distinction made by Martino and Usberti [24] between what we may call a *hypothetical* function f_H that would come into effect once we have a proof of the antecedent to hand, and the *actual* function f_A which maps the antecedent into the consequent when we have the proof of the antecedent. The latter is *actual* since the function is only then an actual map that has come into effect given that a proof of the antecedent of the conditional is available. Since the conditional is intended to "write-into" the object language the relationship defined by the turnstile, and by the functional definition of a proof given at the start of this section, we can generalise this distinction as follows. In general,

---

[21] Kolmogorov's [28] analysis of the conditional also sheds light on this understanding of the role that conditional plays. There, he argues that a solution to the problem posed by a conditional $\alpha \to \beta$ must 'carry the solution of β back to the solution of α' (p. 59). In other words, a solution to the conditional is only partially solved by providing a function, f_H mapping a hypothetical α into β, which becomes a full solution (f_A) when it is shown that this carries back to a solution for α. On this construal, a solution of $\alpha \to \beta$ is a solution of β with premise α, which implies a tacit appeal to a given solution of α, to which a proof of β would "carry back to". See [34] for discussion.

[22] The quotation has been altered slightly to reflect the fact that I am interested in conditional rather than knowability, but the point is theirs.

a proof of $\alpha \vdash \beta$ is just a function that takes α, and maps it into β, hypothetical just in case the required evidence for α is not to hand, and actual just in case there is such evidence. As such, we now have a distinction between a hypothetical proof (in which we assume α), and a valid proof, in which case a proof of α is also given.[23] This, moreover, is a distinction *inside* the notion of a proof, which does not alter the overarching definition given at the begining of this section.

The distinction between hypothetical and actual proofs upturns Prawitz's distinction between closed and open arguments, with hypothetical proofs (open arguments) the norm, which in a subset of cases generate actual proofs (closed arguments). We are also liberalising the notion of proof, since it is here conceived as an activity, which may or may not produce an actual, valid, proof. This is to take seriously the idea that logic is not simply a matter of consequence and the construction of valid proofs, but rather it (also) has to do with the act of proving, reasoning and the construction of judgements. As I say, above, the standard approach to the validity of a proof, stemming from Heyting [27] through Dummett [3], focuses on proof-objects. The central feature that these views share is that a proof of a formula α is a construction π such that π makes α true, and that knowing a proposition is to have a constructive proof of it. In distinction, we may follow Sundholm's [36] account of constructions, which argues for a process/product distinction *inside* constructions, regarded both as processes, or as those processes taken as objects. The idea, is that a proof-object is that which remains *posterior* to the completion of a proof-act, and the trace of a proof is what is written down as the recipe for how to construct that proof.[24] A proof is something that is carried out in time, which then may become an object only subsequently, and in this sense, we also follow Martin-Löf's [37] argument that: '[a] proof is, not an object, but an act [...], and the act is primarily the act as it is being performed, only secondarily, and irrevocably, does it become the act that has been performed.' The process of "proving", then, is more like a process of reasoning that is not required to live up to objective correctness conditions on assertions, which may be thought of as *generating* proof-objects.

The distinction is obscured by the platonistic view of proofs as objects and the constitutive norms approach to assertion, and also by the supposed ease by which it is possible to shift between the hypothetical and the actual by dint of the deduction theorem. On the latter, consider the role of deduction theorem in classifying proofs, which allows us to identify a proof with a deduction from the empty set, and a theorem as the last item in that proof.[25] As is known since Herbrand, to prove that deduction theorem holds intuitionistically requires induction on successive uses of modus ponens from the axioms, so, in effect, this process requires us to shift hypothetical assumptions successively over to the *r.h.s* of the turnstile, removing the hypotheses and treating them as theorems. Perhaps more than other formulations, the categorical version of the single-premise deduction theorem makes this transparent.

[23]The notion of a hypothetical proof bears obvious similarity to Girard's [35] notion of paraproof.

[24]The analogy Sundholm makes is that written proofs are like annotations for a game of chess, as opposed to proof-acts, which are like the game itself.

[25]See [38] for a similar formulation and extended discussion of deduction theorem.

Take 1 to indicate algebraic top, then say that for any propositional formulas α and β, if there exists a derivation of $1 \vdash \beta$ from the assumption $1 \vdash \alpha$, then there exists a derivation of $\alpha \vdash \beta$ (e.g. [39], p. 50). The illusion that this brings with it lies in the ease with which this *appears* to shift from the hypothetical to the actual form, which is surely one of the morals of Carroll's tale: that there is a distinction between the action of an inference, and a relation between propositional contents.[26] One mistake identified in Carroll's story is to think that to infer β from α and $\alpha \rightarrow \beta$ requires an agent to believe "$(\alpha \wedge (\alpha \rightarrow \beta)) \rightarrow \beta$)". The latter may, rather, be that which may be the content of a belief *following* the act of making the inference itself.

3 The Dynamics of Assertions

Let us now consider the relationship between proof and assertion, which is typically taken as stand-in for "truth" in approaches to inferentialist semantics [20, 40]. Importantly, there is an issue with the "proofs as acts" conception that is sometimes taken to rule it out altogether, because of the relationship between proof and assertion that is typical of constructive approaches to meaning. Whilst this is problematic on some accounts of assertion, I will argue that it is consistent with an alternative, social, account of assertions.

On the Prawitz-Dummett approach, in order for a sentence to be asserted, the kind of evidence that is required is a closed proof. The idea being that, to take a mathematical formula as example, its statement is assertible under the condition that there exists an effective proof of that formula (where a proof is defined as object, as above). Open proofs are assertible, only insofar as they can be reduced to closed proofs. That is, if we have an open proof of β from α, then β could not be asserted unless we have some evidence that α holds also. Now, on this view of the relationship between proofs and assertions, there is no room for hypothetical proofs to be asserted: either a proof exists or it doesn't, and whether or not a proof exists is what determines the correctness of assertions. Then, for example, it looks as though for any conditional statement $\alpha \rightarrow \beta$, it could not be asserted by an agent unless the agent also has a proof of α to hand. That is to say, we could not assert conditionals "conditionally", so to speak, and, since we have used this as a means to make a generalised distinction between hypothetical and actual proofs, all such hypotheticals would appear to be "empty" insofar as they are not assertible.

Nonetheless, there is an obvious analogy here to the fact that it seems intuitively incorrect to think a conditional, $\alpha \rightarrow \beta$, is the sort of thing that is simply asserted, especially when the antecedent α is known to be false.[27] The standard treatment of conditionals tells us that $\alpha \rightarrow \beta$ is true whenever α is false. But, in English, when we say that "α, therefore β", or "if α, β", if α were shown to be false, it does not seem

---

[26]Though, unlike the discussion outlined in the introduction, this distinction should not be taken as requiring the reduction of the former to the latter.

[27]There is an excellent discussion of these issues in [41].

correct to say that the conditionals are true. There are well-worn issues with the way in which conditionals are supposed to be understood when they have false antecedents. According to many critics (e.g. [42–44]), this (amongst other issues) suggests that asserting a conditional does not express a "conditional" proposition. Rather, such an assertion is the conditional assertion of the consequent, where the condition of the assertion is that the antecedent holds. On this view, as Humberstone [45] suggests, after Ramsey [43], to assert a conditional is not to be thought of as asserting a conditional proposition, but to make a conditional assertion of the consequent: 'If the latter condition is not satisfied (i.e., if the antecedent is false), then it is as though no assertion had been made. The parallel is with conditional bets, which are void in that no money changes hands unless the condition they are conditional upon obtains' (p. 938).[28] A conditional assertion, then, is a distinctive kind of speech act, as Stalnaker [46] points out, and not a standard speech act (assertion) with distinctive content.

There are different ways of thinking about conditional assertions. For example, some proponents have thought that, whenever the antecedent is false, the condition upon which the assertion of the consequent is made does not hold, and so no assertion is made. Quine puts this as follows:

> An affirmation of the form If α then β is commonly felt less as an affirmation of a conditional than as a conditional affirmation of the consequent. If, after we have made such an affirmation, the antecedent turns out true, then we consider ourselves committed to the consequent, and are ready to acknowledge error if it proves false. If on the other hand the other hand the antecedent turns out to have been false, our conditional affirmation is as if it had never been made [44, Sect. 3].

This approach would suggest, however, that conditional assertions are "empty" until they can be transformed into categorical assetions. But, take a conditional like the following: "If you press that switch, there will be an explosion". When stated by one agent to another, even if the antecedent does not hold (the hearer does not press the switch), it does not seem that nothing has been said, since the hearer has learned some sort of reason to think that, should the antecedent hold, then the consequence will be an explosion![29] This suggests that conditional assertions are not "empty", but that they are distinct from making a categorical assertion of the consequent. As Edgington [47] puts it: 'My hearer understands that if she presses [the switch], my assertion of the consequent has categorical force; and given that she takes me to be trustworthy and reliable, if it does acquire categorical force, it is much more likely to be true than false. So she too acquires reason to think that there will be an explosion if she presses it, and hence a reason not to press it' (p. 178). This way of thinking about the relationship between conditional and categorical assertions coheres, to a great extent, with the account of hypothetical and actual functions given in the previous section. By analogy, we may think that a conditional assertion is like a hypothetical

[28]Edgington [42] goes further still: 'to assert a conditional is to assert that it is true on condition that it has a truth value. To believe a conditional is to believe that it is true on the supposition that it has a truth value. It has a truth value iff its antecedent is true'.

[29]This follows Edgington's [47] argument.

proof insofar as it is a function that takes a categorical assertion (antecedent), and maps it into a categorical assertion (consequent). Whenever no categorical assertion of the antecedent exists this function remains hypothetical insofar as there is nothing yet to map into the consequent. Whenever there is such an assertion, a categorical assertion of the consequent is made. But, again, this would be to make a distinction *inside* assertions (rather than distinguishing different species of speech-acts), where the distinction reflects the *dynamics* of reasoning, with categorical assertion as the objects (sometimes) produced by that process.

It is fairly clear, however, that this kind of approach is incompatible with accounts of assertion that require constitutive norms (such as the existence of proofs), which are supposed to govern the proprieties of assertions. On such accounts, assertions are taken to be an "all or nothing" affair. There are a number of accounts of assertion that require constitutive norms on the making of assertions, such as that one must make an assertion, "c", only in case one knows that "c", as held by Williamson [48]; or that one must make an assertion, "c", only in case it is true that "c", as held by Weiner [49]; or that one must make an assertion, "c", only in case a proof of "c" exists (e.g. [50]). These accounts all share the idea that categorical assertions are taken to be "industry standard". So, even if conditional assertions were allowed some sort of existence, the "all or nothing" status awarded to assertions means that conditional assertions are treated as "empty". In fact, that conditional assertion is ruled out altogether by the Prawitz-Dummett approach on the basis of the definition of the validity of conditionals. This is most clear when we consider negation, which is defined as $\neg\alpha =: \alpha \rightarrow\perp$ (where $\perp$ expresses a constantly false proposition). Now, say that $\alpha \rightarrow\perp$ is categorically assertible on condition that α is categorically assertible, and conditionally assertible otherwise. Then, $\neg\alpha$ could only ever be categorically asserted since, definitionally, neither $\perp$ nor α can be categorically asserted. So, $\neg\alpha$ could never be asserted.[30] I will not discuss the merits, or otherwise, of the constitutive norms approaches to assertions (for this, see the excellent discussion in [52–54]).[31] Rather, I want to point to an alternative account of assertion that coheres much better with the approach to proofs and proving given above.

In [52], this is called the COMMITMENT view, where it is traced to the work of Peirce [55] who suggests that 'to assert a proposition is to make oneself responsible for its truth' (p. 384). The key distinction between this, and the constitutive norms approach, as Macfarlane [52] puts it; '[...] while the constitutive rules approach looks at upstream norms—norms for *making* assertions - the commitment approach looks at downstream norms—the normative *effects* of making assertions'. This view is, perhaps, made most clear in the account of assertion games given by Robert Brandom [40, 56].[32] In [40], Brandom suggests that asserting that "c" is to undertake a commitment to defend "c" when challenged. So, the emphasis here is not on

[30]This is discussed below, and in [51].

[31]Additionally, it may well be the case that these views can be made compatible with the account proposed below in some way.

[32]This also follows Wittgenstein's suggestion that making an assertion is to make a move in a game [57, Sect. 22].

*pre*scribing the conditions under which it is permissible to make an assertion, but rather it is an account of what is *pro*scribed *after* an assertion has been made. Plausibly, then, the key norm on assertion is not simply a commitment to its truth, but rather a commitment to *defend* its truth. In this vein, Pagin [53] also draws attention to the relationship between assertions and promises, as discussed by Watson [58], where Watson notes their similarities. The key distinction between the two, according to Watson, is that the commitment involved is to something that is speaker-independent, which is just the defensibility, rather than the truth, of the assertion. So, again, on this view, the agent making an assertion is obliged to defend the assertion if challenged [58, p. 70]. That is to say, assertoric norms should not be understood to restrict what an agent ought to assert, instead they may be thought of as constraints on how agents respond to challenge in social and dialogical contexts. Furthermore, on this view, the norms on assertions have to do with a willingness to make an attempt to justify those assertions that an agent has brought "into the game of giving and asking for reasons" [59, p. 57].[33] Importantly, then, unlike the constitutive norms approaches to assertion, on which some sort of grounds (presumably justificatory) for making an assertion are required of agents *prior* to making that assertion, the commitment approach requires only that an agent be prepared to make an attempt to justify the assertion *subsequent* to making that assertion.

It is in this sense, then, that making an assertion may be thought of as being akin to making a move in a game of reasons. This makes it available for scrutiny, so that, when asked, then agent should attempt to justify the assertion by way of providing some sort of reasons for it[34]:

> In asserting a claim one not only authorizes further assertions, but commits oneself to vindicate the original claim, showing that one is entitled to make it. Failure to defend one's entitlement to an assertion voids its social significance as inferential warrant for further assertions. It is only assertions one is entitled to make that can serve to entitle others to its inferential consequences. Endorsement is empty unless the commitment can be defended [56, p. 641].

So, on this view, making an assertion is primarily a matter of bringing that assertion into "play". At this point, the assertion is subject to norms involving a commitment to its defense, to providing reasons for it, and allowing it to be "tested" through interaction with other reasons, counterexamples and so on. Assertions do not stand alone, on Brandom's (e.g. [40] p. 167) view, rather, they stand in need of reasons, and it is in the context of language "games" that we ask for, and provide reasons for, our assertions.

This is to place assertions squarely where they belong: in a social setting that involves multiple agents.[35] Further still, what is important in this shift is that these

---

[33] An excellent discussion of these issues can be found in [60].

[34] See also the excellent discussion in [61].

[35] Pagin, in several places [53, 54, 62, e.g.], makes an argument to the effect that the social account of assertions does not, by itself, provide sufficient conditions on the nature of assertions, whilst he accepts that it may be the case that they provide necessary conditions. The discussion in [52] provides a useful rejoinder, though, in any case I do not think that this is an issue for the view espoused

are dialogical norms that are explanatorily prior to constitutive norms of truth or knowledge or justification.[36] If we think of this in relaton to the conditional and categorical assertions distinction, then the commitment view can be thought of as putting conditional assertions first, whilst allowing that categorical assertions are *generated* by the dynamics of assertion games. It is not the case that a conditional assertion that never generates a categorical assertion is merely "empty", it is just that assertion is not an all or nothing kind of affair. So, rather than thinking of the making of an assertion as a fully formed propositional content, which may be thought of as true or false, we rather think of it as "playing" the statement as a kind of "token" in a game.[37] At this point, the statement may be treated hypothetically, and can be challenged and tested by other agents. It may, for example, be defended by the provision of reasons, and it may be contested by other reasons and counterexamples. It is also the case that, at some point during this process, the agents involved might agree that adequate justification has been provided for the original statement to be considered verified, or, indeed, that there is enough reason to think that it is false. At this point, it seems that we would be in a position to evaluate the statement as a kind of propositional content, in the usual way, but this occurs only after this interaction has occurred. The parallel with the distinction that I made above between hypothetical and actual proofs is clear. In fact, as Pagin [53] points out, Dummett makes a similar claim in passing, that 'an assertion is a kind of gamble that the speaker will not be proved wrong' [67, p. 84], and they 'may subsequently be compelled to withdraw it as incorrect' [20, p. 165]. So, we can think of the initial assertion as a hypothetical proof.

(Footnote 35 continued)
here. For example, the kinds of problems usually thought to face commitment approaches involve examples where assertions are made without explicitly making statements, through nonlinguistic signs, for example. I don't think that these are problematic for the account given here, since, it seems perfectly acceptable that one might ask for reasons for such signs, thereby clarifying them, in the same way as linguistic statements. A slightly different example given by Nunberg [63], and discussed in [52], is a waitress who states that "The ham sandwich left without paying". The waitress has made an assertion, though it does not seem correct to say that she has asserted that the ham sandwich left without paying. But, whilst this may seem *prima facie* problematic for a commitment view, I agree with Macfarlane [52], that, to the contrary, this view fares very well in this respect:

> [...] if we wanted to settle, for example, whether Nunberg's waitress had asserted that a *sandwich* had left, or that a *person* who ordered a sandwich had left, we might ask with (if either) of these propositions she meant to commit herself to.

[36] Shieh [64] puts this as follows:

> To be taken as making an assertion, a speaker must acknowledge that the statement she is making is subject to assessment as correct or incorrect, by reference to what she would count as justifying it (cited in [65]).

[37] See also [66] for a similar approach to the relationship between assertion and proving.

But now, whilst the idea of a *hypothetical* proof is somewhat idiosyncratic, it is rendered more transparent if thought of as just a "play", or "move", in the assertion game. This may be transformed into an actual proof by providing a proof for each assumption, and at each stage of the argument providing reasons for the statement in response to "tests". But, now, notice that this would suggest that the the *act* of proving is one which is intrinsically social, rather than just individual, and that, we now have an explanatory structure for this process, which is just that the agent must respond to any *tests* of the initial statement, where these tests are part and parcel of the process of providing a proof of the initial statement. So, the process of *constructing* a proof is just our "game of giving and asking for reasons", which concerns the hypothetical, and that which is "in process", whilst the product which is a *construction* is that which is made explicit (to use Brandom's [40] terminology) over the course of this process, to the point where the initial statement is justified.[38] That is to say, we have provided an explanation of the BHK-style approach that takes a construction to be the construction of a justification, by means of a social account of the nature of assertions and justifications. For example, we might adhere to the conditions on the validity of *actual* proofs as requiring them to end with an instance of an introduction rule (so reflecting the requirements given in the BHK clauses). But, this requirement may now be understood to be the result of a process of interaction, in which the structure of the *process* of proving is rendered explicit by the requirement of canonicity (or reduction to canonicity) holding at the *end* of an assertion game.[39] This is also suggestive of a fuller response to Carroll's regress, since, on this view, it is the fact that agents are engaged in an assertion game of "plays" and "tests", where agents are engaged in an *act* of proving that *gives rise to* the normative force that proofs subsequently have over agents, rather than the other way around.

4 Meanings and "Use"

Let us return, in this context, to the idea that the meanings of statements are determined by their use. On Dummett's view, "use" is construed narrowly since the meaning of a sentence is determined by the *objectively correct* conditions for its assertion,

[38]In Brandom [59] this is related to Socratic method, which, he argues, is:

> [...] a way of bringing our practices under rational control by expressing them explicitly in a form in which they can be confronted with objections and alternatives, a form in which they can be exhibited as the conclusions of inferences seeking to justify them on the basis of premises advanced as reasons, and as premises in further inferences exploring the consequences of accepting them (p. 56).

I shall develop the connection between this approach to logic and the dialogical tradition in the following chapter.

[39]This will be developed in the context of formalism in Chap. 4.

and these are distinct from the conditions under which agents might ordinarily assert the sentence:

> [...] in the general case, we have to consider as primary, in determining the content of an assertion, not the speaker's personal entitlement to make the assertion, but the condition for its objective correctness [67, p. 120].

These objective correctness conditions for sentences involving logical constants (or, at the least, mathematical statements) are equivalent with their proofs, which must be stable, objectively correct, and timeless.[40] In distinction, the commitment approach to assertion clarifies the way in which assertions are understandable only in the context of social interaction, and where those interactions are not determinate from the outset (since, otherwise, agents would be required to foresee every response to an assertion). With this in mind, we can think of establishing the meaning of a proposition as part and parcel with these processes of interaction. In this sense, the meanings of statements are not determined just by an individual's dispositions to follow certain rules, or the way in which assertions and denials are governed by certain constraints. Rather, meaning is established in the broader context of social interaction involving assertions, tests, and reasons amongst multiple agents who are engaged in a common aim.[41] Put together, these two claims require us also to say that the meaning of our statements are not immediately determinate at the outset of such interactions.[42]

Consider, however, what this account would entail for the commonly held distinction between the force and content of speech-acts.[43] This distinction requires there to be something like a pre-existing content, c, of a statement "c", with the assertion of "c" marking the force that is attached to c. This is necessary to ensure that it is possible to have distinct contents c_1 and c_2 which may be both associated with the same force, e.g. assertion; and that it is possible to have a single content c that may be associated with distinct forces, e.g. assertion or denial if we take the bilateralist approach discussed in Sect. 1. This is a view that we may trace to Frege [68], where logical formulas are propositional contents, and the turnstile indicates assertibility. The key message of this view is this: whatever speech acts are, they are not the sort of thing that can have an effect on the content of the associated statements. In distinction, on the view suggested above, this distinction can not be so steadfastly made, since, it is in the context of "assertion games" that the content of a proposition is constructed. As such, it does not seem correct to say that there should be an absolute separation between a prior existing content and the speech act to which it

---

[40] Furthermore, this objectively true notion of the proof of a statement is equivalent with its truth, according to Dummett [67].

[41] So, whilst Restall's work highlights one way in which norms governing the coherence of speech acts can underpin a theory of meaning, devoid of social context, it is difficult to make any sense of how such assertions and denials are supposed to interact with others.

[42] There are certain constraints on the structure of these interactions that we shall discuss in the following chapter, and it is these (I shall argue there and in Chap. 4) that provide a justification for logical rules.

[43] See the excellent discussion in [52, 62].

becomes associated. This may seem controversial, but there are other reasons that we might be pushed in this direction. For example, say that we were to accept the constructivist suggestion that the content of a proposition is equivalent to the set of its proofs. Then, on an objective approach to proofs, it seems sensible to think that this propositional content is unaltered by our interaction with it, since a proof is, on that view, a timeless and platonic entity. However, this not only brings with it issues of epistemic tractability discussed previously in relation to denotational approaches, but it also falls foul of an issue pointed out by Wittgenstein (e.g. [69]). Wittgenstein's issue is that if the meaning of a proposition is determined by a proof of it, then a mathematical conjecture, for example, would seem to have no meaning, or at the least it would seem to have a meaning that is distinct from that which the same statement has subsequent to it being proved true. On the assumption of meaning-stability, this has the unfortunate consequence that the original conjecture could never be proved, since whatever proposition is proved would be distinct from it!

The issue, for the platonist constructivist is just that, whenever agents utter some statement, α, which they are setting out to prove, it is not always clear how to go about proving it, and it is fairly obviously the case that no proof of α exists at that moment *for those agents*. When, on the other hand, the agents do have a proof of α to hand, they would be in a position say that a proposition expressed by α is understood, has a certain meaning, and so on. There are (at least) two responses here. They may want to say that the original conjecture is just meaningless, though this is an unattractive option since the agents can *do* some things with that statement, such as, using it in certain inferences (particularly conditional ones). Another response would just be to bite the Platonic bullet, and say that, since a proof is an objective entity, the content expressed by the conjecture and the proved proposition is the same after all. There are a number of issues with the latter response, however, not least of which is how we are to think of the agents *understanding* of that propositional content. It looks as though we will run straight into the problems that we originally identified with the denotationalist account regarding how agents *understand* the meaning of abstract propositions. Of central importance is that these suggestions void any import to the actual activity of providing the proof. But, it is surely the case that the act of providing a proof does much more than simply establishing the verification of the conjecture. Rather, it also provides an account of the process by which that verification can be established, and in so doing, provides us with a far greater understanding of the statement. That is to say, it provides an account of the *meaning* of the proposition.

The account that I have argued for above, on the other hand, allows us to explain Wittgenstein's concern regarding the propositional contents of assertions made at the start and end of a process of inquiry, or proving activity. Rather than presuppose the stability of meanings, or problematise the original assertion altogether, we can accept that utterances made at different moments during this process do not quite "mean" the same thing, since the process of providing a justification for the statement is part and parcel of the end-proposition's content. It is a mistake to think that the meaning of the statement can be divorced from the way in which it is established, since it is only *a posteriori* that we can evaluate the meaning of a proposition at all. So, the *act* of proving is also an act in which propositional contents are constructed.

On this view, the statement α at the outset of inquiry is just a conjecture whose meaning is in process of construction, and which, subsequently, may be redescribed as a proposition whose meaning is explicable in terms of whatever was required to be *done* in order to provide a proof of that proposition. For example, if we say that A is a proposition, and (c, A) is the condition under which A may be asserted, then both of these would require simultaneous establishment since claiming that A is a proposition is dependent itself upon the conditions under which it is asserted (see [36]). This means that, over the course of a construction (as process), we would subsequently produce an object that we may say is simultaneously both an *actual* proof of A, and a construction of A itself. This is not to say that we need jettison the distinction between force and content, but, rather, that their relationship becomes more complex if we are to say that propositional content is constructed through the process of proving.[44] The content of propositions may be given by the process that is required to construct proofs (or justifications, more broadly), of them. Of course, this requires a non-internalised approach to meanings, and also one in which we do not take meanings to be determinate *prior* to them being established as having justifications. But, this is precisely the point: if we are to take seriously the idea that the meaning of a statement is determined by its use, then we should take the processes and acts of proving in interactions with others to be prior (explanatorily as well as temporally) to the constructed proposition. It is, on this view, the co-ordination of interactions involved in the game of giving and asking for reasons that gives rise to a theory of meaning, in which a propositional content is evaluable whenever adequate justification has been provided (or indeed, some kind of proof) for the assertion in question.

Let us briefly return to the question of logical rules, just in order to introduce the strategy I will outline over the course of the following two chapters. In [70], Tennant suggests that 'harmony for logical operators would be a transcendental constraint on the very possibility of logically structured communication' (p. 629). Above, we pointed out that the approach by harmony fails to provide an adequate justification for logical rules that avoids Carrollian regress, and so, Tennant's transcendental constraint does not even get off the ground. Instead, the approach to proofs and assertions given in this chapter practically reverses the dictum, such that we may say that a structured interaction places transcendental constraints on logic, and logical rules. As such, the approach is suggestive of a slightly different characterization of what is happening with principles of harmony and canonicity. As stated in the previous chapter, a central concern of invertibility in a sequent calculus has to do with the role of cut. Now, if we take the proofs as *acts* viewpoint, then, rather than think of harmony in terms of the self-justification of rules, we might instead think of it as ensuring that an interaction between sequents, as mediated by "cut" will always balance out. That is to say, any sequent that interacts with the premises of a derivation

[44] For example, Wittgenstein's argument in [57] against the force/content distinction is much stronger than the view here.

will also interact with its conclusion in the same way.[45] For example, it is easily seen that the standard conjunction right rule is invertible by looking at the role of "cut":

$$\dfrac{\dfrac{\alpha \vdash \alpha}{\alpha \wedge \beta \vdash \alpha}(\wedge L_1) \qquad \Gamma \vdash \alpha \wedge \beta, \Delta}{\Gamma \vdash \alpha, \Delta}\text{(Cut)}$$

$$\dfrac{\dfrac{\beta \vdash \beta}{\alpha \wedge \beta \vdash \beta}(\wedge L_2) \qquad \Gamma \vdash \alpha \wedge \beta, \Delta}{\Gamma \vdash \beta, \Delta}\text{(Cut)}$$

If we do not take rules to have primacy here, but rather the interactive movement of the proof-act itself, then "cut" effectively ensures balance over the course of the proof, so that it is possible to *end up with* a valid proof.[46] The construction of the conjunction "rules" may then be thought of as following from this prior requirement that, for any proof of $\alpha \wedge \beta$, it must be possible to extract proofs of α, and proofs of β, since if this were not the case, then the premises and conclusion of a derivation would not "balance out". In a similar way, I shall argue in the following two chapters that logical rules may be justified at the outcome of the process of the interactive construction of proofs because of constraints on those interactions.[47]

5 Balancing Assertions and Denials

It is fairly clear that, whilst the above section emphasised assertions alone, the account that we end up with is better understood in a context that is more permissive, such as the bilateralist account with which we began this chapter. A generalised way of putting the issues raised above with Restall's approach is that it is incapable of dealing with neither proof-acts nor proof-objects. Rather it deals only with indirect notions of coherence and incoherence. The above approach to assertions not only provides a way of thinking about the role of assertions that is missing from Restall's account, but also it explains *why* proofs should be taken as central to any such account. This is because the utterance of an assertion brings with it a commitment to its justification in response to requests for reasons, which is tantamount to asking for proofs of

[45] See also [71].

[46] If we take Dummett and Tennant's description of the reasons why harmony is important, then we might say that cut-elimination would therefore ensure that the kind of balance assertions and consequences is maintained across the course of the proof. Of course, it is also the case that cut-elimination allows us to show that a proof-system is consistent, and that it has the sub-formula property - that each formula in the end sequent of a cut-free proof is a sub-formula of one of the premises.

[47] I have briefly suggested the shape that these constraints take, in terms of the relationship between assertions and testing, giving and asking for reasons, but I shall put some meat on these bones in the following chapter.

the assertion to be given. Furthermore, in outlining the approach, we also have an explanation of why we should look towards denial and refutations in addition to assertions and proofs, beyond the idea that their symmetry solves some of the issues raised with formal structures. The above account has provided a way of thinking through the dynamics of a social and dialogical approach to assertions, and, in doing so, we have shown that a dual role for denial is required. If, for example, we take the notion of the "game of giving and asking for reasons" seriously, then, we have a setup that involves both *assertions* and *tests*, which, in a simplified abstraction, we may think of as a interaction between two agents.[48] If we take it that the making of an assertion brings with it a commitment to its defence, then we also require something to defend that assertion *against*. Above, I have mentioned tests, counterexamples, and so on. These *challenges* to the initial assertion may be characterised by means of the bilateralist notion of denial (insofar as denial is understood to be a basic speech-act that is both distinct from, and non-interdefinable with, assertion). That is to say, we can interpret the account given above as a kind of dialogue structure between the roles of prover and refuter, where an assertion of a statement involves a commitment to its defence, and a denial of the statement involves a commitment to its challenge. As such, we can split Brandom's "game" into two parts: the *giving* of reasons on the part of "prover", and the *asking* for reasons on the part of the "refuter".

This would bring with it a very different approach to the one given by Restall, since, in keeping with the above, we would require a separate account of the notion of a valid REFUTATION, in addition to the notion of a valid proof. So, for example, we should be able to provide the conditions under which a proof (or refutation) is valid on the basis of an interaction that provides (or fails to provide) adequate justification for the statement in question. Furthermore, whilst Restall's notions of coherence and incoherence is non-social, we now have a social context that provides an explanatory structure accounting for the interaction between assertions and denials.[49]

There is an additional reason to think that we require a separate account for denial and refutation in the context of providing a constructive approach to proof. As I said above, the asymmetry between proof and refutation in intuitionistic logic is part and parcel with what makes it constructive (for example, that the absence of a proof is inequivalent to a refutation). But, the asymmetry is also problematic when it comes to accounting for non-proved formulae. For example, if non-proved formulae are simply be taken to be false, then falsity is equated with non-truth. But, it does not seem right to say, of some formula α which is not provable at a stage s_1 but may become provable at a later stage s_2, that it is false, and then becomes true. Instead, falsity is typically equated with a reduction to a contradiction (this is the case in Heyting [27], and also Kolmogorov [28]), so that α is false whenever

[48]Though, as mentioned in the introduction (and to be discussed further in the following chapter), it is plausible to think that even individual reasoning involves an internalised dialogue.

[49]This context will be discussed in further detail in the following section, but I note in passing that this would at least require that Restall's CUT rule, which requires coherence to hold unrestrictedly must be jettisoned so that an assertion and a test may be simultaneously present. This alone is enough to think that grounding the account in the notion of incoherence and coherence will not work when the speech-acts are understood within a social context, rather than in terms of an individual.

$\neg\alpha$ is true. This, too, is not unproblematic. As Shramko et al. [26] point out, if we attempt to formalize the definition of falsity in terms of negation, then we are led to a reliance, not only on a syntactic feature (negation), but also on truth, so there is a lack of independence of falsity. As I showed above, this, together with a potentialist approach to proofs is constructively suspect. This is part of the broader issue that intuitionistic logic is incapable of refuting formulae directly. For example, *reductio ad absurdum* is capable only of conditionally proving negative statements, since $\neg\neg\alpha \rightarrow \alpha$ does not hold in general. So, even if we can show that $\neg\alpha$ leads us to a contradiction, we are not in a position to prove α. Say, for example, that we take Kolmogorov's approach, outlined above. There, a negative answer to a problem will not, in general, return a solution. For example, if we answer "is α provable (at some stage s_i)?" negatively, this "no" represents something like a withdrawal that does not involve any commitment to a solution of α or $\neg\alpha$. But, this "no" can not be what is expressed by intuitionistic negation, since that is supposed to record commitment to $\neg\alpha$, whose truth is constructive (i.e. when we have a proof of $\neg\alpha$, this can not change over progressive stages of reasoning). This is problematic, since we do not, for example, want to say even of a mathematical statement that it is false if it is not proved, since then a statement such as "π is transcendental" would be deemed false prior to 1882. Then, to deal with statements which are provable, but not yet proved, requires a weaker than intuitionistic negation. As Dummett puts it:

> Our reluctance to say that π was not transcendental before 1882, or, more significantly, to construe mathematical statements as significantly tensed, is not merely a lingering effect of platonistic misconceptions; it is, rather, because to speak in this way would be to admit into mathematical statements a non-intuitionistic form of negation, as will be apparent if one attempts to assign a truth-value to "pi is not algebraic," considered as a statement made in 1881. This is not because the "not" which occurs in "...is not true" or "...was not true" is non-constructive: we may reasonably view it as decidable whether or not a given statement has been proved at a given time. But, though constructive, this is an empirical type of negation, not the negation that occurs in statements of intuitionistic mathematics. The latter relates to the impossibility of ever carrying out a construction of some fixed type, the former to the outcome, at variable times, of some fixed observation or inquiry [3, p. 337].

Similarly, standard intuitionistic negation is too strong to express statements such that Goldbach's conjecture is undecided at present. What would be required, instead, is a weaker negation, which expresses that the statement lacks a proof at the present stage of reasoning. This weaker negation is precisely the "not" in the statements that Dummett analyses insofar as it records non-provability at a stage of reasoning. One of the properties of this weaker negation must be that, at some future stage of reasoning, the statement may be proved or refuted (and so, the problem solved). It seems, given the above arguments, that this is a key feature of any constructivism, and so, without it, we have an expressive inadequacy regarding the ability to express undeterminedness.

In this sense, typical construals of constructive negation appear as both too weak and too strong. On the one hand, it is not possible to directly establish that some formula α is refuted, rather, it is possible only to conditionally prove that, from assuming α, we can derive a contradiction.[50] On the other hand, we seem to require a kind of "weak negation" where some potential proof of α does not go through, and so, expresses that no proof of α exists at that stage of reasoning, though without also precluding there being a proof of α or $\neg\alpha$ at some later stage. The former issue has to do with a symmetrical means by which it would be possible to directly refute some formula α; the latter, with a means of recording a "withdrawal" or state of non-acceptability for α, which does not preclude later proof or refutation.

Whilst the former requirement is already coherent with the suggestions made above, the latter is not something that we have yet discussed. However, it turns out that, on the view of proofs as *acts*, something like this kind of negation is quite natural. As briefly pointed out above, one of the arguments (e.g. [51]) made against any such view is that it is at odds with typical intuitionistic construals of negation. Simply put, the negation of α could never be asserted, whilst also requiring that in order for the conditional to be asserted, there must also exist a proof of α. So, the *proofs as acts* view, as outlined above, looks as though it is incompatible with intuitionistic negation, as Murzi puts it:

> [...] if functions could only map given proof-tokens, intuitionists couldn't interpret $\neg\alpha$ the way they do, i.e. as $\alpha \to 0$, where 0 expresses a constantly false proposition. For a proof of $\alpha \to 0$ is a function g which evidently maps any proof of α into a proof of 0. Yet there can't be a proof-token of 0! Nor can there be a proof-token of α, if it is false. It follows that g can only map hypothetical proof-tokens [...] [51].

However, whilst this would obviously be an issue for intuitionists, this just provides grist to the mill for the account offered here, in which we want to introduce refutations and denials, alongside and symmetric with proofs and assertions. So, the role ordinarily played by $\neg\alpha$ in intuitionistic logic, such that a proof of α is ruled out, may be played by an actual valid refutation of α. What is more, whilst I do not wish to get into formal details at this stage, it is fairly obvious that, should we define *a form*

---

[50] In fact, Dummett [72] suggests that something like this may be required for non-mathematical statements:

> [A] proof of the negation of any arbitrary statement then consists of an effective method for transforming any proof of that statement into a proof of some false numerical equation. Such an explanation relies on the underlying presumption that, given a proof of a false numerical equation, we can construct a proof of any statement whatsoever. It is not obvious that, when we extend these conceptions to empirical statements, there exists any class of decidable atomic statements for which a similar presumption holds good; and it is therefore not obvious that we have, for the general case, any similar uniform way of explaining negation for arbitrary statements. It would therefore remain well within the spirit of a theory of meaning of this type that we should regard the meaning of each statement as being given by the simultaneous provision of a means for recognizing a verification of it and a means for recognizing a falsification of it, where the only general requirement is that these should be specified in such a way as to make it impossible for any statement to be both verified and falsified (p. 71).

If we translate the latter stipulation into a BHK-style clause for negation, so that the negation of α is verified whenever α is falsified (and vice-versa), then this is analogous to Nelson's strong negation [73], and discussed in a similar context in [74].

of negation by means of conditional and absurdity, 0, that would only ever concern *hypothetical* proofs, and so would provide traction on something like a hypothetical proof that has *failed*. This, weaker than typical negation would then provide a way of dealing with the issues discussed above, such that they can make sense of the status of claims such as "π is transcendental" prior to 1882.[51]

This emphasis on refutation, as playing a role that is symmetric to that of proof, has significant historical connections with the way in which dialogue has been construed historically, in logic, mathematical inquiry, scientific inquiry, and rational argument. It is this to which we turn in the following chapter, where it is shown that refutation is central to a social, and dialogical approach to reasoning. In doing so, I will show how the picture presented above can be clarified and expanded by analysing the ways in which assertion and denial, dialogical response and testing, are central to the dynamics of reasoning and the creation of new knowledge. In particular, it will be argued that these processes should be understood to be *dialectical*, such that the dynamics of reasoning and the construction of meaning is intertwined.

References

1. A. N. Prior. The runabout inference ticket. *Analysis*, 21:38–39, 1960.
2. Gilbert Ryle. 'if','so', and 'because'. *Philosophical analysis*, pages 323–340, 1950.
3. Michael A. E. Dummett. *Elements of Intuitionism*. Oxford University Press, 2000.
4. Dag Prawitz. Meaning approached via proofs. *Synthese*, 148(3):507–524, 2006.
5. Huw Price. 'Not' again. unpublished.
6. Greg Restall. Multiple conclusions. In *Logic, methodology and philosophy of science: Proceedings of the twelfth international congress*, pages 189–205. Kings College Publications, 2005.
7. Greg Restall. Assertion, denial and non-classical theories. In Francesco Berto, Edwin Mares, Koji Tanaka, and Francesco Paoli, editors, *Paraconsistency: Logic and Applications*, pages 81–99. Springer, 2013.
8. Ian Rumfitt. Yes and no. *Mind*, 109(436):781–823, 2000.
9. G. Restall. Multiple conclusions. In Valdes-Villanueva L. Hajek P. and Westerstpahl D., editors, *Logic, Methodology, and Philosophy of Science: Proceedings of the Twelfth International Congress*, pages 189–205. Kings College Publications, 2005.
10. Timothy Smiley. Rejection. *Analysis*, 56(1):1–9, 1996.
11. Jaroslav Peregrin. Meaning as an inferential role. *Erkenntnis*, 64(1):1–35, 2006.
12. Jaroslav Peregrin. Inferentializing semantics. *Journal of Philosophical Logic*, 39(3):255–274, 2010.
13. Greg Restall. What are we to accept, and what are we to reject, while saving truth from paradox? *Philosophical Studies*, 147(3):433–443, 2010.
14. David Ripley. Paradoxes and failures of cut. *Australasian Journal of Philosophy*, 91(1): 139–164, 2013.
15. J Michael Dunn and Gary Hardegree. *Algebraic methods in philosophical logic*. OUP Oxford, 2001.

[51] Whilst such weakly-negated formulas would obviously fail standard monotonicity conditions (i.e. that once α is negated, it is always negated), as [75, Sect. 2.4] points out, adding such an operator to standard intuitionistic logic would be conservative since its existence has no impact on the usual interpretation of all other connectives.

16. Florian Steinberger. Why conclusions should remain single. *Journal of Philosophical Logic*, 40(3):333–355, 2011.
17. Ian Rumfitt. Knowledge by deduction. *Grazer Philosophische Studien*, 77(1):61–84, 2008.
18. Jean-Yves Girard, Yves Lafont, and Paul Taylor. Proofs and types, volume 7 of. *Cambridge tracts in theoretical computer science*, 7, 1989.
19. Reinhard Kahle and Peter Schroeder-Heister. Introduction: Proof-theoretic semantics. *Synthese*, 148(3):503–506, 2006.
20. Michael A. E. Dummett. *The Logical Basis of Metaphysics*. Harvard University Press, 1991.
21. Luca Tranchini. Proof-theoretic semantics, paradoxes and the distinction between sense and denotation. *Journal of Logic and Computation*, 2014.
22. Stephen Read. Proof-theoretic validity. In C. Caret and O. Hjortland, editors, *Foundations of Logical Consequence*. Oxford University Press, 2014.
23. Peter Schroeder-Heister. The categorical and the hypothetical: A critique of some fundamental assumptions of standard semantics. *Synthese*, 187(3):925–942, 2012.
24. Enrico Martino and Gabriele Usberti. Temporal and atemporal truth in intuitionistic mathematics. *Topoi*, 13(2):83–92, 1994.
25. Panu Raatikainen. Conceptions of truth in intuitionism. *History and Philosophy of Logic*, 25(2):131–145, 2004.
26. Yaroslav Shramko, J. Michael Dunn, and Tatsutoshi Takenaka. The trilaticce of constructive truth values. *Journal of Logic and Computation*, 11(1):761–788, 2001.
27. A. Heyting. *Intuitionism*. Amsterdam, North-Holland Pub. Co., 1971.
28. A. N Kolmogorov. Zur deutung der intuitionistischen logik. *Mathematische Zeitschrift*, 35:58–65, (English translation in Mancosu 1998, pp. 328–334) 1932.
29. Paolo Mancosu. *From Brouwer to Hilbert: The Debate on the Foundations of Mathematics in the 1920s*. Oxford University Press, 1998.
30. Jacques Dubucs and Mathieu Marion. Radical anti-realism and substructural logics. In A. Rojszczak, J. Cachro, and G. Kurczewski, editors, *Philosophical Dimensions of Logic and Science*, pages 235–249. Kluwer Academic Publishers, 2003.
31. Jacques Dubucs. Feasibility in logic. *Synthese*, 132(3):213–237, 2002.
32. Michael Dummett. Reply to Dag Prawitz. In Barry Taylor, editor, *Michael Dummett: Contributions to Philosophy*, pages 281–316. Distributors for the United States and Canada, Kluwer Academic Publishers, 1987.
33. LEJ Brouwer. *LEJ Brouwer: Collected Works*. Amsterdam: North-Holland Publishing Company, 1975.
34. Gustavo Fernández Díez. Kolmogorov, Heyting and Gentzen on the intuitionistic logical constants. *Critica*, 32(96):43–57, 2000.
35. Jean-Yves Girard. On the meaning of logical rules i: syntax versus semantics. In *Computational logic*, pages 215–272. Springer, 1999.
36. Göran Sundholm. Constructions, proofs and the meaning of logical constants. *Journal of Philosophical Logic*, 12(2):151–172, 1983.
37. Per Martin-Löf. On the meanings of the logical constants and the justifications of the logical laws. *Nordic Journal of Philosophical Logic*, 1(1):11–60, 1996.
38. Michael Dunn and Greg Restall. Relevance Logic. In D. Gabbay and F. Guenthner, editors, *Handbook of Philosophical Logic*. Kluwer, 2002.
39. Joachim Lambek and Philip J Scott. *Introduction to higher-order categorical logic*, volume 7. Cambridge University Press, 1988.
40. Robert Brandom. *Making It Explicit: Reasoning, Representing, and Discursive Commitment*. Harvard University Press, 1994.
41. David H. Sanford. *If P, Then Q: Conditionals and the Foundations of Reasoning*. Routledge, 2004.
42. Dorothy Edgington. Conditionals, Truth and Assertion. In Ian Ravenscroft, editor, *Minds, Ethics, and Conditionals: Themes From the Philosophy of Frank Jackson*. OUP Oxford, 2009.
43. Frank Ramsey. General propositions and causality. In D. H. Mellor, editor, *Philosophical Papers*, pages 145–163. Cambridge University Press, 1990.

44. W. V. Quine. *Methods of Logic*. Harvard University Press, 1982.
45. Lloyd Humberstone. *The Connectives*. MIT Press, 2011.
46. Robert Stalnaker. Conditional propositions and conditional assertions. In Andy Egan and B. Weatherson, editors, *Epistemic Modality*. Oxford University Press, 2009.
47. Dorothy Edgington. On conditionals. *Mind*, 104(414):235–329, 1995.
48. Timothy Williamson. Knowing and asserting. *Philosophical Review*, 105(4):489, 1996.
49. Matthew Weiner. Must we know what we say? *Philosophical Review*, 114(2):227–251, 2005.
50. Michael Dummett. Truth. *Proceedings of the Aristotelian Society*, 59(1):141–162, 1959.
51. Julien Murzi. Knowability and Bivalence: Intuitionistic Solutions to the Paradox of Knowability. *Philosophical Studies*, 149(2):269–281, 2010.
52. John MacFarlane. What is assertion? In Jessica Brown and Herman Cappelen, editors, *Assertion*. Oup Oxford, 2011.
53. Peter Pagin. Assertion. *Stanford Encyclopedia of Philosophy*, 2015.
54. Peter Pagin. Problems with norms of assertion. *Philosophy and Phenomenological Research*, 91(2):n/a–n/a, 2015.
55. Charles Sanders Peirce. Belief and judgment. *Collected papers*, 5:376–387, 1934.
56. Robert Brandom. Asserting. *Noûs*, 17(4):637–650, 1983.
57. Ludwig Wittgenstein. *Philosophical Investigations, 4th Edition (Trans. Hacker and Schulte)*. Wiley-Blackwell, 2009.
58. Gary Watson. Asserting and promising. *Philosophical Studies*, 117(1-2):57–77, 2004.
59. Robert Brandom. *Articulating Reasons: An Introduction to Inferentialism*. Harvard University Press, 2000.
60. Michael Rescorla. Assertion and its constitutive norms. *Philosophy and Phenomenological Research*, 79(1):98–130, 2009.
61. Mathieu Marion. Why play logical games? In Ondrej Majer, Ahti-Veikko Pietarinen, and Tero Tulenheimo, editors, *Games: Unifying Logic, Language, and Philosophy*, pages 3–26. Springer Verlag, 2009.
62. Peter Pagin. Assertion, inference, and consequence. *Synthese*, 187(3):869–885, 2012.
63. Geoffrey Nunberg. The non-uniqueness of semantic solutions: Polysemy. *Linguistics and Philosophy*, 3(2):143–184, 1979.
64. Sanford Shieh. On the conceptual foundations of anti-realism. *Synthese*, 115(1):33–70, 1998.
65. Mathieu Marion. Game semantics and the manifestation thesis. In *The Realism-Antirealism Debate in the Age of Alternative Logics*, pages 141–168. Springer, 2012.
66. Alain Lecomte. Ludics, dialogue and inferentialism. *Baltic International Yearbook of Cognition, Logic and Communication*, 8(1), 2013.
67. Michael Dummett. What is a theory of meaning? (ii). In Gareth Evans and John McDowell, editors, *Truth and Meaning: Essays in Semantics*. Oxford: Clarendon Press, 1976.
68. Michael Dummett. *Frege: Philosophy of Language*. Duckworth, 1973.
69. Ludwig Wittgenstein. *Philosophical grammar*. Univ of California Press, 2005.
70. Neil Tennant. Rule-circularity and the justification of deduction. *Philosophical Quarterly*, 55(221):625–648, 2005.
71. Peter Schroeder-Heister. Validity Concepts in Proof-Theoretic Semantics. *Synthese*, 148(3):525–571, 2006.
72. Michael A. E. Dummett. *The Seas of Language*. Oxford University Press, 1993.
73. David Nelson. Constructible falsity. *Journal of Symbolic Logic*, 14(1):16–26, 1949.
74. A Pietz. Not quite intuitionism. In *Deutscher Kongress für Philosophie*, 2011.
75. Graham Solomon and David DeVidi. Empirical negation in intuitionistic logic. In *A logical Approach to Philosophy*, pages 151–168. Springer, 2006.

Chapter 4
Dialogue and Refutation

In this chapter, I provide an overview and argument to the effect that reasoning, including scientific and mathematical inquiry, is inherently both dialogical and dynamic. That is to say, reasoning is social and interactive, and requires an account of history that is not captured by an absolutist and monological account of proofs. I begin by considering the role of dialogue in logic, beginning first with a short analysis of Brandom's account of dialogue, before discussing the history of logic, and recent approaches to logical games in the work of Jaako Hintikka and Lorenzen. There, I argue that a distinct approach is required where assertion and denial, dialogical response and testing, are central to the dynamics of reasoning and the creation of new knowledge. It is in this light that I go on to discuss the specific role of dialogue in mathematical proofs, with particular attention to the work of Imre Lakatos. Lakatos' [1] approach to inquiry is understood to operate as a kind of dialogue between "prover" and "refuter", where a conjecture is considered by means of proof attempts and tests in the form of refutation attempts, disproofs and counter-examples. Such "proofs that do not prove" are part and parcel with the development of inquiry, until a conclusive proof or refutation of the conjecture is reached. I go on to provide a discussion of the broader approach to dialogues, with particular reference to the interactional work of Ginzburg (e.g. [2]). I will discuss a number of constraints over dialogues taking interaction, rather than rules, to be a priori, such that co-operative interactions may be understood to form the basis of the kinds of approaches to inquiry previously discussed. I will also briefly turn to considering how this approach fares in relation to providing a way of accounting for "reasons", which does not fall into the trap of important criticisms in feminist and post-colonialist theory. I show that the dialogical approach being advocated here is capable of providing a way of moving beyond such criticisms, with particular attention to the role of emotion in interactive reasoning processes as allowing responses to be "framed" adequately. I finish by drawing together the discussion, arguing that interactions may subsequently give rise to propositional meaning under constraints of coherence, convergence, and symmetry.

J. Trafford, *Meaning in Dialogue*, Studies in Applied Philosophy, Epistemology and Rational Ethics 33, DOI 10.1007/978-3-319-47205-8_4

1 Dialogues and Games

In the kind of "games" introduced in the previous chapter, we consider the activity of proving (and refuting) by means of the reasoned evaluation of statements in dialogical scenarios. This kind of approach to reasoning through dialogue has a long history, from at least Greek approaches to dialectic, or "speech between two" in Aristotle and Plato, through medieval *Ars Obligatoria*, to more recent formal analyses in the work of Lorenzen dialogues [3] and Hintikka's game-theoretic semantics [4], and also the dialogue theory of Hamblin [5] and Walton and Krabbe [6]. Whilst these approaches share many features, they also diverge in numerous ways, such as, for example, the relationship between games and inquiry; the way in which "rules" of games are conceived; the relationship between games and actual dialogues; the goals of games, and so on.[1]

For example, in Chap. 1 a number of reasons were given against thinking that agents' activities in a game are prescribed by certain rules of inference. In Chap. 2, I discussed a way of thinking about proof and assertion that liberalises the standard inferentialist account, and places both in the context of interactive games. The account developed there leant upon Brandom's philosophical approach, which requires games to be part-and-parcel with assertions. However, Brandom's formal account follows that of Dummett and Prawitz (discussed in the previous chapter). So, his formal approach to assertion is given in terms of natural deduction rules, which are understood as norms governing correct moves in a game in a similar vein to the dispositionalist approach discussed in Chap. 1. In the previous chapter, I argued that we could reinterpret these games as having the goal of providing proofs, and possibly also refutations, of statements. In addition, on the view outlined there, I said that the meaning of statements should not be divorced from the means by which they are established, in which case, the process of constructing propositional content is also intertwined with these games of giving and asking for reasons. As such, we might look to more minimal normative constraints than rules, over these dialogue games, and through which the meaning of logical statements can be constructed. For example, this way of approaching "games" requires that agents are open to reasons and tests towards the construction of propositional content. This requires that such content is not circumscribed by the individual agents' presuppositions, but rather it is, to some extent, generated by that interaction, and without requiring that the rules of the game are fixed in advance.

An additional reason why Brandom's approach is unsatisfactory in this respect is that it requires that the rules of the game are fixed in advance, but also, that there is a lack of symmetry between agents involved in the games. This is perhaps shown most clearly by the fact that, the interpretation of speech-acts in Brandom's theory makes

[1]There is also the issue of what counts as a legitimate question in a dialogue, and whether or not all statements allow for legitimate questioning. For example, Brandom [7] suggests that: '[c]laims such as *There have been black dogs* and *I have ten fingers* are ones to which interlocutors are treated as prima facie entitled. They are not immune to doubt in the form of questions about entitlement, but such questions themselves stand in need of some sort of warrant or justification' (p. 177).

central use of a "community" involved in the assessment of utterances made by a speaker, rather than by: '[...] an addressee who is expected to give the speaker an answer. Every round of a new discourse opens with an ascription that the interpreter undertakes from the observer's perspective of a third person' [8, p. 161].[2] In [7], Brandom argues that agents involved in this process also act as "scorekeepers", who keep track of each others' commitments and entitlements, where commitments are sentences to which an agent is committed (though perhaps unknowingly as consequence of other commitments), and entitlements are sentences to which an agent is justified after having defended them successfully (e.g. Sect. 3). These ideas come together, in [7], with the idea that agents' have "deontic scoreboards", instituting what Brandom calls, the 'communal assessment theory' (p. 37). The issue with this emphasis on the community of assessment, over the interaction of two agents, is that it introduces an external assessment process, which takes precedence over the internal dynamics of the dialogue itself. For example, in [10], Brandom states that: '[u]nderstanding a speech act—grasping its discursive significance—is being able to attribute the right commitments in response. This is knowing how it changes the score of what the performer and the audience are committed and entitled to' (p. 165). That is, rather than see the game of giving and asking for reasons as emerging from the testing relationship that we discussed in the previous chapter, we are now asked to introduce a kind of checking-mechanism in the form of the third-person. It is my supposition that this is why Brandom's formalism is close to Dummett-Prawitz. The natural deduction formalism privileges a monological approach to proofs, and the requirement of third-person assessment plays a role that is, perhaps, more analagous to the role of model-theoretic checking in modest approaches to inferentialism, as discussed in Chap. 1.

The kind of approach discussed in the previous chapter, in distinction to Brandom, takes the dialogue game between two agents to be essential to the process of giving and asking for reasons, as an activity of the construction of proofs and refutations for statements (that is to say, where a proof may be identified with certain dialogical interactions a posteriori). In this respect, the assessment of utterances takes place immanently to each other, with no external assessment process. This assessment occurs through a relationship of testing, working towards the point where the two agents *agree* upon what would count as justifying the proof, or refutation, of the statement in question.[3] As such, it is necessary that, internal to these interactions between agents, there is a goal-directed norm such that the two "players" aim towards the construction of a proof, and also that they interact in order to attempt to justify the statement that forms the focus of the interaction. In this vein, Brandom's [11] account of entitlement, where;

[2]See also the discussion of this issue in [9]. It is worth pointing out that Habermas' theory actually coheres with Brandom's to a great extent, but with a much greater emphasis on political action and democracy.

[3]In [7], Brandom discusses what would count as a legitimate challenge to an assertion (this is analogous to what we are calling a TEST), suggesting that this may simply be some reason to doubt the statement, or that it may take the form of an assertion of its negation (or what we have in the form of a direct denial).

> Giving reasons for a claim is producing other assertions that *license* or *entitle* one to it, that *justify* it. Asking for reasons for a claim is asking for its *warrant* for what entitles one to that commitment. Such a practice presupposes a distinction between assertional commitments to which one *is* entitled and those to which one is *not* entitled. (p. 114)

may be reinterpreted in terms of agreement upon what counts as justification for a statement, which both players would then become *entitled* to. This would ensure that the approach to interaction does not require any external arbiter, or assessment, of the dynamics of the game itself. So, we can think of the construction of a proposition over the course of an interaction by its co-operative nature between agents who are directly involved in that interaction (rather than as external assessors).[4] That is to say, the construction of a proposition over the course of an interaction would be a collective process that requires the symmetrical involvement of agents engaged in cooperative activity. The meaning of a statement, then, is inextricable from agreement on what counts as justifications for a statement *by the agents involved in the interaction.* Indeed, it is in this regard that this approach to dialogue games is most radical in distinction to more standard approaches to formal dialogue and games, so it is this to which we now turn.

2 Logical Games

By considering inquiry as a game constructed as a dialogue, we are explicitly calling to mind the dialectical understanding of dialogue in the Socratic elenchos. The dialogues are well-known to be central to the philosophical activity of Plato, and also Aristotle considers different forms of argument: didactic, dialectical, examination and contentious (eristic) arguments (discussed in [13]).[5] I do not intend to dwell on this, but note that the typical interpretation of such dialogue is that there exists a thesis whose proponent must reply to a set of questions from an opponent. The opponent's aim is to drive the proponent to an inconsistency involving the thesis, so as to construct an elenchos, (taken to be a refutation of the thesis) (e.g. [15]). In this vein, we might follow [16], where it is argued that the aim of such dialogue games is to ensure that the thesis passes a test of consistency:

> α has to fit into a set of beliefs Γ held by its proponent in such a way that the addition of α to Γ does not render it inconsistent. If and when the opponent unveils an inconsistency, it is shameful for the proponent.

As such, the elenchus provides a way of testing the initial assertion by means of 'consistency management' [16].[6] When it happens that the proponent is driven into an

---

[4]See also [12], where a formal structure for identifying propositions as types with successful games is discussed in detail.

[5]This is prior to the consolidation of argument in the form of the syllogistic by Aristotle in the *Prior Analytics*, see [14] for a discussion of this process.

[6]There is an interesting connection, then, between this notion of the maintenance of consistency and the maintenance of coherence in Restall's work, discussed in the previous chapter.

elenchus, it is required of them that they retract their initial assertion, and reconsider their understanding of the concepts involved (see the discussion in [16]).[7]

In [18], an overview is given, of how these sort of dialectical games resurface in the Medieval *Ars Obligatoria*, where, again, a dialogue is modelled in terms of a commitment to avoid falling into contradiction once the proponent has accepted or denied an initial sentence. In these dialogues, a timeframe is specified, and either proponent or opponent is said to "win" if, either the opponent is unsuccessful in driving the proponent to a contradiction, or is successful in driving the proponent to a contradiction.[8] In Hodges' [19] discussion of the relationship between logic and games, he characterises *obligations* slightly differently. There, he says that both *opponents* and *respondents* agree on a "positum", which is usually a clearly false statement. Then, respondents answers questions from opponents, which assume that the positum is true. Again, it is the task of opponents to drive respondents into contradictions. It is interesting, in the context of the discussion above, that, as Hodges notes, these games tend not to prescribe rules that must be followed by the players, and also that they may be open-ended. For this reason, Hodges argues that they can not constitute logical games at all (which would then count against our view also). However, this view, is argued against in detail in the work of Dutilh Novaes (e.g. [20]), and, in particular, she shows how similar obligations is to contemporary developments in logical semantics.

Nonetheless, it is common to both approaches, that proponent defends a thesis, whilst opponent attempts to lead proponent to make a series of claims that end with the proponent contradicting themself. So, the idea is that an initial thesis may be refuted by means of its leading to a contradiction, and, in particular, this is often used in the form of an indirect proof. This equates to the idea that, for a thesis α, if $\neg\alpha$ leads to a contradiction, then α must be correct (or true) (e.g. [16]). As such, the practice of dialectics, and obligations, looks to the logical principle of non-contradiction to provide the driving force of the dialectic, but also to the principle of excluded middle to provide a means by which these dialectics can be transformed into proofs. As such, these approaches neither capture the co-operative interactions that we are interested in, nor do they adequately provide a *constructive* approach to the activity of proving (or refuting). This latter point is not merely logical, rather, it also reflects the fact that, for formal dialogues of this kind, we end up with something that looks much more like a formula-checking operation. As such, the role of opponent appears to be just an externalisation of the standard logical practice of proof-checking. For example, if we could make explicit the set of initial commitments that proponent holds, together with the thesis, then we merely have to check that list to see whether or not there exists α and $\neg\alpha$, for some α, within it, to know whether or not the initial thesis may be refuted. In addition, in the previous chapter, I have suggested that there are good

[7] See also the discussion in [17].

[8] A distinct approach is discussed in [18], which he calls the "argument method", which is more similar to the approach argued for here, since it proceeds until both proponent and opponent agree on some statements that both accept.

reasons to think that a primitive notion of denial is required in addition to negation, and that this should be treated as symmetrical with assertion.[9]

These criticisms also cohere with issues that are raised against more recent dialogical approaches to logic, particularly in the work of Lorenzen, and Hintikka, by Wilfred Hodges [19, 21]. Let us briefly rehearse these approaches before considering Hodges' criticisms. Both approaches, whilst differing in detail and motivation, interpret formulas by identifying them with winning strategies in a particular game. These games are asymmetric, and they are governed by a set of rules that a player must follow in making each move subsequent to the first. In Lorenzen games, a game ends when a player loses because they have no moves left to make, whilst in Hintikka games, a winning strategy requires a player to end up with the assertion of an atomic formula that is true in a model.

Lorenzen dialogue games begin with proponent making an assertion, before opponent challenges it, taking turns to make moves thereafter until the game has ended. "Particle" rules prescribe the way in which a statement may be attacked, or defended, so that, at any stage over the course of a dialogue, there are rules that clearly specify how each player is to proceed, until no further moves are available. So, for example, if proponent asserts $\alpha \wedge \beta$, then opponent must choose one of the conjuncts to attack, so proponent then defends that conjunct; if proponent asserts $\alpha \vee \beta$, then opponent asks that proponent choose one of the disjuncts to defend; if proponent asserts $\alpha \rightarrow \beta$, then opponent concedes α and attacks β; if proponent asserts $\neg\alpha$, then opponent is forced to assert α which also requires that the roles of proponent and opponent are switched; if proponent asserts $\forall x F(x)$, opponent takes a value for x to be attacked; if proponent asserts $\exists x F(x)$, then opponent requests that proponent produce an x that has property F.[10] More formally, take P and O as constant symbols representing Proponent and Opponent, then, for any expression α we let $P : \alpha$ and $O : \alpha$ denote that α is signed by P and O. X and Y are two variables taking their valies in P, O, where $X \neq Y$, and let A and D denote Attack and Defence.

Definition 11 (*Lorenzen dialogue* [23]): A dialogue is a couple (γ, θ) such that:

- γ is a sequence of signed expressions
- θ is a function on the ranks of the signed expressions in γ such that $\theta(n)$ is a couple $[m, Z]$ where $m \leq n$ and $Z = A$ or D, which satisfies the following three conditions:
- $\gamma(n)$ is P-signed if n is even and O-signed if n is odd; $\gamma(n)$ is a compound formula;
- If $\theta(n) = [m, A]$, then $\gamma(m)$ is a compound formula and $\gamma(n)$ is an attack on $\gamma(m)$ which corresponds to a regular argumentation form;
- If $\theta(p) = [n, D]$, then $\theta(n) = [m, A]$, and $\gamma(p)$ is an answer to the attack $\gamma(n)$ which corresponds to a regular argumentation form.

---

[9]Further still, in order that we treat proofs as constructive, we have said that $\neg\alpha$ may not bring with it the refutation of α, since, as long as $\neg\alpha$ is defined as $\alpha \rightarrow 0$ (which we may assume to cohere with this view from the description of a proponent accepting α being driven to an elenchus), then it is impossible to assert $\neg\alpha$ at the termination of a dialogue (such that the dialogue can be identified with a valid proof).

[10]See also [22] for an overview of both Lorenzen and Hintikka games.

"Structural" rules, in distinction to particle rules, provide a set of stipulations that are taken to determine the overarching structure of dialogues. For example, it is typical to take as structural rules that all dialogues must end in a finite number of moves, and that whoever can not move has lost, and their adversary has won.[11] An additional rule is the FORMAL RULE, which states that proponent may not utter an atomic sentence, unless it has already been uttered by opponent. This rule, together with the particle rule for conditionals, ensures that there is an asymmetry between proponent and opponent, which is built-in to the structure of the dialogue.[12] In fact, this is exactly analogous to the asymmetry discussed in relation to intuitionistic logic in the previous chapter. If anything, this is made more explicit here, since, if proponent were allowed to assert atomics, then this would bring in the meaning of atomic formulas, which would lead to a requirement of a primitive (or model-theoretic) account of true statements, which are not proved true.

It is in this latter domain that Hintikka's [4] game-theoretical semantics (GTS) may be understood, since GTS is more aligned with model theory than proof-theoretical approaches to semantics. Hintikka's GTS is primarily concerned with providing an analysis of quantifiers, explaining these, and other logical rules, by means of a semantic game whose task is to find a correlation between utterances and objects (as described by a model-theory). As such, as Tulenheimo [22] points out, '[t]he rules giving rise to semantic games are not argumentative or discursive [...] they are *correlative* or *projective*. A move in a semantic game consists of creating a correlation between a word and an object; it is not any sort of communicative act' (Sect. 3). It follows that, in distinction with Lorenzen's formal rule, players are actively encouraged to search for true atomic formulas in a model in which the truth-values of atomic formulae are fixed. So, in this sense, GTS can not be considered to cohere with the aims of an inferentialist approach to logic and meaning, and, rather, may be better understood as attempting to provide a means of elucidating the model-theoretic approach to meaning by means of winning strategies in a game.[13] For example, universal and existential quantifiers are differentiated from each other in terms of which player is allowed to choose a value from the model that the variable x ranges over, to see whether or not the formula $F(x)$ is satisfied in that model; similarly, conjunction and disjunction are distinguished by means of whether or not every conjunct or some disjunct is true in the relevant model; negation switches positions, much like Lorenzen dialogues; conditionals are unnecessary because $\alpha \rightarrow \beta$ is defined as $\neg\alpha \vee \beta$, since

[11]It is also common to take as structural rules certain constraints on which moves must be attacked or defended, and in which order. For example, if each player may attack, or respond to any attack, including those that have already been defended against, then the dialogues will have the same provable formulas as classical logic, whereas if these are restricted so that it is not possible to revise a defense, they will be equivalent with intuitionistic provability, see [23].

[12]On the asymmetry induced by the rules for conditionals, see the discussion in [22, Sect. 2] where it is argued that any explanation of the particle rules will make use of semantic assumptions involving, for example, role distribution of players.

[13]There are connections between GTS and Lorenzen's "material" games, however, which replace the formal rule with a MATERIAL rule that allows true atomics to be asserted. Further discussion may be found in [22, 24].

the two are identical according to classical logic, for which Hintikka is intent upon providing a game-theoretic explanation. As pointed out in [22], this has the result that, once a set of rules for play, and winning conditions, are specified, all strategic properties of the game are fixed.

The key distinction between GTS and Lorenzen dialogues, then, is that GTS requires an external model to fix the meaning of atomic propositions, assuming that players have access to this model, whilst Lorenzen dialogues are much more "internal", in that only previously conceded atomic propositions may be asserted. However, there is a problematic circularity in both approaches, due to the relationship between a winning strategy, and the plays themselves. In Lorenzen games, for example, the particle rules are supposed to be both normative over agents' commitments, depending upon which role they are playing, and also descriptive of the possible ways in which utterances may be challenged and defended in ordinary dialogue, whilst structural rules are supposed to govern overarching and semantic attributes of dialogues in terms of attribution of strategy. As pointed out in [22, Sect. 4], agents' commitments must already be strategic in order to make sense of dialogue games at all. So, there is a circularity where the semantics specified at the level of strategy must already be presupposed at the level of plays, since, if the strategic notions were absent, then there would be no justification for the particle rules.[14] A similar circularity to the one identified in the introductory chapter with respect to model-theoretic attempts to justify logical rules, is also argued to hold for GTS by [22, Sect. 4]. There, it is argued that the fact that the meanings of logical expressions relies upon sentential context also requires a presupposition of their truth in that context. This, again is a strategy-level notion, that can not be built-in at the foundation of an attempt to justify the game rules that are, in turn, supposed to provide its definition.

Even if these issues could be dealt with, there exists a much greater challenge. For example, in [19, Sect. 2], Hodges argues that, whilst these approaches to logic purport to capture some sort of pre-existing practice such as argumentation and debate, in fact the formal account provides no traction on these at all, and we end up with a system that is of little explanatory value:

> [i]n many applications of logical games, the central notion is that of a winning strategy for the [proponent]. Often these strategies (or their existence) turn out to be equivalent to something of logical importance that could have been defined without using games: for example a proof. But games are felt to give a better definition because they quite literally supply some motivation: [proponent] is trying to win. This raises a question that is not of much interest mathematically, but it should concern philosophers who use logical games. If we want [proponent's] motivation in a game [...] to have any explanatory value, then we need to understand what is achieved if [proponent] does win. In particular we should be able to tell a realistic story of a situation in which some agent called [proponent] is trying to do something intelligible, and doing it is the same thing as winning in the game. [19]

Hodges calls the latter question the "Dawkins question" (this is also explored in [24]). There are two inter-related issues raised by the Dawkins question. The first has to do with mathematical interest, and whether or not the formal structures provided by

[14]There is an obvious analogy with the discussion of the dispositionalist approach in Chap. 1.

Lorenzen and Hitikka achieve something that it is not just of philosophical interest. This issue is somewhat tricky, since it is certainly not beyond mathematical interest to show that distinct systems are mathematically identical. So, for example, the winning proponent strategies in Lorenzen dialogues can be shown to be identical with intuitionistic (and classical, depending upon the restrictions upon moves) provability, by an equivalence theorem between them, is not without interest in itself. Furthermore, in [25], for example, it is shown that winning proponent strategies correspond with sequent calculus proofs, that do not require connective rules, but rather parameterize left and right rules corresponding to the way in which attacks and defenses are specified in Lorenzen dialogues. Since it may be presumed that Hodges is well aware of these mathematically interesting results, I think that what Hodges must be getting at is that there is little that is explanatorily substantial about these identifications. For example, we may ask what has been achieved by the dialogical explanation of logic if we end up with a fairly trivial correspondence between standard intuitionistic proofs and winning strategies in a Lorenzen dialogue. This "feeling" of triviality is then directly connected to with the second issue raised by Hodges concerning whether it is plausible to provide a realistic story of what a proponent is doing, or attempting to do, which is identifiable with the provision of a winning strategy in a logical game.[15] For example, it may be of obvious mathematical interest to plot, formally, plays within an interaction that do not end up corresponding to proofs (or winning strategies), and that are representative of moves within ordinary dialogue games that *attempt* to provide support for some statement or other, and yet fail to do so.

That is to say, both dialogue games and GTS build-in an obvious divergence between actual dialogical activities and the formal structure that is meant to capture them. For instance, for all that Lorenzen dialogues purport to be motivated by their foundation in debate, argumentation, and dialogue, what we end up with is little more than a formula-checking procedure, where we simply take a complex formula, and decompose it, step-wise, in a manner similar to the tableaux method of proof-checking.[16] Both approaches require predetermined criteria for deciding on the existence of a winning strategy, whether a model for atomic formulas in GTS, or a requirement that a player "runs out" of moves to play in Lorenzen dialogues. As such, these predetermined criteria can be better understood by means of traditional game-theoretic notions such as payoff functions. For example, in game theory, there are rules that prescribe players' actions which are also dependent upon their opponents actions, whilst the goal of a game is the payoff that each player aims towards (for example, winning the game of checkers). Moreover, the aim of game theory is to figure out strategies that, given certain specified payoff functions, and certain contexts, will enable players to move in response to previous moves by the opponent, such that a winning strategy may be found that guarantees the payoff function regardless of the way that the game proceeds. Of course, this requires that a game is both finite, and also that it is deterministic, both of which are also the case for

[15]The question of whether or not the formal account given in the following chapter is of mathematical interest will be dealt with (and hopefully answered in the affirmative) in Chap. 6.

[16]On the latter, see (e.g. [26]).

GTS and Lorenzen dialogues. Whilst it is certainly the case that some dialogues, and arguments, end with what may be interpreted as a payoff function, it is surely not the case that all do. Indeed, even the notion of a winning strategy is alien to many forms of dialogue, and linguistic activity, which may be both non-deterministic and infinite. Furthermore, whilst these approaches to dialogue require an antagonistic relationship between players, many forms of dialogical inquiry (as discussed above) are co-operative in the sense that both agents base their *disagreement* upon a more fundamental *agreement* regarding the focus and coherence of the dialogue.

As such, there is a substantial discrepancy between these approaches to games and dialogue, and the kind of activity that we are interested in here. For example, rather than prescribe the set of possible moves available to players from the outset of the game, we need to leave open the possibility that a game is open-ended, and also that it is not possible to predetermine winning strategies by proof-checking. Furthermore, whilst standard approaches require that either proponent or opponent "win", just in case there exists a winning strategy, if we think of games as more co-operative than antagonistic practices, then a game may be "won" where there exists an agreement between players that adequate justification has been given either to prove, or to refute, an utterance. This is an important distinction if we would like interactions to allow agents to construct something novel (a proof, for example), which is not merely unpacking what is already contained in the formal structure (by decomposing a complex formula, for example). From a game-theoretic point of view, this kind of flexibility is not possible. We could, for example, just say that whatever assumptions regarding atoms are required for a winning strategy should be adopted, and then allow players to run through games altering assumptions until a winning strategy is found. But, this does not adequately reflect the interaction of agents in dialogue, which may involve unexpected information, or unexpected "plays" in the game, that can not be predetermined at the outset by the strategy players have decided to follow. Furthermore, as Başkent [27] points out, this would also be at odds with the very notion of a strategy, which is suppose to be a function determining which move a player ought to make at each state, as based upon the moves which other players have made. If the existence of a winning strategy is determined simply by a posteriori success (depending upon which assumptions, and which moves work), then this calls into question the very idea of an a priori strategy existing prior to the playing of the game.

3 Mathematical Games

Let us now turn to the notion of mathematical dialogue games, particularly in the work of Imre Lakatos. In [1], Lakatos develops a methodological framework for mathematical reasoning that is in sharp distinction to a formalist (or rigorist) program (e.g. p. 137ff). The latter is understood to be the reconstruction of conclusive proofs by means of a set of inferential steps from axioms. According to Lakatos, this fails to reflect both the actual process by which mathematical theorems are constructed, and also the centrality of informal and fallible "proofs that do not prove" (e.g. p. 13). Instead, Lakatos' suggested framework operates as a kind of dialogue between "prover" and "refuter", where a conjecture is considered by means of proof attempts, and tests, in the form of refutation attempts, disproofs and counter-examples, either to the conjecture itself or to a lemma of the proof attempt. For Lakatos, then, a "proof" is not required to be conclusive (something alien to the rigorists (p. 137)), rather, it may be a rough set of arguments in support of a conjecture that admits reexamination and revision. In this sense, Lakatos' approach to philosophy of mathematics may be understood as an extension of Karl Popper's [28, 29] approach to philosophy of science. So, just as Popper argues that scientific claims are open to refutation, so too mathematical theorems ought not to be seen as irrefutably true, but rather conjectures that may come to be refuted at some later stage of inquiry.[17] However, whereas Popper's framework is better understood to be FALSIFICATIONIST, in the sense that statements may never be truly verified, only refuted, Lakatos' is more dialectically oriented. That is, in Lakatos' framework it must be possible to deal with the simultaneous consideration of proofs and refutations, in a dialectical process of discovery that is "self-correcting".[18]

Central to Lakatos' argument is his understanding of the notion of "proof", which is in distinction, as I say, to the standard logical definitions (though not necessarily out of keeping with the informal level of rigour of actual mathematical practice). Near the beginning of [1], we are presented with the following statement by "Teacher":

> I propose to retain the time-honoured technical term "proof" for a thought-experiment - or "quasi-experiment" - which suggests a decomposition of the original conjecture into subconjectures or lemmas, thus embedding it in a possibly quite distant body of knowledge.

So, a proof is understood in distinction to the formalist definition, which according to Lakatos identifies mathematics with its "formal mathematical abstraction", (it is also worth noting the distinction between this and the approach of Dummett-Prawitz that was argued against in the previous chapter). As Lakatos has it:

> The proof is only a stage of the mathematician's work which has to be followed by proof-analysis and refutations and concluded by the rigorous theorem. [1]

A proof, then, is seen as something more like an experiment at the beginning of the process of inquiry, through which we may come to find some justification for the theorem proposed, but also by which we may come to find ourselves justifying something that is quite distinct from it. For, as Lakatos argues, formalism removes historical and social processes from the domain of mathematical knowledge, by restricting proofs to a small set of written texts that follow the canons of logical deduction. These, are a posteriori records of the *actual act of proving*, on Lakatos' view, which have rendered it possible to forget that '[c]orroborations never compare with counterexamples, or even exceptions, as catalysts for the growth of concepts' [1]. So, rather than take proofs to simply guarantee the validity of a mathematical

---

[17] See the discussion in (e.g. [30, 31]).

[18] On the relation between Lakatosian methodology and dialectics, see [32, 33].

statement, they are more like the starting point for a dialogue game in which we engage in a process of testing, looking for counter-examples, providing reasons, and so on, until some conclusion is established. There is, therefore, a central role in this account played by counter-examples, and *incorrect* proofs and unsuccessful arguments over the course of reasoning in mathematical inquiry.[19] Without allowing that these play an integral role in the process of mathematical inquiry, we would be both obscuring actual historical practice, and also selling ourselves short insofar as mathematical inquiry is capable of giving rise to novel theorems and new realms of knowledge.

Let us dwell on a more explicit reconstruction of Lakatos' [1] framework a little.[20] There, as we have described it above, we begin with a conjecture that is typically of the form that some mathematical objects obey certain principles, or have certain properties. The proponent of the conjecture, "prover", may attempt to provide a proof of that conjecture, presumably from a set of premises that are either accepted already, or are, at least, not immediately false. It is then incumbent upon the opponent, "refuter", to test the attempted proof, by attempting to provide a refutation of it. This process, as Lakatos points out, is usually considered to be part and parcel of an informal kind of mathematical thought, which is "lost" when we eventually formalise a proof. However, in ([34] p. 156), Lakatos suggests that this is simply incorrect:

> To call this sort of proof an informal proof is a misnomer and a misleading one. It may perhaps be a quasi-formal proof, or a 'formal proof with gaps', but to suggest that an informal proof is just an incomplete formal proof seems to me to be to make the same mistake as early educationalists did when, assuming that a child was merely a miniature grown up; they neglected the direct study of child behaviour in favour of theorizing based on simple analogy with adult behaviour.

So, in this setting, it is possible that a "proof" of a conjecture may yet come to be refuted. For example, there may exist a successful test of one or more of the premises, which refutes the introduction of the conjecture altogether. Or, there may exist a successful test of one or more of the premises, which doesn't refute the conjecture, but refutes the proof-attempt given for it at that time (these are called global and local counterexamples, respectively (e.g. [1] p. 42ff)). We can provide a little more precision by considering quantified mathematical statements. Take a statement of form $\forall x \phi(x)$, for which, over the course of our inquiry we come across some mathematical object p, for which it is the case that $\neg\phi(p)$. This p would provide a refutation of the original statement, so in Lakatos' terms it is a global counterexample since it shows us that the universal statement is incorrect. The provision of such a refutation does not require that we look at the actual structure of the proof that has been provided for the mathematical statement, whilst a local counterexample is directed towards the proof of the statement itself. So, for example, we may look at a proof of $\forall x \phi(x)$, and find a refutation of some statement used to provide argument for it. Then, supposing we find an object such that, for some statement p used in the proof of $\forall x \phi(x)$, we find a good argument for $\neg a$, then we have a local counterexample for the statement

[19] See also the discussion in [27, 31–33].

[20] See also [33].

by way of providing a refutation of that specific attempt to prove it. In this vein, Lakatos provides four "rules" for the general method of proofs and refutations:

> Rule 1. If you have a conjecture, set out to prove it and to refute it. Inspect the proof carefully to prepare a list of non-trivial lemmas (proof-analysis); find counterexamples both to the conjecture (global counterexamples) and to the suspect lemmas (local counterexamples).
>
> Rule 2. If you have a global counterexample discard the conjecture, add to your proof-analysis a suitable lemma that will be refuted by the counterexample, and replace the discarded conjecture by an improved one that incorporates that lemma as a condition. Do not allow a refutation to be dismissed as a monster. Try to make all "hidden lemmas" explicit.
>
> Rule 3. If you have a local counterexample, check to see whether it is not also a global counterexample. If it is, you can easily apply Rule 2.
>
> Rule 4. If you have a counterexample which is local but not global, try to improve your proof analysis by replacing the refuted lemma by an unfalsified one.

With this, following [35], we can reconstruct the *process* through which inquiry is supposed to be carried out, according to Lakatos as follows:

- Primitive conjecture.
- Proof (a rough thought experiment or argument, decomposing the primitive conjecture into subconjectures and lemmas).
- Global counterexamples.
- Proof re-examined. The guilty lemma is spotted. The guilty lemma may have previously remained hidden or may have been misidentified.
- Proofs of the other theorems are examined to see if the newly found lemma occurs in them.
- Hitherto accepted consequences of the original and now refuted conjecture are checked.
- Counterexamples are turned into new examples, and new fields of inquiry open up.

Throughout the text, Lakatos illustrates his approach by an analysis of the process by which Euler's theorem is constructed. The theorem, properly called the polyhedron formula, is a statement that, for any polyhedron p, we have $V - E + F = 2$ where V, E, and F are, respectively, the number of vertices, edges, and faces of p. Through analysis of this example, he argues that the history of mathematical inquiry moves through the kind of reconstruction that we suggest above, to the point where the theorem evolved into its final form. This approach brings to light many novel concepts and ideas over the course of the inquiry, even as the original statements of the theorem are found wanting. It is through acknowledgement of these concepts and ideas, which gives rise to the increase in mathematical knowledge.

There is, as we have described them here, a clear structural analogy between Lakatos' approach to inquiry and the assertion games that we discussed in the previous chapter. Further still, the distinction that we made there *inside* proofs between the act of proving as a social and dynamic process, and the object of the proof that we may evaluate subsequent to termination of this process, may also map onto Lakatos' distinction between a valid proof that is justified and written down a posteriori, and proves are not required to "prove". In the previous chapter, I argued that in constructive logic there is an issue with certain negated sentences indicating that which is "not" proved. Here, this distinction inside proofs that Lakatos provides also allows us to make meaningful sense of statements such as "Wiles's proof of Fermat's Last Theorem was incorrect".

Let us end our discussion of Lakatos' work with a brief analysis of the role of logic in this approach to mathematical inquiry. In the work of both Lakatos and Popper, it is argued that the dialectics of inquiry is not formalisable. We might, however, conjecture, that this is, at least, in part due to the way in which logic was conceived to operate in the sort of foundationalist way that we objected to in the introductory chapter. Whilst both writers have a metalogical commitment to consistency as an *aim* of inquiry, this is obviously not something that is built in at the outset.[21] Indeed, there is always, in the process of inquiry, interplay between proof attempts, counterexamples, and the revision of the conjectures and proofs in question. This is in tension with classical logic, insofar as it has a strict form of bivalence, as bedrock, and, further still, admits non-constructive proofs. Then, the key issue becomes whether or not there is some way in which it is possible to maintain logical consistency, which Lakatos (and also Popper [29]) take to be central to rationality, whilst also constructing a formal structure for the consideration of the process of inquiry itself where this does not hold.[22] Classical logic, then, is not capable of dealing with the possibility of a process of inquiry that maintains the *force* of dialectical development containing both α and $\neg\alpha$, for some formula α.[23] Furthermore, as pointed out in the previous chapter, intuitionistic logic alone is incapable of dealing with refutation on a par with proof, since falsity is a non-constructive notion in intuitionistic logic (e.g. [36]). An alternative suggestion made in [32] considers a paraconsistent logic as formalizing Lakatos' framework. Any logic with access to negation, it can be defined to be paraconsistent if the logic is not trivialized under the presence of a formula and its negation. That is, where there exists some α, and some β, for which $\alpha, \neg\alpha \nvdash \beta$.[24] It is important to distinguish this from certain philosophical construals of paraconsistent logics as dialetheic.[25] For example, paraconsistent logics have, at times, been dismissed outright (e.g. [38]). One reason for this has to do with a confusion between paraconsistency and dialetheism. The latter is typically understood as the view that there exists true contradictions (e.g. [39]). There are a number of arguments given in favor of dialetheism in the literature (e.g. [39, 40]) that I will not discuss these here.[26] Rather, I will take it that dialetheism

[21] There is a useful discussion of this is in [32].

[22] [33] identifies this as the crucial tension in Lakatos' work.

[23] This is particularly obvious in [28].

[24] This is also explored in [31].

[25] The distinction is made clearly in [37].

[26] I will discuss this a little further in Chap. 6. Amus [41] argues that the relationship between paraconsistency and dialetheism is complicated by differing views on consequence, and is therefore dependent upon the motivations for the logic.

is ruled out in this context, since Lakatos is concerned with consistency at the aim of inquiry, so admitting "true" contradictions, whatever these are supposed to be, is not possible.

Instead, we might pursue the much more widely accepted view that inconsistencies are prevalent in our epistemic grip on the subject of inquiry (e.g. [42]). The consideration of a paraconsistent logic as adequate to scientific falsification has also been explored in [43], following some of Popper's [29] remarks. There are several issues, nonetheless, with the consideration of a paraconsistent logic in this context. First, paraconsistent logic is inadequate as a logic for constructive proof, since, for example, it can not handle constructive truth, nor does it have a detachable conditional, so modus ponens fails to be universally valid.[27] However, of particular note is Shramko's [45] work, which directly connects scientific inquiry with paraconsistent co-intuitionistic logic. Shramko considers a logic of refutation that extends Nelson's [46] logic, by taking co-intuitionistic logic to account for constructive falsity. Just as intuitionistic logic may express constructive truth, but can not deal with falsity constructively, the opposite is the case for Shramko's interpretation of co-intuitionistic logic. What results is a logic that favors falsification as central to the progression of scientific inquiry:

> It seems that not only constructive falsity appears a natural semantic counterpart for falsification, but also the non-constructive truth expresses the idea of empirical verification much better than [*constructive truth*] does. Indeed, within empirical science the requirement for constructive provability of all the statements is much too strong. In some minimal sense a hypothesis (conjecture) can be considered verified if our attempts to refute it do not succeed, i.e. if we could not find any empirical (factual) counterexample for this hypothesis. In this case we have sufficient reasons for accepting it, at least until convincing evidence against the hypothesis are found. [45, p. 353]

To a degree, this view coheres with Lakatos', since it promises to allow for the acceptability of conjectures without assuming that they are conclusively proven. Nonetheless, this is lop-sided in exactly the opposite way as intuitionistic logic, since it is now truth that is non-constructive. For Popper, we can never be sure that a theorem is true since it is open to future counterexample. This asymmetry of proof and refutation is, however, anathema to Lakatos, where it is always possible that there is exists a conclusive proof or refutation for some theorem, at the end of a process of inquiry. Furthermore, Lakatos, whilst directly influenced by Popper, considers the progress of inquiry to allow for theoretical revision without unconditional rejection of a conjecture under the existence of attempted counterexamples.[28]

In line with Lakatos, we are also interested in the consideration of proofs and refutations on a par with each other, whilst also ensuring that the logical structure is capable of dealing with them simultaneously, both in process, and at the termination of inquiry. In this respect, in the following chapter, we shall consider a formal

[27] See [44] for discussion and relevant references to the literature.

[28] An additional issue with Shramko's [45] falsificationist logic is that, through a focus on the role of negation, he fails to consider conditional operators for either proofs, or refutations. But, given the centrality of the conditional in reasoning and logic, this is a notable omission.

structure that, combines the benefits of dialogical approaches to logic with a form of BI- INTUITIONISTIC logic, through which we shall provide a structure ensuring the symmetry of "prover" and "refuter", with interactions not pre-determined by a set of logical rules.

4 Dialogical Constraints

The above discussion of dialogues, both logical and mathematical suggests that we should look to a more minimal set of normative requirements on players than is required by, for example, Lorenzen dialogue rules. Rather than stating each rule that is permitted at each stage of a dialogue, the most important norm is just that each player attempts to keep the dialogue coherent so that it converges towards agreement where possible. This approach, in contrast with that of GTS and Lorenzen dialogues, gains support from research in dialogue theory (e.g. [5, 6, 13]).

In contemporary dialogue theory, whilst heavily influenced by Wittgenstein (e.g. [47]), it is perhaps Grice's [48] work, which looms large. There, an argument is understood to be part and parcel of collaborative conversation between two agents in dialogue with each other. As Walton [13] notes, for Grice, arguments and speech acts are to be evaluated in terms of their contribution to a conversation, at that specific stage of the conversation. In this regard, Grice considers a series of maxims that are supposed to represent guiding norms over "polite" dialogue. In [5], Hamblin documents this strand of thought, whilst also arguing that traditional logic is incapable of dealing with the way in which distinct tokens of statements may express the same propositional content. This is also developed in the argumentation theory of Walton and Krabbe (e.g. [6]), which to some extent, may be understood as a continuation of Hamblin's WHY- BECAUSE SYSTEM WITH QUESTIONS [5, pp. 265–276], which gives a formal characterisation of two person dialogues in terms of the way in which alternating questions affect the players' commitments. To follow Walton [13], we can identify, in Hamblin's and Walton and Krabbe's systems, four fundamental features of collaborative conversation:

> (1) the two participants, called the proponent and the respondent; (2) the types of moves (taking the form of various speech acts) that the two participants are allowed to make, as each takes his or her turn to speak; (3) the sequence of moves, in which the appropriateness of each move depends on the type of preceding move made by the other party, (4) the goal of the dialogue as a whole.

In addition, four kinds of moves are especially important:

> (1) the asking of questions; (2) the making of assertions; (3) the retracting of assertions; (4) the putting forward of arguments.

A detailed discussion of these kinds of dialogues is given in the work of Walton and Krabbe [6], where conversational games are used to analyse dialogues, and, as with Brandom's account of dialogues, the central requirement for such games is that agents involved agree on the focus of the interaction.[29] These approaches to dialogue and conversation suggest that we should look at communication as a cooperative activity, made possible through intentional acts involving language. In recent work on dialogue, however, it is often argued that this can be developed further, and that language itself is predicated upon more fundamental interactions. For example, as the work of Ginzburg [2, 50–52] makes clear, it is not possible to adequately consider many notions of grammar and conversation outside of the context of interaction. In [2], Ginzburg argues that grammatical competence cannot be described without reference to conversational and social interactions. This means that grammar should be viewed as characterizing 'talk in interaction', and study of grammar requires studying information from the interactions, events, gestures, and so on, that comprise the actual data of language use:

> the conventional meaning of a word or a construction involves notions that irreducibly involve reference to interaction - notions such as "current issue under discussion", "disengagement from conversation", "acknowledgement of understanding" and "ask intended reference of other's utterance". [2, p. 5]

Following from this, in [51], Ginzburg and Fernandez articulate a model of dialogue called QUESTION UNDER DISCUSSION (QUD). This defines a question as constituting a "live issue", in the sense that it has been introduced into the dialogue as a focus for inquiry, and also that discussion of that question has not yet terminated. As such, QUD offers a way of considering a question as a focus of the dialogue, which also offers a form of "scorekeeping" that is distinct from the collective assessment process that we are given in Brandom's work. On this approach, dialogues are to be understood in terms of a 'collection of dynamically changing, coupled information states, one per conversational participant' [51].

As such, central to these approaches, is Ginzburg's [53] development of Hamblin's dictum that 'to know the meaning of a query is to understand what counts as a relevant response to that query'. Moreover, rather than look to a set of predetermined "rules" to determine what counts as a relevant response, Ginzburg argues that we need to consider what counts as relevant in this context, with particular attention to the notions of that which is optimal at that point in the conversation, and that which is coherent. So, in [2, Chap. 1], Ginzburg argues that there are several features of natural dialogues that must be accounted for:

- Coherence: each conversational move seems to cohere smoothly with its predecessor: questions are followed by answers which, in turn, raise new questions;
- Conciseness: conversation is, by comparison with a text, a highly efficient medium;
- Radical context dependence: isolated from their occurrence in a dialogue many utterances lose most of their import.
- Equal access to context: as a conversation proceeds a shared context (the common ground) emerges: A has her turn; Then either A proceeds or B takes over *from the common ground point at which A spoke*.

---

[29] See also [49] for a discussion of the central role of coherence in dialogical situations. It is also worth noting that the view provided in [6] may be understood as analogous to Brandom's account, but in the context of dialogue modelling.

Though note the the last feature is captured better, according to Ginzburg, to be understood in terms of a QUD, where the QUD is something like the maximal element of an interaction (or the central focus). This coheres, both with the idea that dialogue moves come in move/counter-move pairs, and that the action of a move sets up a restricted set of actions by the other participant, together with the idea that this should not be so strict as to prevent discussion regarding a players' utterance, nor of its challenges. So, on this approach, it is not suggested that an agent's assertion removes certain options for the other players' response, from some pre-defined set of possibilities. Neither is it the case that "anything goes". Rather, the idea is that certain ways of responding are rendered more viable, and so are normatively acceptable in that context, given that the interaction is focused around the QUD.[30] As such, whilst this approach has much in common with the question-structuring approach to dialogue in GTS and Lorenzen dialogues (as well as Hamblin's), it also has a much more flexible attitude towards the structure of those dialogues.[31]

This latter point is important, since it is the case that the norms upon conversational participants are incredibly minimal, and they are imposed *internally* to thos conversations insofar as they are coherent, concise, and symmetric. In this sense, we can look towards these minimal norms on interactions, rather than a predetermined strategy, or a prior notion of "rules", as central to keeping dialogue coherent. The range of an agents' possible responses to the utterance of a statement are wide and varied, yet, there are normatively correct actions *in the specific context* of that utterance determined by the coherence, conciseness, and symmetry of the interaction. That is to say, what is explanatorily basic in attempting to understand what it is that keeps interactions together so that they are capable of achieving some goal or other, is not logical rules, but more basic principles. To an extent, this approach bears an obvious resemblance to Grice's CO- OPERATIVE PRINCIPLE. Take Grice's [48] example:

A: I am out of petrol.
B: There is a garage around the corner

B's response has the implicature that A may purchase petrol at the garage, which may be interpreted only on the basis that B is obeying the co-operative principle in that conversational context. It is this principle, according to Grice, which ensures that the convergence of the dialogue is maintained. Note also that it is the utterance by A that sets out the context of the dialogue in a similar way to Ginzburg's analysis of QUD. The norm at work here, according to Grice [48] is just this: 'Make your conversational contribution such as is required, at the stage at which it occurs, by the accepted purpose or direction of the talk exchange in which you are engaged'.

According to Ginzburg [52], though;

> The big gaping hole in Grice's account, one which he was fully aware of, was the lack of substance concerning the maxim of relevance. Two crucial ingredients are missing: an explicit

[30] This flexibility also has the benefit that it is possible to engage in discussion regarding an utterance, with that discussion leading to further questions prior to the original discussion being resolved.

[31] Ginzburg also notes that this bears an obvious connection with work in artificial intelligence.

notion of current conversational state and a means for generating the range of potentially relevant contributions.

It is precisely this point that Ginzburg's QUD is supposed to answer. For example, whilst co-operation makes central the notion of an internal coherence between agents' statements and responses, it is central to the notion of convergence that dialogues have some sort of shared aim, or goal. It is important, nonetheless, that this goal may not be specified at the outset of an interaction between agents, but rather requires a less prescriptive norm that the agents involved are fully *collaborative* in their inquiry. So, for example, whilst it may be the case that an inquiry begins with a statement that prover asserts, and refuter tests, the original statement may be successfully refuted whilst the dialogue continues to determine a justification of a distinct, yet related statement (as we saw with Lakatos, above).

Furthermore, and unlike Grice, this mode of collaboration may be better understood in terms of non-intentional cooperation underlying interaction. For example, in [54], Gregoromichelaki and Kempson provide evidence accounting for this kind of collaboration towards a shared goal by means of lower-level cognitive phenomena (rather than subject-level Gricean principles):

> We should distinguish here between the terms "coordination" and "cooperation": cooperation is taken as involving a defined shared goal between interlocutors whereas coordination is the dynamically matched behaviour of two or more agents so that it might appear that there is a joint purpose, whether there is one or not [...] In this respect, psycholinguistic studies on dialogue have demonstrated that when individuals engage in a joint activity, such as conversation, they become "aligned", i.e. they (unconsciously) synchronise their behaviour at a variety of different levels, e.g. bodily movements, speech patterns.

There are, of course, many forms of dialogue, not all of which are cooperative. However, in [55], it is argued that non-cooperative dialogues may be modelled only on a prior understanding of cooperative behaviour. Furthermore, Gregoromichelaki and Kempson go on to argue that *meaning* should be understood by means of intentionally underspecified, yet incrementally goal-directed, dialogue:

> [J]oint-construal of meaning in dialogue is fundamentally based on the same mechanisms underlying language structure: structure is built through incremental procedures, that integrate context in every step, and this provides principled explanations for the syntactic properties of linguistic signals; but, in addition, since the grammar licenses partial, incrementally constructed structures, speakers can start an utterance without a fully formed intention/plan as to how it will develop relying on feedback from the hearer to shape their utterance and its construal and this provides the basis for the joint derivation of structures, meaning and action in dialogue. [54, p. 192]

Accordingly, it is these prior mechanisms of cooperation and coordination, which enable the sustaining of interactions without either external influence, or an abstract structure of a language supposed to be fixed prior to interaction. This approach, therefore, corresponds well with Ginzburg's QUD, whilst also developing its basis in sub-personal mechanisms of cooperation. Of central importance here is that agents are required neither to possess a plan or strategy, nor even a clear goal at the outset of an interaction. So, in this sense, we are able to provide an explication of the results of the intention-based view of meaning that Grice provides, without also requiring these norms to be part of the *intentions* of the agents involved.

An additional concern, which is discussed in detail in [6] is that of the "retraction" of commitments. If we are to construe interactions along the lines suggested here, then we are aiming towards some kind of resolution in which one of the agents retracts attempts to test a statement further, and, rather, accepts, or is entitled to that statement (to use Brandom's terminology). Over the course of a dialogue, prover may retract the initial statement, or refuter may give up its testing, and so the *termination* of a collaborative dialogue ends up with agreement. The point, as I say, is not to "win" the game, or to possess a winning strategy, but rather to agree whether or not agents are entitled to the statements in question. As such, it is required that commitment states must be defeasible so that they can be revised over the course of an interaction, such that agents may change in accord with the provision of arguments, information, and tests. Of course, as Walton and Krabbe [6] point out, the ideal of the kind of interactions that we are interested in here is to have cumulative movement towards agreement. We might, therefore, identify two forms of "retraction". The first, a retraction of a move made in support of a statement, which failed to uphold that statement which may yet be supported in some other way (similar to a local counterexample in Lakatos' terminology). The second, a resolution (similar to a global counterexample in Lakatos' terminology), by which an interaction terminates in agreement, and where any further discussion regarding the QUD is put aside. As such, in addition to the four fundamental features of interactions summarised above, we should also consider DENIALS, alongside assertions, as argued both above and in the previous chapter. Then, for example, we should have precisely the same expectations on the retraction of denials and assertions, such that the two are symmetric. That is, agents' retractions and resolutions should be taken to be the result of agreement regarding the failures, and successes, of justifications.

5 Reasoning, Emotion, and Co-operation

In the context of the discussion above, I have outlined a rather different approach to dialogue from that which is offered both in Brandom's work and in standard logical approaches. Furthermore, I have also offered a way of thinking about inquiry and interaction that is out of keeping with standard formal approaches to reasoning. That is to say, reasoning is social and interactive, and requires an account of history that is not captured by an absolutist and monological account of proofs. As such, we are also in a position to briefly deal with two, interrelated issues arising in relation to accounts of reasoning. The first has to do with the relationship between reasoning and emotion, which have often been seen to be antagonistic; the second concern has to do with criticisms of reasoning from the point of view of feminist and post-colonial theory.

There are long-held, and significant, criticisms of rationality and rationalism, particularly within feminist and post-colonial theory. In the main, these criticisms are aimed at the supposed aperspectivalness and absolutism of reason, particularly insofar as this has been used to exclude groups of agents from the domain of reason, and exercise oppression over them. According to this criticism, this view of reason also has its roots in an idealised account of rationality, which is dissociated from the actual exercise of reasoning insofar as it is forms a prescriptive set of norms by which reasons may be held up as adequate, or otherwise. Moreover, the approach to dialogue that is supposed to underlie GTS and Lorenzen dialogues given above, is also considered to be problematic due to the adversarial nature of these arguments, in contrast to sharing common goals.[32]

For example, a key criterion of standard approaches to rationality, according to the feminist critique, is that it allow for some form of detached objectivity, where this objectivity is often assumed to be an overhang from Cartesian approaches to rationalism. For example, in [57], Prokhovnik argues that emotion, which has typically been both identified as a dichotomous feminine trait in contrast to masculine rationality, should rather be brought under the rubric of reasoning, along with social practices.[33] This emphasis on reasoning, rather than objective rationality, is clearly coherent with the position that is argued for here. For example, the aperspectival nature of rationality is obviously no longer plausible on the view that is set out here, since reasoning is always rooted in a specific social context. As such, it is possible to reconceive reasoning in ways that preclude the possibility of prescriptive notions of epistemic authority.[34]

Let us consider, for example, the roles played by emotion in reasoning. These are often overlooked, particularly regarding the relationship between emotions and reasoning according to logic. This has obvious roots in classical theories of rationalism, whilst in the contemporary literature, it is arguable that a kind of suppression of the role of emotion in reasoning is primarily found in the prevalent construal of reasoning in terms of DUAL- PROCESS theories. Specifically, in the extant relevant literature, reasoning is construed as the outcome of two separate, often antagonistic reasoning systems. One system (Type-1) is fast, automatic, emotional, and subconscious; whereas the other (Type-2) is rule-based, analytical, deliberative, and explicit [60–63].[35]

A key motivation for this move lies in what has been identified over the past fifty years or so in cognitive science as human doxastic conservativeness.[36] The study of human reasoning has led, in essence, to a pretty solid eradication of Piaget's paradigm that reasoning, broadly speaking, follows the rules of classical logic. The experimental psychology of human reasoning suggests that humans have a fundamental bias in "the tendency to automatically bring prior knowledge to bear when solving problems" [69]. This is seen in matching biases [70]; and deductive biases [71]. The latter

[32] See, (e.g. [56]).

[33] See [58, p. 149].

[34] See the work of Miranda Fricker [59].

[35] Whilst this view is dominant throughout cognitive science, the idea that reasoning results from the competition between dual systems has also been questioned in fields as diverse as psychology [64–66], philosophy [67], and neuroscience [68].

[36] I take this phrase from the work of Catarina Dutilh Novaes [67].

are exemplified by the now infamous "Linda problem". Here, subjects are presented with the following piece of fiction:

> Linda is 31 years old, single, outspoken, and very bright. She majored in philosophy. As a student, she was deeply concerned with issues of discrimination and social justice, and also participated in anti-nuclear demonstrations.

before being asked:

> Which is more probable?
> 1. Linda is a bank teller.
> 2. Linda is a bank teller and is active in the feminist movement.

In such cases, around 90 % of participants routinely and systematically choose option (2), suggesting that the probability of the conjunction is greater than one of its conjuncts. One key response to such evidence is to suggest that a "representative heuristic" is involved in the participant's judgment, which results from a fundamental "belief bias" across human reasoning (e.g. [63]).

It appears from these results, that humans have the tendency towards doxastic conservativeness, in that subjects routinely seek to confirm their extant beliefs. One explanation for this, which is now widely accepted in the literature is that there is distinction made between the two dual processes, both of which underlie cognition, and often conflict with each other. However, in recent work with Alex Tillas [72], we have argued against dual-process theory, by providing a philosophical account of how a unitary reasoning system operates. We showed that what are often called intuitions (allegedly Type-1) are sub-activated associations, which are in turn influenced by the weightings of the connections between different representations. In that paper, we explain standard biases by appealing to the role of attention in thinking processes and argue that selectively attending to different aspects of a perceived stimulus, inevitably engages different associations, which in turn ground different cognitive strategies. In this sense, reasoning is not underpinned by two antagonistic systems but rather by a unitary one.

Our suggestions about a unitary system gathers pace from recent psychological work advancing a view of reasoning as the outcome of a compound process consisting in cognitive and affective features (e.g. [73]). More specifically, Phelps et. al. [68] show that the alleged antagonistic relationship between the two systems is rather shallow and simplistic, and, instead, there is a clear modulatory relationship between the two which more accurately reflects the current research in affective neuroscience. Specifically, they show that that there are two potential mechanisms by which affective states modulate the computation of subjective values and decisions. Briefly these mechanisms concern (a) incidental affective states that may carry over to the assessment of subjective value and the decision, while (b) emotional reactions to the choice may be incorporated into the value calculation. Crucially, for our purposes, [68] also shows that changing affective states can result in changing the choices or inferences made in that context.

It is in this light that it is also possible to see that emotions may play a much larger role in reasoning than previously thought. Whilst approaches such as Brandom's leave little role for emotion in reasoning, we think that it is clearly the case that inferential relations involve emotions from the outset. Indeed, there are good arguments suggesting that emotions are a necessary requirement for reasoned inference (e.g. [74]). Emotions generally contribute positively to reasoning by enhancing the formation of associations and modes of attention to features in the world.[37] Moreover, as pointed out in [75], emotions also provide focus on what is important to an agent, so usually the inferences that the subject makes concern goal-relevant information rather than diversions and trivialities, providing in this way a direct connection between beliefs and actions. For example, we are clearly capable of dealing with incredibly complex sets of associations without computational collapse. This is due mainly to an ability to identify those associations and inferences which have immediate valence. As such, emotions are one of the most prevalent means by which attention is both constrained and directed.[38] Emotions both constrain the framework within which reasoning operates, and 'they render salient only a tiny proportion of the available alternatives and of the conceivably relevant facts'. In addition, and since we are considering reasoning in a social context, Greenspan [77, 78] argues that emotions are particularly important in determining the context of actions in such settings.

The omission of emotion from the account of reasoning in Brandom's work mirrors that in dual-process theory, where, for example, it is often suggested that overcoming biased (emotion-led) associations requires slow, deliberative and rational thought. On the suggested view this cannot be the case, since emotions shape and construct reasoning across all levels.[39] In particular, we are considering the role of dialogue in which an agent's assertions are open to challenge to provide reasons for their commitment to that which is asserted. This makes room for the explicitation of inferential relations such that they are open to question and challenge in the public domain. It may well be the case that the explicitation of certain inferences in public discourse allows us to reconsider our commitments and their relations. Nonetheless, this process takes place, not through the dissociation of inference and emotion, but rather through a better understanding of the role that emotions play in rational thought.

As such, far from adhering to Enlightenment rationality's excising of emotion from the domain of the rational, we have reason to think that emotion is part and parcel with the framing of reasoning activities so that they remain convergent and coherent from the ground up. It is important, nevertheless, that the overarching picture that is given here is a reconceptualisation of rationality which does not simply jettison all claims of "correctness". Without this, as Heikes [58] forcefully points out:

> [I]f feminists are always and necessarily bound by their own point of view, with no opportunity to transcend that perspective, then they are giving up more than a claim to speak the truth: they are giving up a claim to speak any truth. (p. 150)

[37] It is also the case, of course, that emotions hinder reasoning by bypassing existing thinking patterns and the routine deployment of associations. However, these mechanisms are more complex than often assumed by standard accounts.

[38] See also the discussion in [76].

[39] See also [79, 80].

The problem is that, without an account of the correctness of reasons, any attempt to argue against oppression and supposed epistemic authority would be given up on insofar as we also give up any account of the normativity of meaning and also of truth that is not *just* perspectival.[40] It is important, then, that we do not give up the ability to engage in a struggle for truth that may have universal consequences, whilst also not regressing towards the kind of foundationalist impulse through which oppression is made possible. We require, for example, some means of justifying the statement that "the oppression of women is wrong". As Heikes puts it:

> How does one establish a sufficient objectivity to assert the truth of feminism's claims without introducing an oppressively totalising rationality? An obvious path is through a concept of rationality that integrates some regulatory structures with the content governed by those structures. In other words, rationality should unite the transcendental and the immanent since this brings together the universality and the perspectivity necessary to provide constraints on diversity. [58, p. 151]

This path is, then, somewhere between the Scylla of totalising epistemic authority, and the Charybdis of perspectival immanentism, such that a normative role is played by reason that is somehow beyond the actual instantiations of reasoning activity. The issue that Heikes points out, is that androcentrism clarifies the way in which many forms of reasoning are subject to norms that are held up by a small group of people in privileged positions of authority. Of course, in the domain of logic, this kind of appeal to foundations is already undone, as explained in the introductory chapter, since any attempt to provide ultimate grounds for logical reasoning in semantics is problematically circular. Furthermore, the idea of a foundational condition upon which reasons may be judged to be correct, or otherwise, is not plausible on the view of assertion games given in the previous chapter.

6 Internalised Dialogues Without Rules

Let us finish this chapter by returning to the question of the specific kind of interactions that we are interested in, paying particular attention to the way in which we approach them.

This is, to some extent, similar to recent work by Catarina Dutilh Novaes [67, 81] on dialogical approaches to logic. In particular, she takes a view of logic in which "proponent" and "opponent" are not working *against* each other, but rather *with* each other in co-operative inquiry:

> Proponent's job is not only to beat Opponent; she also seeks to persuade Opponent of the truth of the conclusion, if he has granted the truth of the premises. In fact, the goal is not only to show that the conclusion follows from the premises, but also why it does; this corresponds to the idea that deductive arguments ought to have explanatory value. In this sense, Proponent and Opponent are cooperating in a common inquiry to establish what follows from the premises, and thus to further investigate the topic in question.

---

[40] We shall return to these issues in Chap. 7.

This is somewhat more restricted than the kinds of interactions that we have discussed in the above, but, nonetheless, her work importantly surpasses GTS and Lorenzen dialogues in two major ways. First, rather than see dialogues as purely adversarial, there is an important cooperative component to these interactions that coheres with the approach developed here. Second, more standard frameworks, according to Dutilh-Novaes, overlook the often internalised, and implicit, role that is played by one of the dialogical participants.[41] In this setting, a proof is to be understood as a kind of discourse which is aimed at persuading an audience that, they ought to accept the truth of a conclusion, when they accept the truth of the premises. This gives rise to Dutilh Novaes' "built-in opponent" conception of proofs [67], which argues that ordinary proofs in deductive logic have internalised the opponent in proof-construction. The idea is that, by appeal to the origin of proofs in the Greek dialogues discussed earlier in this chapter, when constructing a proof each step should have no counterexamples, and it is the role of the opponent to find such counterexamples. In the ordinary presentation of proofs in contemporary deductive logic, the role of the opponent is built directly into the framework, hence the built-in opponent conception of deduction. The inferential steps that are deductively correct are only those that are indefeasible, and so do not admit counter-examples. As such, Dutilh-Novaes [67] argues that these build-in the role of opponent to search for counterexamples since, by excising the possibility of counterexamples altogether by definition, a deductive argument is guaranteed to be a winning strategy for the proponent since it is a strategy that must yield a win regardless of what move the other (internalised player) makes. This reflects the distinction made in the previous chapter between the dynamics of the interaction and the resulting proof-object that is constructed over the course of that interaction. Deductive logic removes the possibility of accounting for the dynamics of the process of the interaction, by requiring that each step of an inference be immune to counter-examples, and so on. As such, the internalisation of the role of opponent is analogous to the formalist, or rigorist, in Lakatos' discussion.

There is an interesting connection between these arguments regarding the prevalence of the monological approach, and the role of power and control over the provision of reasons and meaningful dialogue, to follow Goffman's argument that;

[41]On the former, Dutilh Novaes draws upon the work of cognitive scientist Keith Stenning, who says, for example:

> Our job, as speaker or hearer of a discourse that purports to be a deduction, is to test the drawing of inferences to destruction, to ensure that what is inferred is true in all models of the premises, not just the intended one. It is in this technical sense that logic models adversarial discourse. We may actually be socially co-operating in testing a deduction for validity, and certainly we have to co-operate a great deal to be sure that we are assuming the same things about the range of interpretations which are intended, but there is a core of seeking out all possible assignments of things, not simply finding one intended one. This is perhaps not accidentally related to the fact that logic arose as a model of legal and political debate. [82, p. 138]

> the proper study of interaction is not the individual and his psychology, but rather the syntactical relations among the acts of different persons mutually present to one another. [83, p. 2]

For example, the field of conversational analysis (e.g. [84–86]), provides a structure for understanding conversation as organised into turns, where the passing of turns between agents is also a matter of passing the "right to speak" [84]. In this field, the passing of a turn does not rely upon certain rules, but is rather a manifestation of the alternation of speaking and listening over the organisation of interactions. As Sacks and Schegloff have it, a turn-constructional unit (TCU) is the smallest complete unit constituting a turn, and a transition-relevance place (TRP) is the natural break in conversation where a transition to a next speaker becomes relevant (though it may not actually occur). In the specific case of question-answer pairs, they argue that the agent who asked the question has an obligation, or at least right, to respond subsequent to the provision of an answer to that question by another agent, giving rise to the kind of question-answer dialogues discussed above in the work of Ginzburg. Central to conversational analysis, in distinction to the latter, however, is the role of power and control, which operates via the domination of speaking and the taking of turns, topic-refusal, and so on. For example, understanding the mechanisms of TCU's and TRP's requires accounting for the obligations and rights to speak. Control, in this setting operates through the blocking of these rights such that agents are *denied the right* to have equal status, or symmetry. These blocks include denying access to common ground, being refused the right to set the topic of conversation (or QUD in Ginzburg's system), being refused the symmetry of TRP, by interruption, and by delimiting the range of possible (or acceptable) responses. Complete control, then, involves the blocking of interaction altogether such that we are left with a monologue, rather than a dialogue. The monological, in effect, removes the possibility for interaction through the prevention of other agents' rights to speak, and the constriction of topic choice, preventing both coherence and symmetry as possibilities for interaction.

In most of the approaches discussed above, interactions are required to have an explicit set of rules that prescribe agents' roles and responses. Even in Ginzburg's work a set of conversational rules is required to make sense of the way in which common ground is affected by a change in context and the latest move, for example. In distinction, and for the reasons discussed in Chap. 1, I have proposed that a set of conventional rules can not be are required a priori in order for successful interaction to take place. Rather, I take it that successful interactions do occur without recourse to such rules, and that such interactions can be taken to be the foundation through which any "rules" may be constructed a posteriori. This is because we can not determine in advance of the process of reasoning, either what counts as a reason, or the meaning of propositions involved. In general, the idea is that a successful interaction requires only that we respond to challenges in a specific context such that the interaction between agents is kept coherent. That is, we do not say that the meaning of statements can be fixed by a general set of rules that are given in advance. Instead, meaning is determined only over the course of our linguistic activities, and reasoning correctly in these contexts is just a matter of keeping our dialogues convergent in the sense that we respond when challenged, and that we challenge existing assertions that are made. Since these activities are, by definition dependent upon the actual context of the dialogue (the arguments, reasons, and speech-acts made by other agents), they can not be determined *prior* to the process of reasoning itself. Nor, can these activities be given as some abstract set of features of a language, or features of a hypothesised rational faculty.[42]

We require that the coherence and convergence of interactions provides a retrospective mechanism for consideration of how mutual understanding is constructed as immediate and radically local to the interaction, rather than requiring access to abstract or generic aspects of language. Furthermore, in contrast to standard foundationalist accounts requiring a basis of propositional knowledge for deductive practice, this account provides a clear way of thinking about reasoning as practices that also involve heuristic associations and tacit knowledge. It is through the process of dialogue that we are able to bring reasons into the domain of the conscious through language, providing the scaffolding for further expansion of reasoned inquiry, and the construction of propositional content.

In this light, what is required for the process of an interaction to be one that is capable of aiming towards truth, is a minimal set of requirements over interactions ensuring that an interaction is coherent in the sense that it converges towards the same goal with symmetric access to common ground.[43] It is not the case, then, that we can adopt the notions of winning strategies, or of formula checking, from standard accounts of game semantics and dialogue. Instead, what is important about interactions is that they are coherent, and that whenever they terminate, the reason for their termination is that one of the agents involved endorses a statement and gives up any further tests of that statement by *agreement through co-operative activity*. As Brandom puts it, in this context:

> [u]nderstanding a speech act - grasping its discursive significance - is being able to attribute the right commitments in response. This is knowing how it changes [...] what the performer and the audience are committed and entitled to. [10, p. 165]

In our terms, this is to say, it is *subsequent* to interactions that assertions can be evaluated to have normative force, and it is, therefore, only through these processes that we are capable of "struggling" for truth.

References

1. Imre Lakatos. *Proofs and Refutations: The Logic of Mathematical Discovery*. Cambridge University Press, 1976.
2. Jonathan Ginzburg.*The Interactive Stance: Meaning for Conversation*. Oxford University Press, 2015.
3. Paul Lorenzen and Kuno Lorenz. *Dialogische logik*. Wissenschaftliche Buchgesellschaft, 1978.

---

[42]We have also noted difficulties in appealing to a common dispositional core of abilities in Chap. 1.

[43]For example, Brandom argues that an interaction requires an agreement to disagree, so that we do not simply talk past one another, see [7, Sect. 8] for a fuller discussion.

4. Jaakko Hintikka. *Logic, Language-Games and Information: Kantian Themes in the Philosophy of Logic*. Oxford,Clarendon Press, 1973.
5. C. L. Hamblin. *Fallacies*. Vale Press, 1970.
6. Douglas Walton and Erik CW Krabbe. *Commitment in dialogue: Basic concepts of interpersonal reasoning*. SUNY press, 1995.
7. Robert Brandom. *Making It Explicit: Reasoning, Representing, and Discursive Commitment*. Harvard University Press, 1994.
8. Jürgen Habermas. From Kant to Hegel: On Robert Brandom's pragmatic philosophy of language. *European Journal of Philosophy*, 8(3):322–355, 2000.
9. Mathieu Marion. Game semantics and the manifestation thesis. In *The Realism-Antirealism Debate in the Age of Alternative Logics*, pages 141–168. Springer, 2012.
10. Robert Brandom. *Articulating Reasons: An Introduction to Inferentialism*. Harvard University Press, 2000.
11. Robert Brandom. *Between Saying and Doing: Towards an Analytic Pragmatism*. Oxford University Press, 2008.
12. Aarne Ranta. Propositions as games as types. *Synthese*, 76(3):377–395, 1988.
13. Douglas Walton. The place of dialogue theory in logic, computer science and communication studies. *Synthese*, 123(3):327–346, 2000.
14. Lorenzo Magnani. The eco-cognitive model of abduction. *Journal of Applied Logic*, 13(3): 285–315, 2015.
15. Walter Carnielli. Meeting hintikka's challenge to paraconsistentism. *Principia*, 13(3):283–297, 2009.
16. B. Castelnérac and M. Marion. Arguing for inconsistency: Dialectical games in the academy. In Giuseppe Primiero, editor, *Acts of Knowledge: History, Philosophy and Logic*. College Publications, 2009.
17. Matthew Duncombe and Catarina Dutilh Novaes. Dialectic and logic in Aristotle and his tradition. *History and Philosophy of Logic*, 37(1):1–8, 2016.
18. Ignacio Angelelli. The techniques of disputation in the history of logic. *Journal of Philosophy*, 67(20):800–815, 1970.
19. Wilfrid Hodges. Logic and games. In *Stanford Encyclopedia of Philosophy*. 2008.
20. Catarina Dutilh Novaes. Medieval theories of consequence. *Stanford Encyclopedia of Philosophy*, pages 1–21, 2012.
21. Wilfrid Hodges. Dialogue foundations: A sceptical look. *Aristotelian Society Supplementary Volume*, 75(1):17–32, 2001.
22. Tero Tulenheimo. On some logic games in their philosophical context. In *Ludics, dialogue and interaction*, pages 88–113. Springer, 2011.
23. Walter Felscher. Dialogues as a foundation for intuitionistic logic. In *Handbook of philosophical logic*, pages 341–372. Springer, 1986.
24. Mathieu Marion. Why play logical games? In Ondrej Majer, Ahti-Veikko Pietarinen, and Tero Tulenheimo, editors, *Games: Unifying Logic, Language, and Philosophy*, pages 3–26. Springer, 2009.
25. Morten Heine Sørensen and Pawel Urzyczyn. *Lectures on the Curry-Howard isomorphism*, volume 149. Elsevier, 2006.
26. Graham Priest. *An Introduction to Non-Classical Logic: From If to Is*. Cambridge University Press, 2008.
27. Can Başkent. *Perspectives on Interrogative Models of Inquiry: Developments in Inquiry and Questions*, volume 8. Springer, 2015.
28. Karl R. Popper. What is dialectic? *Mind*, 49(196):403–426, 1940.
29. Karl R. Popper. *Conjectures and Refutations: The Growth of Scientific Knowledge*, volume 15. Routledge, 1989.
30. David Miller. *Critical Rationalism: A Restatement and Defence*. Open Court, 1994.
31. Graham Priest and Neil Thomason. 60 *Australasian Journal of Logic*, 5:89–100, 2007.
32. C. Başkent. Inquiry, refutations and the inconsistent. In C. Baskent, editor, *Perspectives on Interrogative Models of Inquiry*. Springer, 2015.

33. Ladislav Kvasz. Lakatos' methodology between logic and dialectic. In *Appraising Lakatos: Mathematics, Methodology and the Man*. Kluwer, 2002.
34. Imre Lakatos. What does a mathematical proof prove. In Thomas Tymoczko, editor, *New directions in the philosophy of mathematics*, pages 153–162. Princeton University Press, 1998.
35. D. Corfield. Assaying Lakatos' philosophy of mathematics. *Studies in History and Philosophy of Science Part A*, 28(1):99–121, 1997.
36. Yaroslav Shramko, J. Michael Dunn, and Tatsutoshi Takenaka. The trilaticce of constructive truth values. *Journal of Logic and Computation*, 11(1):761–788, 2001.
37. Bryson Brown. Yes, Virginia, there really are paraconsistent logics. *Journal of Philosophical Logic*, 28(5):489–500, 1999.
38. B. H. Slater. Paraconsistent logics? *Journal of Philosophical Logic*, 24(4):451–454, 1995.
39. Graham Priest. *In Contradiction: A Study of the Transconsistent*. Oxford University Press, 2006.
40. Graham Priest. *Doubt Truth to Be a Liar*. Oxford University Press, 2006.
41. C. M. Amus. Paraconsistency on the rocks of dialetheism. *Logique Et Analyse*, 55(217):3, 2012.
42. Newton Da Costa and Steven French. *Inconsistency in science: A partial perspective*. Springer, 2002.
43. David Miller. Paraconsistent logic for falsificationists. In *Proceedings of the First Workshop on Logic and Language (Universidad de Sevilla)*, pages 197–204, 2000.
44. J. C. Beall. Free of detachment: Logic, rationality, and gluts. *Noûs*, 49(1), 2013.
45. Yaroslav Shramko. Dual intuitionistic logic and a variety of negations: The logic of scientific research. *Studia Logica*, 80(2-3):347–367, 2005.
46. David Nelson. Constructible falsity. *Journal of Symbolic Logic*, 14(1):16–26, 1949.
47. Ludwig Wittgenstein. *Philosophical Investigations, 4th Edition (Trans. Hacker and Schulte)*. Wiley-Blackwell, 2009.
48. H Paul Grice. Logic and conversation, in syntax and semantics, vol 3, speech acts. *ed. Peter Cole and Jerry Morgan*, pages 41–58, 1975.
49. Paul Piwek. Meaning and dialogue coherence: A proof-theoretic investigation. *Journal of Logic, Language and Information*, 16(4):403–421, 2007.
50. Jonathan Ginzburg, Ivan A Sag, and Matthew Purver. Integrating conversational move types in the grammar of conversation. *Pragmatics and Beyond*, pages 25–42, 2003.
51. Jonathan Ginzburg, Raquel Fernández, et al. Computational models of dialogue. *Handbook of Computational Linguistics and Natural Language, Oxford. Blackwell*, 2010.
52. Jonathan Ginzburg. The semantics of dialogue. In Maria Aloni and Paul Dekker, editors, *Cambridge Handbook of Semantics*. Cambridge University Press, 2016.
53. Jonathan Ginzburg. Questions: logic and interactions. *Handbook of logic and language*, pages 1133–1146, 2011.
54. Eleni Gregoromichelaki and Ruth Kempson. Grammars as processes for interactive language use: incrementality and the emergence of joint intentionality. In *Perspectives on linguistic pragmatics*, pages 185–216. Springer, 2013.
55. Brian Plüss, Paul Piwek, and Richard Power. Modelling non-cooperative dialogue: the role of conversational games and discourse obligations. In *Proceedings of SemDial 2011, the 15th Workshop on the Semantics and Pragmatics of Dialogue*, September 2011.
56. Catherine E. Hundleby. Feminist perspectives on argumentation. *Stanford Encyclopedia of Philosophy*, 2015.
57. Raia Prokhovnik. *Rational woman: A feminist critique of dichotomy*. Manchester University Press, 2002.
58. Deborah K. Heikes. *Rationality and Feminist Philosophy*. Continuum, 2010.
59. Miranda Fricker. *Epistemic Injustice: Power and the Ethics of Knowing*. Oxford University Press, 2009.
60. Seymour Epstein. Integration of the cognitive and the psychodynamic unconscious. *American Psychologist*, 49(8):709, 1994.

61. Jonathan Evans. *Bias in human reasoning: Causes and consequences.* Lawrence Erlbaum Associates, Inc, 1989.
62. Steven A Sloman. The empirical case for two systems of reasoning. *Psychological bulletin*, 119(1):3, 1996.
63. Keith E Stanovich. *Who is rational?: Studies of individual differences in reasoning*. Psychology Press, 1999.
64. Wim De Neys. Dual processing in reasoning two systems but one reasoner. *Psychological science*, 17(5):428–433, 2006.
65. Magda Osman. An evaluation of dual-process theories of reasoning. *Psychonomic bulletin & review*, 11(6):988–1010, 2004.
66. Magda Osman. A case study dual-process theories of higher cognition—commentary on Evans & Stanovich (2013). *Perspectives on Psychological Science*, 8(3):248–252, 2013.
67. Catarina Dutilh Novaes. A dialogical, multiagent account of the normativity of logic. *Dialectica*, 69(4):587–609, 2015.
68. Elizabeth A Phelps, Karolina M Lempert, and Peter Sokol-Hessner. Emotion and decision making: multiple modulatory neural circuits. *Annual Review of Neuroscience*, 37:263–287, 2014.
69. Keith E. Stanovich and Richard F. West. Advancing the rationality debate. *Behavioral and Brain Sciences*, 23(5):701–717, 2000.
70. P. C. Wason and J. Evans. Dual processes in reasoning? *Cognition*, 3(2):141–154, 1974.
71. Daniel Kahneman, Paul Slovic, and Amos Tversky. *Judgment Under Uncertainty: Heuristics and Biases*. Cambridge University Press, 1982.
72. J. Trafford and A. Tillas. Intuition and reason: Re-assessing dual-process theories of cognition. *Teorema: Revista internacional de filosofía*, XXXIV(3), 2015.
73. P. Thagard and B. AuBie. Emotional consciousness: A neural model of how cognitive appraisal and somatic perception interact to produce qualitative experience. *Consciousness and Cognition*, 17(3):811–834, 2008.
74. Antonio R Damasio. *Descartes' error*. Random House, 2006.
75. Adrian Wells and Gerald Matthews. *Attention and Emotion (Classic Edition): A Clinical Perspective*. Psychology Press, 2014.
76. Ronald De Sousa. Emotion. *Stanford Encyclopedia of Philosophy*, 2007.
77. Patricia Greenspan. *Emotions and Reasons: An Inquiry Into Emotional Justification*. Routledge, Chapman and Hall, 1988.
78. Patricia Greenspan. Emotional strategies and rationality. *Ethics*, 110(3):469–487, 2000.
79. Georg Brun, Ulvi Dogluoglu, and Dominique Kuenzle. *Epistemology and Emotions*. Ashgate Publishing Company, 2008.
80. Ronald B. de Sousa. *The Rationality of Emotion*. MIT Press, 1987.
81. Catarina Dutilh Novaes. A dialogical account of deductive reasoning as a case study for how culture shapes cognition. *Journal of cognition and culture*, 13(5):459–482, 2013.
82. Keith Stenning. *Seeing Reason: Image and Language in Learning to Think*. OUP Oxford, 2002.
83. Erving Goffman. *Interaction ritual: Essays in face to face behavior*. AldineTransaction, 2005.
84. Harvey Sacks, Emanuel A Schegloff, and Gail Jefferson. A simplest systematics for the organization of turn-taking for conversation. *language*, pages 696–735, 1974.
85. Emanuel A Schegloff and Harvey Sacks. Opening up closings. *Semiotica*, 8(4):289–327, 1973.
86. Emanuel A Schegloff, Gail Jefferson, and Harvey Sacks. The preference for self-correction in the organization of repair in conversation. *Language*, pages 361–382, 1977.

Chapter 5
Rules in Dialogue

This chapter returns to some of the foundational issues discussed in the introduction and first chapter, particularly regarding the justification of logical rules. In Chaps. 1 and 2, I argued that standard inferentialist accounts do not suffice to provide justification of basic logical rules. In response, an analysis of logical inference as acts taking place in dialogical situations is provided, by taking interactions to be structured around moves that may be defined as coherent under certain circumstances. This is used to underpin a novel account of the proof-theoretic notion of harmony as a way of balancing moves. I show that this leads to a constructive logic over both proofs and refutations, where logical rules are justfied internally, at the termination of dialogue. This means that, rather than start with logical rules a priori, the rules themselves are derivative of dialogical balance, so we can reconstruct certain logical rules subsequent to interaction. I finish by discussing the relationships between the formal structure presented here, and related structures.

1 Dialogue, Interaction, and Coherence

We turn now to consider the way in which we might formalise interactions whilst also requiring that our formalism is both non-deterministic, and also not based on autocratic rules. Rather, we want to get to the heart of what logic is, such that the justification of what we typically think of as "rules" is generated from a more primary relationship between proofs and refutations. As previously discussed, typically, an inferential rule is defined so that it is supposed to have normative traction over agents' reasoning *and* that reasoning in accord with an inferential rule will always be valid, in the sense that reasoning with a rule from valid premises will always lead the agent to a valid conclusion. If we take these two aspects to be definitional of an inferential rule, then our formalism should contain no "rules" as such. Rather, we want interactions of a certain kind to give rise to what may be subsequently taken to be rules. So, the MOVES that agents make are not here to be thought of as correct, or

J. Trafford, *Meaning in Dialogue*, Studies in Applied Philosophy, Epistemology and Rational Ethics 33, DOI 10.1007/978-3-319-47205-8_5

valid, in the usual sense. Instead, we need to think of proof and refutation-attempts that may not "prove" or "refute", insofar as the the latter clauses are understood to be formally valid constructions of proof (and refutation) *objects*. As discussed in the previous two chapters, we are taking the notion of the "game of giving and asking for reasons" seriously, where an assertion of a statement involves a commitment to its defence, and a denial of the statement involves a commitment to its challenge. As such, we are relaxing what counts as a proof or refutation, insofar as moves made in interactions are more like reasons, or arguments, in support of a possible assertion, or denial.

With this in mind, we shall take as central notions, positive and negative moves, rather than proofs or refutations, with "tests" as the relationship between them.[1] The primary relationship through which this interaction is constructed, then, is between a positive move, α^+, and negative move, α^-, and it is through this testing relationship that an interpretation of a statement is constructed. That is to say, we are building stepwise refutations (or counter-proofs) that cohere with proofs (whilst relaxing both of these notions), and where the relationship between the two replaces the role of counter-models discussed in Chap. 2. Since we are thinking of this structure, not purely formally, but also as an inferentialist structure that is rooted in the cooperative dialogues discussed in the previous chapter, we need to consider a number of key desiderata for the construction of the relationship between moves:

- Shared focus
- Symmetry between agents
- Coherence
- Agreement and resolution

Each of these will be examined in turn.

1.1 Shared Focus

First, consider the focus of an interaction, and how this may be formalised. Say that an interaction concerns some utterance α, between prover $\mathcal{P}$, and refuter $\mathcal{R}$. Then, we allow $\mathcal{P}$ to offer some reasons for that utterance, which we formalise in terms of a sequent $\Gamma^+ \vdash \alpha^+$. In turn, $\mathcal{R}$ offers some reasons against that utterance, directed towards α^+ itself, or directed to one of the reasons given in support of α^+, that is, some $\beta^+ \in \Gamma^+$. To indicate this, let us use a "negative" turnstile, so we have in the first case, $\Delta^- \dashv \alpha^-$. We are using the superscripts to indicate simply that formulas are involved in positive or negative "moves", where, for example, Γ^+ indicates that each $\beta_i \in \Gamma$ is involved in a positive move, so Γ^+ is just $\{\beta_1^+, \beta_2^+, \ldots, \beta_n^+\}$. Then, in this setting, as is obvious, the two moves have the same FOCUS, and so they agree upon what it is that constitutes the QUD to use Ginzburg's terminology. In a different

[1] I adopt the notion of "test" from Girard (e.g. [1], Sect. 12.5).

case, we may have that $\mathcal{P}$ makes the move represented by $\Gamma^+ \vdash \alpha^+$, whilst $\mathcal{R}$ makes a move $\Delta^- \dashv \beta^-$, with $\beta^+ \in \Gamma^+$. Now, to ensure that the focus of these moves is coherent, we need to be able to show that the move made by $\mathcal{R}$ can be transformed into a move that has the same focus as that made by $\mathcal{P}$ (i.e. α^-).

For the moment, let us think of an interaction as two, dual, derivations where prover attempts to prove a formula α, and refuter attempts to refute α. Let us treat these individually for now, since the requirement that it is possible to transform moves into other moves sharing the same focus can provide us with an explanatory framework for cut-elimination procedures in each *separate* derivation. Recall that, in Chap. 2, we said that the admissibility of "cut" effectively ensures balance over the course of derivations, so that it is always possible to extract a canonical object subsequent to the process of constructing a proof. The transformation procedure of one move into another more direct move can then be understood as analogous to cut-reduction procedures as a way of ensuring that all "detours", where a formula is introduced into a move and then eliminated immediately, can be eliminated. Of course, the point of our approaching logic by way of interaction is that, ordinarily, proofs are not constructed in the most direct manner possible, and that many "detours" of this kind may arise. The requirement that these interactions are focused, or can be *brought into focus* with each other, can then be understood in terms of a reduction procedure ensuring that derivations involving detours *could* be transformed into derivations that do not, but instead are more direct characterisations of the same argument.

Let us consider an abstract schema that we can apply to formulae across moves in interactions (this is ordinarily called a rule since it is typically understood to allow derivations to move from valid formulae to valid formulae).[2] An n-premise schema is simply an $n+1$-ary relation on moves (represented as sequents). Where a schema, #, is associated with an operation on moves, the set of formulas occurring in that move may be said to be closed under the operation, i.e. $\#(\alpha_1, \ldots, \alpha_n)$.[3] This is to be called an insertion move, since it brings an operation with a certain shape into the interaction, where the operation may be inserted into the focus of a move (focal insertion), or not (non-focal insertion). Although in a different context, I adopt the the definition of a COMBINING schema from [3], as having more than one premise. To be more precise, a combining schema with n premises can be given as follows.

Definition 12 (*Combining schema*) A combining operation $\#_n^c$ has $m \geq 1$ focal insertion schema of the form:

$$\frac{\Gamma_1 \vdash \beta_1 \qquad \ldots \qquad \Gamma_n \vdash \beta_n}{\Gamma_1 \ldots \Gamma_n \vdash \#_n^c(\beta_1, \ldots, \beta_n)} \text{(Combining-I}^+\text{)}\ \#_n^c$$

Where $n \geq 2$.

[2]See [2] for a detailed discussion of inference rules that share much in common with the notion of schema in the following.

[3]I am gliding over some formal niceties here, which we will deal with later, including, for example, the parameterization of schema that depend upon free variables.

So, for example, the schema usually taken as the form of standard conjunction, $\wedge$, is just a form of this rule where $n = 2$. We can form a schema that inserts the combining operation $\#_n^c$ into the reasons taken as support for a move, that is to say, a non-focal insertion schema. This schema should be in balance with the original schema so that the latter schema is neither too strong, nor too weak with respect to the focal insertion schema. This can be formalised by a requirement that any schema introducing an operation into the focus of a schema that is followed by a schema using that operation as reasons supporting a move may be eliminated (i.e. removed from the derivation without the overarching derivation being altered).[4] So, for example, for any positive combining schema with focal insertion $\#_n^c$ of the form

$$\frac{\Gamma_1 \vdash \beta_1 \qquad \ldots \qquad \Gamma_n \vdash \beta_n}{\Gamma_1 \ldots \Gamma_n \vdash \#_n^c(\beta_1, \ldots, \beta_n)} \text{ (Combining-I}^+\text{) } \#_n^c$$

(where $n \geq 2$), there exist n non-focal insertion schema of the form:

$$\frac{\Gamma_i, \beta_i \vdash \sigma_i}{\Gamma_i, \#_n^c(\beta_1, \ldots, \beta_n) \vdash \sigma_i} \text{ (Combining-N}^+\text{) } \#_n^c$$

This allows us to transform any set of moves that takes a set of reasons Γ to support the focal insertion of $\#_2^c(\alpha_1, \alpha_2)$ (more clearly, $(\alpha_1 \#_2 \alpha_2)$, then uses $\#_2^c(\alpha_1, \alpha_2)$ as reasons to support the focal insertion of, say, α_1, into a direct move from Γ to α_1. Schematically:

$$\frac{\dfrac{\begin{matrix}[A_1] \\ \Gamma \vdash \alpha_1\end{matrix} \qquad \begin{matrix}[A_2] \\ \Gamma \vdash \alpha_2\end{matrix}}{\Gamma \vdash \alpha_1 \#_2^c \alpha_2} \text{ (Combining-I}^+\text{) } \#_2^c \qquad \begin{matrix}[A_3] \\ \Gamma, \alpha_1 \#_2^c \alpha_2 \vdash \gamma_1\end{matrix}}{\Gamma \vdash \gamma_1}$$

Transforms to:

$$\Gamma \vdash \gamma_1$$

By cutting away the detour from inserting the combining operation into the focus, and using the combining schema as reasons in support of a further move. Since there is no more information hidden in the focal and non-focal insertion rules for $\#_2^c$, they can be eliminated as a detour from the derivation. That is, if, by hypothesis $[A_1]$ and $[A_3]$ are accepted arguments, then the argument from Γ to α_1 must also be acceptable. In other words, the conditions upon which it is possible to argue for α_1 at the end of the former derivation are already present at the beginning of that derivation. The transformation procedure is properly understood as a condition upon cut-elimination such that, cut-reductions on a formula α must yield cuts on formulae β that are lower complexity than α.[5]

---

[4] This is, obviously, just a form of cut-admissibility, as discussed below.

[5] As noted in [4], this condition is only plausible in the presence of the structural rule of contraction, though this is not an issue for the system here since we are dealing with sets of formulae.

This form of balance is called intrinsic harmony in [5], which is defined in terms of local soundness and completeness for the schema involved:

Definition 13 (*Local Soundness*) The schema inserting the operation into the non-focus are locally sound iff derivations containing a formula introducing the operation into the focus can be transformed by cut-reductions to derivations of the same conclusion avoiding the detour through that formula.

Definition 14 (*Local completeness*) This operates in the other direction, and says that a schema inserting the operation into the non-focus is locally complete iff 'every derivation of a formula α with principle operator $\#^x$ can be expanded to one containing an application of an non-focus insertion rule of $\#^x$, and applications of all focal insertion rules of $\#^x$ each with conclusion α' ([5], p. 9).

The latter ensures that the schema inserting the operation into the non-focus is not too weak, allowing us to reconstruct the original argument. For example, given the combining schema outlined above, we should be able to construct schema that removes $\#^c_n$ from the focus such that there is one such schema for each n. If not, then the schema that do exist would not be capable of expansion so would not be locally complete.

As discussed in Chap. 2, this process requires a rerouting of a derivation that may only occur subsequent to the process by which that derivation is made (and this will be complicated further still when we consider both sides of an interaction, momentarily). Nonetheless, the fact that it is possible to transform formulae in this way is a requirement of the idea that it is possible to maintain the focus of an interaction; so that, when one player makes a move testing one of the assumptions made by the other player in support of some formula α this may be understood as testing the formula α, albeit indirectly.[6]

[6]Below, I suggest that the two separate schematic structures which may be taken to represent moves on behalf of prover and refuter, are sometimes equivalent with the sequent calculus, *LJ* for intuitionistic logic, and the sequent calculus *LDJ* (e.g. [6]) for co-intuitionistic logic, respectively (though the latter is altered by taking it to string together negative moves corresponding to refutations). It is worth noting, in this context, the following theorems proved in Urbas.

Theorem 15 (Cut Elimination) *The cut rule is eliminable from LDJ; every sequent which is derivable in this system has a cut-free derivation* ([6], *p. 447).*

(Subformula Property): Every derivable sequent of the sentential fragment of LDJ has a cut-free derivation with the Subformula Property ([6], *p. 448).*

(Decidability): Sentential LDJ is decidable ([6], *p. 448).*

Bear in mind, however, that the structure of interaction outlined can not be reduced to derivability in one or both sequent calculi. Furthermore, as I highlight below, even subsequent to the construction of a valid proof or refutation, the structure of interaction differs from the characterisation of valid proofs and refutations as formal derivations in sequent calculi.

1.2 Symmetry over Agents

Let us consider the shapes of the schema which agents might use over the course of an interaction. In doing so, we can also begin to shed light on the approach to justification that we are pursuing here. The shape of schematic moves made in response to other moves can be understood in terms of internal constraints on the way in which syntax is utilised by agents in response to one another.[7] It is these constraints that give rise, subsequent to the termination of an interaction, to what are typically known as inference rules. The central notion of symmetry at work in the following is simply an elaboration of the primitive relationship between α^+ and α^-, where moving between the two involves just the reversal of the direction of a move.

Combining and Splitting Schema

We have already mentioned a combining schema with n-premises. Let us consider this intuitively from the point of view of a dialogical interaction. Say that $\mathcal{P}$ makes a move introducing the assertion of a statement $\#_n(\alpha_1 \ldots \alpha_n)$. Then, in response to this assertion, $\mathcal{R}$ may test this assertion. How would they go about this? Since in $\#_n(\alpha_1 \ldots \alpha_n)$ we have the combination of the formulas $(\alpha_1 \ldots \alpha_n)$ by definition, it would be appropriate to say that the formula may be tested by testing any of the formulas so combined. In other words, $\mathcal{R}$ might test any $\alpha_i \in \{\alpha_1 \ldots \alpha_n\}$. If their test of α_2 (where $n \geq 2$) were successful, then this would have the result that the combined formula $\#_n(\alpha_1 \ldots \alpha_n)$ would also be successfully tested, since each formula closed under $\#_n$ is combined into a single formula. As such, it is simple to construct a schema that is DUAL to the combining schema, which (after Francez) we call SPLITTING, which is negative since we are testing the utterances for their possible refutation.

Definition 16 (*Splitting Schema*) A splitting operation $\#^s_n$ has n focal insertion schema of the form:

$$\frac{\Gamma \dashv \beta_j}{\Gamma \dashv \#^s_n(\beta_1, \ldots, \beta_n)} \text{ (Splitting-I}^-\text{) } \#^s_n$$

Where $1 \leq j \leq n$

This is the j'th schema over any formula, and can be intuitively understood as a negative move by player $\mathcal{R}$ that tests any formula β_j where $\beta_j \in \{\beta_1, \ldots, \beta_n\}$, may also be understood to test the combining formula $\#^c_n(\beta_1, \ldots, \beta_n)$. Say, for example, that $n = 2$, then the combining schema introducing $\#^s_2$ allows us to introduce the formula $\alpha_1\#_n\alpha_2$ as a POSITIVE move. Dually, the splitting schema introducing $\#^s_2$ allows us to introduce the formula $\alpha_1\#_n\alpha_2$ as a NEGATIVE move. These are exactly dual to each other, so, in case a positive move is made that uses a positive version of the splitting schema (simply swap the turnstile for a positive one), then the corresponding negative move is a negative version of the combining schema (simply swap the turnstile for a negative one).

[7] These are schematic in the sense that any formula may be substituted into them.

Conditional Schema

We are using the positive and negative turnstiles to indicate positive and negative moves, on behalf of agents $\mathcal{P}$ and $\mathcal{R}$, respectively. We also allow, as is typical, that these may involve a series of assumptions (on the left), which are supposed to support the formula introduced into the interaction (on the right).[8] As such, we would also like to express this form of reasoning in the object language itself. It is typical to do so using the connective $\rightarrow$ as a detachable operation that allows, for example, for $\alpha \vdash \beta$ to be expressed in the object language as $\vdash \alpha \rightarrow \beta$. Since we are treating $\mathcal{P}$ and $\mathcal{R}$ symmetrically, we would then also require a connective, which we denote $\leftarrow$, which expresses a negative move in the object language. Dually, we will also require that $\leftarrow$ is detachable, allowing that, for example, $\alpha \dashv \beta$ can be expressed in the object language as $\dashv \beta \leftarrow \alpha$. More generally, this gives us the following dual schema:

Definition 17 (*Positive Conditional Schema*) A conditional operation $\#^h_n$ has $m \geq 1$ conditional insertion schema of the form[9]:

$$\frac{\Gamma_1, \beta'_1 \vdash \beta_1 \qquad \ldots \qquad \Gamma_n, \beta'_n \vdash \beta_n}{\Gamma_1 \ldots \Gamma_n \vdash \#^h_n(\beta_1, \ldots, \beta_n)} \text{(Conditional-I}^+) \#^h_n$$

(Negative Conditional Schema) A conditional operation $\#^h_n$ has $m \geq 1$ insertion schema of the form:

$$\frac{\Gamma_1, \beta'_1 \dashv \beta_1 \qquad \ldots \qquad \Gamma_n, \beta'_n \dashv \beta_n}{\Gamma_1 \ldots \Gamma_n \dashv \#^h_n(\beta_1, \ldots, \beta_n)} \text{(Conditional-I}^-) \#^h_n$$

Where $n \geq 2$.

The symmetry of the roles played by $\leftarrow$ and $\rightarrow$ (which are just cases of the above where $m = 1$ and $n = 2$) is clearly accounted for by these schema.[10] Yet, if we consider the intuitive roles of the players as *testing* the other player's utterances, then it may seem less intuitive to think of a "test" of a positive move of $\alpha \rightarrow \beta$ as equivalent to a negative move of $\beta \leftarrow \alpha$, which inserts the co-conditional operation $\leftarrow$. We are following, however, the general idea that switching the direction of a move ought to be symmetric to the original move. This is out of keeping with, for example, the strong negation of a conditional $\alpha \rightarrow \beta$ in Nelson's [8] falsification logic, which combines the verification of α with the falsification of β.[11] But, in order to maintain symmetry between players, this is not plausible. Specifically, I shall suggest that, in

[8] Note, however, that the sequent structure combining the two will differ in direction (see below).

[9] The subscript stands for hypothetical.

[10] This also follows Mellies [7] suggestion that the 'permutation Prover $\leftrightarrow$ Denier should be understood as an internal symmetry of logic' (p. 1).

[11] As discussed below. Similarly, in [3], which also offers an account of the duality of rules involving positive and negative formulas (though in a very different context to the one offered here), the dual of our positive conditional schema involves the combination of positive and negative formulas. Furthermore, whilst a definition of duality is given there (which differs from that presented below) it is not clear what the justification of the definition is supposed to be.

cases where there exists a successful move in support of α^+, and also of β^-, we only have the result that the initial hypothesis expressed by a conditional $\alpha \rightarrow \beta$ fails. Instead, what we are interested in is whether or not it is possible to categorically assert (or deny) β, on condition α.[12]

The testing relationship between $\leftarrow$ and $\rightarrow$ is understandable in the context of the distinction made in Chap. 2 *inside proofs* (and refutations). Recall that the way we are approaching proofs and refutations as activities alters the standard account of the validity of a conditional, since it requires that $\alpha \rightarrow \beta$ may be valid only in case an agent has an actual proof of α to hand. That is, a proof of a conditional is a function that maps actual proofs of the antecedent into actual proofs of the consequent since: 'as long as no proof of α is known, [the function] f has nothing to map. So we can still define f as the constant function which, once a proof π of α is known, maps every proof of α into the proof of β' [10]. So, a valid proof of a conditional $\alpha \rightarrow \beta$ requires a proof of α in addition to a proof that there exists a function mapping every proof of α into a proof of β. We may think analogously for the co-conditional. We also need to remember that the reasoning that we are attempting to capture in the object language with these schema is just this: "α provides reason to think that β is assertible" (positively); "α provides reason to think that β is deniable" (negatively). In this setting, it is the principle formula, β that is of importance, since this denotes the focus around which the interaction takes place.

In this sense, we may construe the formula upon which the principle formula is supposed to rest (as reason for it) as a kind of hypothetical *placeholder* for justifying the principle formula. In effect, we may think of interactions of this form as constituting attempts to settle whether or not the principle formula, β may be justified positively, or negatively. Then, a conditional move is just a move made inside a hypothesis, which initiates a search for the principle formula under the minor assumption. So, when we say that $\alpha^+ \rightarrow \beta^+$ is valid, what we *mean* is just that β is categorically assertible on condition α. On this basis, if we test $\alpha^+ \rightarrow \beta^+$ by means of testing α^+, then no categorical assertion can be made, since the condition on which β is categorically assertible fails to hold.[13] So, we can test a conditional $(\alpha^+ \rightarrow \beta^+)$ by simply refuting α^+ since $(\alpha^+ \rightarrow \beta^+)$ *says* that α^+ is reason to take β^+ to be assertible. Then, if refuted, this means that, either β^- is deniable, so it can not be assertible; or, it means that α^+ does not give us reason to take β^+ to be assertible. If α^- is refuted, then α^+ can not stand as reason in support of β^+. So, whilst

[12]To a large extent, this also coheres with Kapsner's [9] excellent analysis of constructive falsification, and particularly his suggestion that providing the conditions under which a conditional is falsified (read "tested") should be a method of turning a falsification of the consequent into a falsification of the antecedent.

[13]This follows Edgington's [11] argument considering the conditional "If you press that switch, there will be an explosion": 'My hearer understands that if she presses it, my assertion of the consequent has categorical force; and given that she takes me to be trustworthy and reliable, if it does acquire categorical force, it is much more likely to be true than false. So she too acquires reason to think that there will be an explosion if she presses it, and hence a reason not to press it' (p. 178).

this (successful) test would mean that no categorical assertion of β is possible on condition α, it does not rule out such a categorical assertion on a different condition.

To test this, much stronger claim, requires a test of β^+. Furthermore, since we are working under the hypothesis that $\alpha^+ \rightarrow \beta^+$, if the test of β^+ were successful, then $\mathcal{P}$ has no option but to agree that this test "carries back" to α^+. Note that, at this point the tables must turn, since, whilst a sucessful test of β^+ would allow $\mathcal{R}$ to insert $\beta^- \leftarrow \alpha^-$, it is open to $\mathcal{P}$ to test α^-. Thus, whilst a successful test of β^+ carries back to α^+ by hypothesis, this means only that α^+ is part of an unsuccessful move, and not that α^+ itself has been refuted. So, in order for $\beta^- \leftarrow \alpha^-$ to pass "all" of the tests, that is, to ensure that there exists a refutation of α^-, $\mathcal{P}$ may go on to test α^-.[14] The point here is that there must be a categorical assertion or denial at the termination of interaction, and that this is made on condition of the minor assumption. So, whenever β^- is refuted, *then* $\mathcal{R}$ may insert $\beta^- \leftarrow \alpha^-$, to make transparent that any refutation of the principle formula also carries back to the condition upon which the hypothesis was originally made.

Consider that the conditional moves place into the object language the form of a positive and negative move. By the local soundness and completeness conditions given above, we know that, for every positive conditional operation $\#^h_n$, there exists n non-focal insertion schema of the form:

$$\frac{\Gamma_i \vdash \beta_i \qquad \Gamma_i, \beta_i' \vdash \alpha}{;\Gamma_i \ldots \Gamma_n, \#^h_n(\beta_1, \ldots, \beta_n) \vdash \alpha}\ (\text{Conditional-N}^+)\ \#^h_n$$

(where $1 \leq i \leq n$).

And, for every negative conditional operation $\#^h_n$, there exists n non-focal insertion schema of the form:

$$\frac{\Gamma_i \dashv \beta_i \qquad \Gamma_i, \beta_i' \dashv \alpha}{\Gamma_i \ldots \Gamma_n, \#^h_n(\beta_1, \ldots, \beta_n) \dashv \alpha}\ (\text{Conditional-N}^+)\ \#^h_n$$

(where $1 \leq i \leq n$).

These schema give shape to the way in which players may use conditional schema as reasons for the insertion of other formulae. These non-focal insertion schema also make perspicuous that a positive move of form $\alpha \vdash \beta$ may lead to further interactions regarding reasons for α, and reasons for β (similarly for the dual negative move). For example, the following move:

$$\frac{\vdash \alpha \qquad \vdash \beta}{\alpha \vdash \beta}$$

is just a non-focal positive insertion when contexts contain zero formulae. In a similar vein, to make a positive move of form $\Gamma \vdash \alpha$ we should take reasons for each formula in Γ and transform them into a reasons in support of α. That is, we can decompose moves like this as follows:

[14] Of course, if we are left in a position where β^- and α^+ successfully pass tests, this just tells us that the initial hypothesis fails.

$$\frac{\begin{array}{cc}[\alpha^+] & \\ \Gamma \vdash \beta & \vdash \alpha\end{array}}{\Gamma, \alpha \vdash \beta}$$

Think of these moves in relation to their tests. As the above clarifies, for any $\Gamma \vdash \alpha$, we may test α directly, or we may test a formula $\beta \in \Gamma$. Say, for example, that $\mathcal{P}$ is looking for is a proof of β from a proof of α. Then, $\mathcal{R}$ may test the proof of α, and see if they can transform that into a refutation of β. So, take $\alpha \vdash \beta$, and a negative move introducing α as test against it, such that α can be transformed into a test of β:

$$\frac{\dfrac{\begin{array}{c}[\alpha^-] \\ \Gamma \dashv \beta \leftarrow \alpha\end{array}}{\Gamma, \alpha \dashv \beta} \qquad \dashv \alpha}{\Gamma \dashv \beta}$$

Now, let us take a test of β, where that test should be *transformable* into a test of $\Gamma, \beta \vdash \alpha$. Echoing the case with empty contexts above,

$$\frac{\dashv \beta}{\dashv \beta \leftarrow \alpha}$$

is admissible. Now, say, for example, that $\mathcal{P}$ is attempting to support β^+, and that they proceed as follows:

$$\frac{\dfrac{\begin{array}{c}[\alpha^+] \\ \Gamma, \alpha \vdash \beta\end{array}}{\Gamma \vdash \alpha \to \beta} \qquad \Gamma \vdash \alpha}{\Gamma \vdash \beta}$$

That is, $\mathcal{P}$ gives as reason, $\Gamma, \alpha \vdash \beta$, to insert $\Gamma \vdash \alpha \to \beta$, going on to use α^+ to insert β. Here, α^+ is used (twice) in such a proof-attempt, and so, $\mathcal{R}$ may test this minor assumption, α^+, where this test of α^+ can be used to test β^+. In other words, an attempted refutation of α^- is transformable into an attempted refutation of $\Gamma, \alpha \dashv \beta$, which is, therefore, a test of $\Gamma \vdash \alpha \to \beta$. Now, say, instead, that $\mathcal{R}$ tests β^+ by attempting to provide *some* reason, σ^- in support of β^-. Then, take $\Gamma, \sigma \dashv \beta$ to insert $\Gamma \dashv \beta \leftarrow \sigma$, where $\mathcal{R}$ then requires a refutation of σ to complete the test of β. So in this case, $\mathcal{P}$ will no doubt test σ, with this test being transformable into a test of $\beta \leftarrow \sigma$. Then, an attempt to provide a categorical assertion of β under this hypothesis may proceed by decomposing $\Gamma \vdash \alpha \to \beta$ into a subproof of $\Gamma, \alpha \vdash \beta$ as follows:

$$\dfrac{\dfrac{\dfrac{\Gamma, \alpha \vdash \sigma \qquad \Gamma, \sigma \vdash \beta}{\Gamma, \alpha \vdash \beta}}{\Gamma \vdash \alpha \rightarrow \beta} \qquad \dfrac{\vdash \gamma \qquad \gamma \vdash \alpha}{\Gamma \vdash \alpha}}{\Gamma \vdash \beta}$$

The point is just that that, if $\Gamma \vdash \alpha \rightarrow \beta$ is to *pass* every possible test, then a test of $\beta \leftarrow \alpha$ will have been constructed along the way.

Think of this is by analogy with Popper's [12] falsification conditions of a theory.[15] In general, take a theory, Γ, where a claim, α follows from Γ, by hypothesis. This initial conjecture (on behalf of either player) sets the hypothetical "mood" of the interaction. Say that the hypothesis is made by $\mathcal{P}$. Then, to refute the theory, we need only refute α, unless the hypothesis is incorrect, since according to the hypothesis, α follows from Γ. That is to say, *under the hypotheses*, a test of β^+ *by any means*, carries back to a test of α^+. So, this means that what is happening with the refutation-as-test in this setting is essentially that refuter accepts the underlying ambit. Nevertheless, this initial conceit is also tested by the way in which formulae are manoeuvered across tests in an interaction. Take $\alpha^+ \rightarrow \beta^+$ made as hypothesis on behalf of $\mathcal{P}$. Given the above, we should imagine that $\mathcal{R}$ responds with a test of β^+ in the form of an attempt to refute β^- (transformable into a refutation of $\beta^- \leftarrow \alpha^-$). Consider how $\mathcal{P}$ responds to this. Given that they have already suggested, by hypothesis, that $\alpha^+ \rightarrow \beta^+$, it is natural for $\mathcal{P}$ to support a possible categorical assertion of β by appeal to a proof of α^+. Now, $\mathcal{R}$ responds by also testing α^+ through an attempted refutation of α^-. That is, an attempted proof of $\alpha^+ \rightarrow \beta^+$ is met with an attempted refutation of $\beta^- \leftarrow \alpha^-$.

What the conditional expresses is something like the *possibility* of a categorical assertion, or a categorical denial. This is clarified by these non-focal insertion schema (where $n = 2$), which say (positively or negatively): if we take reason for $\alpha \rightarrow \beta$, and for α, then this provides reason in support of β. That is, if we have a valid proof (or refutation) of $\alpha \#_n^h \beta$, then we should be able to extract from it enough reason to provide us with a categorical assertion (denial) of β. As such, there is an interesting connection here between this account and approaches to relevance in logical conditionals (e.g. [14]), which require there to be some sort of relevance (this is cashed out in different ways) between antecedent and succedent. An additional advantage of the approach is that we automatically ensure that some form of relevance holds between premises and conclusions in terminating interactions (where there is some agreement between agents). For example, the paradox of relevance results from the standard fact that, whenever some α is a theorem, it is derivable from any other statement, so, for example, from any α, where α is assumed, it is possible to infer $\beta \rightarrow \alpha$. The paradox holds in, for example, Prawitz's natural deduction rules for

[15] Though, as is well-known, Popper's suggestion is that a falsificationism of a conditional claim is based on classical modus ponens, as discussed in ([13], p. 123ff).

the conditional. Here this is not the case, since, for any interaction to *terminate* with agreement upon $\beta \rightarrow \alpha$, this requires it to have passed *all* the relevant tests, such that we have also justification for β, together with a function mapping each instance of β into α.

Parameterised Schema

Let us also consider schema that take a domain of variables and make claims over that domain. Such as, for example, that some such predicate $\alpha(x)$ exists, or that, for all x_n in a domain $\mathcal{D}$, $\alpha(x)$ holds. Schematically, these are just PARAMETERISED forms of combining or splitting schema, where a schema is parameterised if at least one of its premises depends upon a free variable.[16] So, for example, let a domain $\mathcal{D} = \{x_1, x_2, \ldots, x_n\}$. Then, a parameterised combining operation, $\#^{cp}$, may be defined as equivalent to a generalised combining operation as follows: $\#^{cp}_{x\in\mathcal{D}}\alpha(x) = \#^{c}_{n}(\alpha(x_1), \alpha(x_2), \ldots \alpha(x_n))$. Similarly, a parameterised splitting operation, $\#^{sp}$, may be defined as equivalent to a generalised splitting operation as follows: $\#^{sp}_{x\in\mathcal{D}}\alpha(x) = \#^{s}_{n}(\alpha(x_1), \alpha(x_2), \ldots \alpha(x_n))$. Schematically, these can be written as follows.

Definition 18 (*Positive Combining Parameterised Schema*) A combining parameterised operation $\#^{cp_{x_n}}_{x\in\mathcal{D}}$ has $m \geq 1$ insertion schema of the form:

$$\frac{\Gamma_n \vdash \alpha[x_i/x]}{\Gamma_n \vdash \#^{cp_{x_n}}_{x\in\mathcal{D}} x.\alpha(x)} \text{ (Combining Paramaterised}-I^{+}\text{) } \#^{cp_{x_n}}_{x\in\mathcal{D}}$$

(Negative Combining Parameterised Schema) A combining parameterised operation $\#^{cp_{x_n}}_{x\in\mathcal{D}}$ has $m \geq 1$ insertion schema of the form:

$$\frac{\Gamma_n \dashv \alpha[x_i/x]}{\Gamma_n \dashv \#^{cp_{x_n}}_{x\in\mathcal{D}} x.\alpha(x)} \text{ (Combining Paramaterised}-I^{-}\text{) } \#^{cp_{x_n}}_{x\in\mathcal{D}}$$

Where the upper label on $\#^{cp_{x_n}}_{x\in\mathcal{D}}$ indicates that each variable x_i must be fresh such that it can not occur in Γ_n or in $\#^{cp_{x_n}}_{x\in\mathcal{D}} x.\alpha(x)$. The box $[x_i/x]$ indicates that a fresh variable $x_i \in \mathcal{D}$ is substituted for x.

As is clear, the combining operation is essentially embedded within the requirement to substitute arbitrary fresh variables into the first line of the schema.[17] Similarly, for splitting parameterised schema as follows.

Definition 19 (*Positive Splitting Parameterised Schema*) A splitting parameterised operation $\#^{sp_{x_n}}_{x\in\mathcal{D}}$ has $m \geq 1$ insertion schema of the form:

$$\frac{\Gamma_n \vdash \alpha[x_i/x]}{\Gamma_n \vdash \#^{sp_{x_n}}_{x\in\mathcal{D}} x.\alpha(x)} \text{ (Splitting Paramaterised}-I^{+}\text{) } \#^{sp_{x_n}}_{x\in\mathcal{D}}$$

---

[16] Again, I adopt the terminology of parameterised rules discussed in [5] though there is no discussion there of the relationship between these and combining/splitting rules.

[17] Freshness simply means that the variable has not been used previously.

(Negative Splitting Parameterised Schema) A splitting parameterised operation $\#^{sp_{x_n}}_{x \in \mathcal{D}}$ has $m \geq 1$ insertion schema of the form:

$$\frac{\Gamma_n \dashv \alpha[x_i/x]}{\Gamma_n \dashv \#^{sp_{x_n}}_{x \in \mathcal{D}} x.\alpha(x)} \text{ (Splitting Paramaterised}-I^-\text{) } \#^{sp_{x_n}}_{x \in \mathcal{D}}$$

Where the variable x is free in the conclusion, and there is no restriction upon freshness of variables substituted in $[x_i/x]$ since we require only one variable to be substituted in order to introduce the operations.

1.3 *Excursus: Duality as Symmetric Reversal*

The symmetry between conditionals and (parameterized and non-parameterized) combining and splitting schema is borne out by the idea introduced above, that a symmetric duality between players may be understood in terms of the reversal of the direction of a move. Let us briefly discuss this notion of duality, since it is often used without adequate explanation. Duality in logic is typically understood by means of a De Morgan mapping by the negation operator in classical logic. For example $\neg$ in classical logic is an involutive mapping that obeys the following:

$$\neg(\alpha \wedge \beta) =: \neg\alpha \vee \neg\beta$$
$$\neg(\alpha \vee \beta) =: \neg\alpha \wedge \neg\beta$$

This duality between conjunction and disjunction is also mirrored by the analogous duality in set-theory where intersection and union are interchangeable under complementation. That is, where complementation is denoted by $\overline{(-)}$ it is, again, an involutive mapping that obeys the following:

$$\overline{\alpha \cup \beta} =: \overline{\alpha} \cap \overline{\beta}$$
$$\overline{\alpha \cap \beta} =: \overline{\alpha} \cup \overline{\beta}$$

However, a more direct approach to duality is possible, and one that does not require the existence of the logical operation of negation to be defined in advance of giving its definition. This way of considering it is suggestive of a more elementary duality than negation, considered in the abstract, by not invoking classical negation, but rather, by inverting the direction of entailment. At the category-theoretic level, duality is well-understood, where, for example, it is always possible to construct the dual category ${\mathbb{C}_{\mathbb{H}}}^{op}$ for any cartesian closed category $\mathbb{C}_{\mathbb{H}}$, by inverting all morphisms in $\mathbb{C}_{\mathbb{H}}$. So, there, duality is not mediated by negation, but by reversing the direction of entailment (that is, if we interpret canonical morphisms as entailment). So, for example, there exists a duality between conjunction and disjunction, which is constructed by reversing the direction of the morphism. Let us just give a few definitions so that we may highlight the way in which this approach to duality operates.[18]

Definition 20 (*Cartesian closed category (CCC)*) A CCC is a category $\mathbb{C}_{\mathbb{H}}$ with a terminal object, finite products and exponentials such that, for every object A of $\mathbb{C}_{\mathbb{H}}$, the functor $- \times A : \mathbb{C}_{\mathbb{H}} \to \mathbb{C}_{\mathbb{H}}$ has a right adjoint (e.g. [15, 16]).

Now, consider the typical definition of product in C_H, typically taken to model the logical operation of conjunction.

Definition 21 (*Product*) A product of A and B is an object $A \times B$, with morphisms:
$i_{A,B} : A \times B \to A$;
$i'_{A,B} : A \times B \to B$;
such that, for any object C, and any morphisms $f : C \to A$, and $g : C \to B$, there is a unique arrow $h : C \to A \times B$ making the following diagram commutative:

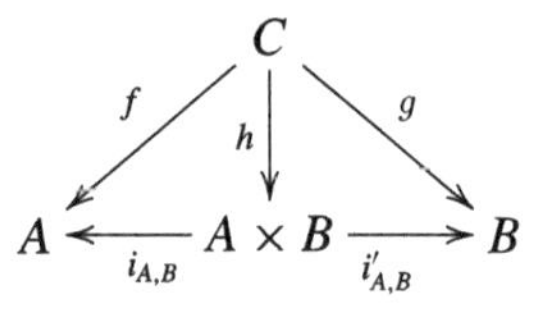

Now, consider co-product, taken to model disjunction.

Definition 22 (*Co-product*) A co-product of A and B is an object $A \oplus B$, with morphisms:
$i_{A,B} : A \to A \oplus B$;
$i'_{A,B} : B \to A \oplus B$;
such that, for any C, and any morphisms $f : A \to C$, and $g : B \to C$, there is a unique arrow $h : A \oplus B \to C$ as in the following diagram:

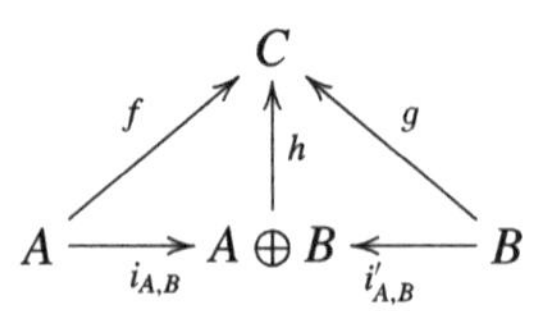

As is clear, disjunction and conjunction are dual to each other by inverting the direction of each morphism.[19] Let us continue by defining initial and terminal objects in $\mathbb{C}_{\mathbb{H}}$.

Definition 23 An object 1 is terminal in $\mathbb{C}_{\mathbb{H}}$ if, for any object A of $\mathbb{C}_{\mathbb{H}}$, there is a unique arrow $\Box_A : A \to 1$. An object 0 is initial if for any object A of $\mathbb{C}_{\mathbb{H}}$, there is a unique arrow $\circ_A : 0 \to A$.

---

[18] These will be utilised further in the following chapter.

[19] Note also that, in category theoretic terms, the dual of conjunction is only disjunction (co-product) when it is intuitionistic, rather than classical since morphisms $i_{A,B}$ and $i'_{A,B}$ are "injections", which ensure the disjunction property.

We can then define exponentials, which are taken to model logical conditional as follows.

Definition 24 (*Exponential*) An exponential of objects A and B, denoted B^A, is defined by the arrow $\in_{A,B}$: $B^A \times A \to B$ which satisfies the following property: for any object C, and any arrow $f : C \times A \to B$, there is a unique arrow $g : C \to B^A$ as in the following diagram:

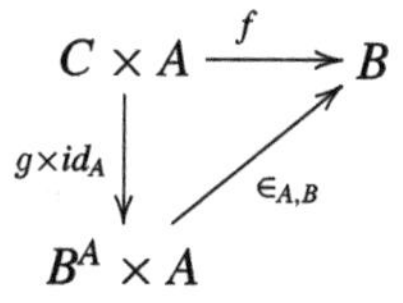

The arrow $\in_{A,B}$ is usually called evaluation, which clearly models what modus ponens. The operation dual to exponential is co-exponential, which may be directly constructed by means of dualizing the above definition, such that every $- \oplus A : \mathbb{C}_{\mathbb{H}} \to \mathbb{C}_{\mathbb{H}}$ has a left adjoint.

Definition 25 (*Co-exponential* [17]) A co-exponential of objects A and B, denoted B_A, is defined by the arrow $\ni_{A,B}$: $B \to B_A \oplus A$ which satisfies the following property: for any object C, and any arrow $f : B \to C \oplus A$, there is a unique arrow $g : B_A \to C$ as in the following diagram:

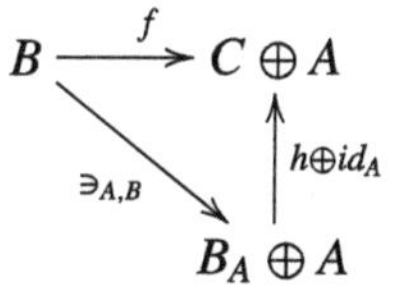

Let us note that the morphism $\ni_{A,B}$, which we call co-evaluation, if we allow B_A to interpret $(\beta \leftarrow \alpha)$, gives us $\beta \leq (\beta \leftarrow \alpha) \vee \alpha$. As such, it is possible to consider duality much more directly than is typically the case when invoking negation, or complementation, operations. Moreover, existential and universal quantification may also be understood as dual (left and right) adjoints to a certain form of substitution.[20]

The schema introduced above ensure that there is symmetry between the two agents such that the moves made by corresponding agents are equally capable of introducing the same operation of the relevant formulae. This is distinct from the *interpretation* of the above duality-formation in category theory by reversing the direction of morphisms in the above, since, rather than consider, for example, disjunction as dual to conjunction in the same structure, we are thinking of the construction

---

[20]In fact, without invoking any category theory, this can be constructed algebraically by means of reversing the direction of an ordering relation over a set of formulas. I shall discuss this in far more detail in this chapter, where I also note proofs that whenever co-exponential is added to a CCC, it collapses to a single pre-order (the proof is given in [17]), which I strengthen in the context of topos-theoretic semantics to provide collapse results to a semantics whose algebraic structure is Boolean. This is grist to the mill for reasoning to the syntactic separation discussed below.

of separate derivation structures by means of this notion of duality using the reversal of the relation $\vdash$ to $\dashv$. That is, rather than consider duality within the same derivation structure, we shall consider duality mapping as constructing a dual *structure*. So, rather than take disjunction to be the dual of conjunction in a derivation structure, we may rather think of conjunction as *combining* in the positive structure as dual to conjunction as *splitting* in the negative structure, and vice-versa for disjunction. We can extend this mapping between positive and negative structures to all of the above schema as follows:

Definition 26 (*Duality*) Let us denote by α^+ a formula that is an element of a positive move, and α^- a formula that is an element of a negative move, and note that all schema must be of the same arity.

- α^+ is dual to α^-
- A positive combining schema is dual to a negative splitting schema
- A positive splitting schema is dual to a negative combining schema
- A positive conditional schema is dual to a negative conditional schema
- A positive combining parameterised schema is dual to a negative splitting paramterised schema
- A positive splitting paramterised schema is dual to a negative combining parameterised schema

Duality is defined to be idempotent (so all maps are symmetrical).

As such, we can use this duality mapping to define a formula's "counterpart", so that we can identify formulae across positive and negative moves. For example, above, we said that a formula introduced using a combining schema in a positive move will be tested by a splitting schema in a negative move. So, in order to formally account for this idea that the formulae inserted by positive combining and negative splitting are the same formula as seen from the differing point of view of $\mathcal{P}$, and $\mathcal{R}$, we use the notion of counterpart formulas.

Definition 27 (*Counterpart formulas*) For any formula p, α with superscript $^{+,-}$, the counterpart formula, denoted $\alpha_\circ$, is simply the same formula, rewritten with:

- (a) dual superscripts;
- (b) dual schema.

So, for example, where $n = 2$, this ensures that the counterpart of $\alpha^+ \wedge \beta^+$ is just $\alpha^- \wedge \beta^-$, where the former is inserted using a positive combining schema, and the latter a negative splitting schema.

With this in mind, we can begin to consider the reconstruction of inference rules by means of the symmetry of positive and negative moves, in terms of a TESTING relationship. Let us consider the standard forms of the above schema (e.g. the positive combining schema where $n = 2$ is analagous to conjunction). Then, take the following interaction. "Prover" asserts a conjunction, $\alpha \wedge \beta$, putting $\alpha \wedge \beta$ "into the game", and it is incumbent upon prover, to provide a move that serve as some sort of reason or evidence in support of both α and β. "Refuter", on the other hand, challenges $\alpha \wedge \beta$ providing a move that tests either α or β. As such, a "test" for an assertion of a conjunction is disjunctive, and so is more fine-grained than a counter-model since we test each formula used as a premise for the initial assertion. Exactly the reverse is the case if the formula in question is a disjunction. With this in mind, we can give a BHK-style informal interpretation of the testing relationship between positive and negative moves, which we denote with superscripts $(-)^+$ and $(-)^-$, respectively.

Definition 28 (*Informal testing interpretation*) Where a test of an atomic positive move α^+ is just a negative move of form α^-, and a test of α^- is just a positive move of form α^+, and $\mathcal{D}$ is a domain of variables x_n, so $\mathcal{D} = \{x_1, x_2 \ldots, x_n\}$:

- Testing $\alpha^+ \wedge \beta^+$ involves testing α^+ or testing β^+.
- Testing $\alpha^+ \vee \beta^+$ involves testing α^+ and testing β^+.
- Testing $\alpha^- \wedge \beta^-$ involves testing α^- and testing β^-.
- Testing $\alpha^- \vee \beta^-$ involves testing α^- or testing β^-.
- Testing $\alpha^+ \rightarrow \beta^+$ involves testing for a function that maps each test of α_i^+ into a test of β^+ (where α_i may shift in the testing process).
- Testing $\beta^- \leftarrow \alpha^-$ involves testing for a function that maps each test of α_i^- into a test of β^- (where α_i may shift in the testing process).
- Testing $\exists_{x \in \mathcal{D}} x(\beta^+)$ involves testing each x in the domain $\mathcal{D}$ such that the test is also a test of $\beta^+(x)$.
- Testing $\exists_{x \in \mathcal{D}} x(\beta^-)$ involves testing each x in the domain $\mathcal{D}$ such that the test is also a test of $\beta^-(x)$.
- Testing $\forall_{x \in \mathcal{D}} x(\beta^+)$ involves testing any x in the domain $\mathcal{D}$ such that the test is also a test of $(\beta^+(x)$.
- Testing $\forall_{x \in \mathcal{D}} x(\beta^-)$ involves testing any x in the domain $\mathcal{D}$ such that the test is also a test of $(\beta^-)(x)$.[21]

[21] If we consider just the standard forms of the schema, as suggested, then the inference structures that are appropriate to the above informal testing interpretation are intuitionistic logic, L_I, for the positive structure, and a negative, refutation, form of co-intuitionistic logic, L_C, for the negative structure. I have argued elsewhere [18, 19] that L_C, which is formed by a duality-mapping from the sequent calculus *LJ* into the sequent calculus *DLJ* in [6] (which may be seen as an instance of the above duality), is better understood as a logic for "refutation". This is, in part, due to the fact that the co-conditional connective, which is dual to conditional, detaches on the left, and also that the dual to a sequent of *LJ* has multiple formulas on the *r.h.s*, and a single formula on the left. As such, in essence, the idea is just to read sequents of *DLJ* from right to left, and the above turnstile $\dashv$ simply reflects this reading. Moreover, it is well known that, just as discussed above, the duality mapping between *LJ* and *DLJ* is a form of De Morgan duality, without negation. Taken together, I suggest that these facts give reason to think that the interpretation of duality between the two is better understood by means of the duality definition above, and that they may be taken as separate structures, which are combined via the notion of interaction between prover (operating in L_I) and refuter (operating in L_C).

1.4 Coherence, Convergence and Resolution

With the duality of the testing relationship between schematic moves in hand, we can now consider the notion of coherence of an interaction. In the previous chapter, the coherence of a dialogue is defined such that each conversational move seems to cohere smoothly with its predecessor with questions followed by answers which, in turn, raise new questions. Since the testing relatonship between schematic moves is perfectly symmetrical on the definitions provided above, in this setting, this amounts to a requirement that both symmetry and shared focus hold over the course of an interaction.[22] We may think of coherence, then, as that which underlies interactions in which symmetry and shared focus feature. Of course, it is also the case that neither symmetry nor shared focus are constant over all interactions. So, in this sense, we can also utilise coherence as a way of classifying interactions, allowing for interactions to be classified as coherent under certain circumstances, rather than being written-in to the definition of an interaction.

I will formally define coherence in the following section, but for now, take coherence to be defined in terms of both the duality relationship between schematic moves, and the requirement of shared focus. An interaction shares focus when a positive move in support of α^+ is opposed to a negative move in support of α^- (or that we may transform one or the other to such a move as discussed above), and these moves are dual whenever $\alpha^+ = \alpha^-_\circ$ and vice-versa. So, rather than think of the schematic shape of moves as being constrained by logical rules, it is underlying constraints upon coherence of schematic moves via which logical rules may be constructed *over the course of an interaction*. That is to say, in order that an interaction is coherent, it is a requirement that, through subsequent moves, players respond to a move with a schematic move of a certain shape.

When both conditions (symmetry and shared focus) hold, we should, over the course of an interaction, be able to construct sets of formulae that are coherent in this way, where this ensures that an interaction terminating with a proof of α^+ is always coherent with an attempted refutation of α^-. That is, if an interaction terminates such that a formula is considered proved (or refuted), then coherence should ensure that this formula has passed the relevant tests for that formula (in the form of an attempted refutation (or proof) of that formula). This coheres with the notion, discussed in Chaps. 2 and 3, that justification of a formula requires that players *agree* that adequate justification has been provided for that statement. When players agree in this way, the interaction terminates either with refuter endorsing proof of the statement and marking their attempt to refute it as failing, or prover endorsing the refutation of the statement marking their attempt to prove it as failing. As such, we also allow either player to make a move that we shall call RESOLUTION,

[22]Note that this approach to coherence is at odds with Restall's (discussed in Chap. 2), where it is incoherent for an agent to simultaneously assert and deny the same formula. I return to this in Chap. 6, where I suggest that Restall's form of coherence may be generated at the termination of interaction by this prior form of coherence.

which signals the termination of the interaction through agreement, given that the interaction is coherent, and the relevant tests have been successfully passed. Let us turn, now, to consider the formalisation of these requirements.

2 Interactions and Coherence Conditions

Let us now provide an overview of the formalism for this interactive logic for proofs and refutations (IPR).

Definition 29 Define a language S for IPR, over a denumerable set of atomic formulas F, in Backus-Naur form (p is any atomic formula of F, β a formula generated from them, and x is a variable)[23]:

$$S\ \beta ::= p \,|\, \#^c_n(\beta, \ldots, \beta) \,|\, \#^s_n(\beta, \ldots, \beta) \,|\, \#^h_n(\beta, \ldots, \beta) \,|\, \#^{cp_{x_n}}_{x \in \mathcal{D}} x.\beta(x) \,|\, \#^{sp_{x_n}}_{x \in \mathcal{D}} x.\beta(x) \,|\, 0 \,|\, 1$$

For every atomic formula p of S, we may form p^+, p^-, where the labelling $(-)^{+,-}$ is distributive over compound formulas. For any compound formula, $\beta^{+,-}$, its counterpart, $\beta^{-,+}$, is defined above.

Note that syntax is not formed unrestrictedly, since whether or not a formula is well-formed is parameterized by its role in an interaction. This also reflects the fact that we are not thinking of formulae as propositions, since they represent moves in an interaction, and, therefore, may be better understood as "placeholders" for propositions subject to the termination of the interaction. For example, $\#^h_n(\beta, \ldots, \beta)$ is an operation that embeds into the object language a "positive play", or a "negative play", depending upon context. As such, it is not possible to form a negative conditional insertion schema with β^+, nor a positive conditional insertion schema with β^-. Consider, also, that it is possible to form negation-like operations by means of $\#^h_n$ and units 0 and 1. So, for example, $\neg_+\beta^+$ may be defined as $\beta^+\#^h_2 0^+$, and $\neg_-\beta^-$ is defined as $1^-\#^h_2\beta^-$. As such, it follows that $\neg_+\beta$ can never be formed with β^-, nor can $\neg_-\beta$ be formed with β^+.[24] As stated above, on our approach, the validity of a conditional move of the form $\beta^+\#^h_2 0^+$ would require that we also provide a valid proof of β^+, which is impossible by definition. This means that $\neg_+\beta^+$ can never be valid at the termination of an interaction. Rather, $\neg_+\beta^+$ indicates that an attempted positive move involving β^+ as focus is unsuccessful. But, it is not the case that a positive move of $\neg_+\beta^+$ also brings with it a negative move of β^-, so that β is successfully refuted. Instead, we have a distinction between a successful refutation of β, and a positive move involving an attempt to provide a reason for β that is unsuccessful.[25]

[23] For simplicity, the only terms that we discuss will be variables.

[24] This also has the result that, for no β is it possible to have a *wff* with $\neg_+\neg_-\beta$, or $\neg_-\neg_+\beta$.

[25] This allows us to formally address the distinction between, for example, the use of the word "not" in "π is not algebraic" as a statement made in 1881, and as a statement made subsequent to a proof justifying that claim. See also the related discussion of empirical negation in [20].

As such, the units 0 and 1 are not to be interpreted as "false" and "true" here, but rather, they perform the role of a "dead-end", bringing with them the retraction of anything that was intended to map into them (positively or negatively).[26] Moreover, formulae decorated with $(-)^+$ do not directly interact with formulae decorated with $(-)^-$, since they are played by distinct roles in an interaction, which is defined as follows.

Definition 30 (*Interaction*) For any set of positively labelled formulae A^+, and negatively labelled formulae A^-, where $\alpha^+ \in A^+$, and $\alpha^- \in A^-$ (a subset of positive formulae $\alpha_1^+, \alpha_2^+ \dots \alpha_n^+$ are denoted Γ^+, and negative formulas $\alpha_1^-, \alpha_2^- \dots \alpha_n$ are denoted Δ^-), and PLAYERS $\mathcal{P}$, $\mathcal{R}$, an INTERACTION is a pair, (A^+, A^-), which consists of a possibly infinite alternating sequence of MOVES. A move is a sequence of signed formulas $a^+, \dots, a^+$ or $a^-, \dots, a^-$, so a POSITIVE MOVE is a just a pair (Γ^+, α^+), and a NEGATIVE MOVE is a pair (Δ^-, α^-), (each may be empty), where the second formula is the FOCUS of the move, and a RESOLUTION is a move with an empty focus (effectively it just "marks" a formula). Moves will be denoted by sequents, which are quadruples, composed of two sets of formulae, Γ^+ and Δ^-, and two distinguished sets (stoups) containing zero or one formula, Π^+ and Λ^-.[27] This may be noted $\Lambda^-; \Gamma^+ \dashv\vdash \Delta^-; \Pi^+$, though ordinarily we will rewrite this as $\alpha^-; \Gamma^+ \dashv\vdash \Delta^-; \beta^+$, making transparent that α^- and β^+ are single formulae; where empty, we will write this as $; \Gamma^+ \dashv\vdash \Delta^-;$. The dual direction of the turnstile, $\dashv\vdash$ indicates that it can be read in either direction: from left to right, as a positive move, or from right to left, as a negative move, though a sequent must be positive or negative as employed in an interaction. A player's being positive (*resp.* negative) is determined simply by the player who begins the interaction, and $\mathcal{P}$ may only make positive moves, whilst $\mathcal{R}$ may only make negative moves. So, if an assertion is brought into the game by a player, then that player becomes $\mathcal{P}$; if a denial is brought into the game by a player, then that player becomes $\mathcal{R}$. Moves are alternating, so, whenever $\mathcal{P}$ plays some $\alpha^+ \in A+$, $\mathcal{R}$ plays some $\alpha^- \in A^-$, and a stage is just a pair of moves.

This does not define a game semantics or any kind of semantics at all (though see following chapter) it rather lays down the structure of interaction space). An interaction, so defined, is unstructured, and rather than proscribe structure by permissible moves (as in Lorenzen dialogues, for example), we will provide structure by means of the coherence of some interactions.[28] Note also that we allow tests of atomics by either of $\mathcal{P}$, $\mathcal{R}$ (in distinction with Lorenzen dialogues), since the meaning of atomic formulae is not fixed at the outset of an interaction and also so that we are able to handle assumptions appropriately. We may also think of a specific formula, on which the interaction may be thought to converge.

---

[26] See below for further discussion.

[27] We are using a technique similar to that used in [21], and in a similar context to this one in [22] (there are also marked differences, particularly regarding the notion of positive and negative rules in those systems).

[28] That is, I shall define when an interaction is coherent, allowing that some (possibly many) interactions do not meet those criteria.

Definition 31 Say that, for any interation, there exists a CONVERGENT FORMULA, which is just the formula that the interaction converges upon (if it does so converge), noting that this convergent formula may be inserted into the interaction *at any stage*.

It is the convergent formula that an interaction may be considered as reasoning *about*, such that it may be justified and its meaning constructed. But, in keeping with the unstructured nature of interactions, we shall leave this imprecise, to be made exact by the specificities on an individual interaction. We should also keep in mind that interactions may converge on more than one formula, or that convergent formulae may be used in subsequent interactions, and so on. As such, we shall formally define convergent formulae by means of coherence. To define coherence, we first define a condition that is always admissible for terminated interactions, which we call TERMINAL- CUT, in terms of the notion of counterpart formulae (defined above).

Definition 32 $$\frac{\alpha_{\circ}^{-}; \Gamma^{+} \vdash\!\dashv \Delta^{-}; \qquad ; \Gamma^{+} \vdash\!\dashv \Delta^{-}; \alpha_{\circ}^{+}}{; \Gamma^{+} \vdash\!\dashv \Delta^{-};} \text{ (Terminal-cut)}$$

Terminal-cut formally specifies the dynamics of an interaction, ensuring that the testing relationship between positive and negative moves keeps the interaction coherent, which we can now define.

Definition 33 An interaction is coherent iff it is the case that *if terminal-cut were to be applied*, then taking moves pair-wise, either: both negative and positive *foci* would be derived to be empty in the course of an interaction, or if one or both is non-empty then the formula(s) in the foci may be transformed by a single insertion move so that they become empty when so transformed; or a focal insertion of the convergent formula is derived in either the positive or negative foci, which is identified by the interaction terminating with a resolution move marking the counterpart formula.

We can think of this in terms of *in process*, and *after process*. On the former, coherence ensures that the interaction converges such that prover and refuter are engaged in common inquiry over the course of the interaction, so that the testing relationship between the two players is "balanced".[29] That is to say, within one move of each other, coherence requires that both agents are "talking about the same thing". On the latter, for an interaction to be coherent requires that terminal cut is admissible for dialogues terminating with either prover or refuter "successful", and so may be understood as inducing a kind of harmony between the two logics by means of the duality mapping between them. Furthermore, as discussed above, in this respect, it is possible to see the construction of logical rules as secondary to the balance that is required to hold between positive and negative moves in order for the interaction to be coherent, so the form that schema take for inserting operations on formulas into the focus of moves follows from these prior conditions. The fact that terminal-cut must be admissible at the termination of dialogue requires that a convergent formula can not be inserted into the focus of a move as both a proof and a refutation *at*

[29]In the following chapter, I shall provide a slightly liberalised version of the definition, which does not require that moves are always made subsequent to each other.

the termination of dialogue. This is also held together by the ability to transform schema by means of a more standard notion of balance, (as discussed above, and in Chaps. 1 and 2), between focal insertion schema and non-focal insertion schema. This is defined by local soundness and local completeness requirements, which can be formalised by means of the admissibility of positive and negative cut conditions, which are admissible over positive and negative schema, respectively:

Definition 34 (*Positive cut*) A positive cut condition on positive formulae has the form:

$$\frac{;\Gamma_i^+ \vdash\!\dashv ;\alpha_i^+ \qquad ;\Delta_i^+,\alpha_i^+ \vdash\!\dashv ;\beta_i^+}{;\Delta_i^+,\Gamma_i^+ \vdash\!\dashv ;\beta_i^+}\ (\text{Cut}^+)$$

Where $1 \leq i \leq n$.

(Negative cut) A negative cut condition on negative formulae has the form:

$$\frac{\beta_i^-; \vdash\!\dashv \Delta_i^-,\alpha_i^-; \qquad \alpha_i^-; \vdash\!\dashv \Gamma_i^-;}{\beta_i^-; \vdash\!\dashv \Delta_i^-,\Gamma_i^-;}\ (\text{Cut}^-)$$

Where $1 \leq i \leq n$.

That these conditions are admissible (such that they may be eliminated over the course of a set of positive (negative) plays) ensures that the focal and non-focal insertion schemas are balanced such that they may be transformed in accord with local soundness and completeness definitions. Essentially, what is required is just that, for any derivation in which a formula is inserted into the reasons providing support for a move, and also as the focus of a move (where these moves are confined to one agent), then it is permissible to use cut to "cut away" the unnecessary detour. Then, the construction of non-focal schema from focal schema requires that such detours may be directly eliminated by the balance between those schema.

The shape of the schema discussed above, is presented in full as follows.

Definition 35 (*Schematic moves for IPR*) There are no identity axioms since we are not assuming that the meaning of atomic formulae is fixed prior to interaction, nonetheless, whilst terminal-cut is admissible at the termination of interaction, identity for atoms (without superscripts) are also admissible at termination. Similarly, since we are operating with sets of positive and negative formulae as supports for (at most) single formulae in the foci, we avoid structural rules.[30]

$$\frac{}{;\Delta^+, 0^+ \vdash\!\dashv ;\gamma^+}\ (\text{0-N}^+)$$

$$\frac{}{\gamma^-; \vdash\!\dashv \Delta^-, 1^-;}\ (\text{1-N}^-)$$

$$\frac{}{; \vdash\!\dashv ; \not\alpha^+}\ (\text{Resolve}^+) \qquad \frac{}{\not\alpha^-; \vdash\!\dashv ;}\ (\text{Resolve}^-)$$

where the counterpart of $\alpha^{+,-}$ is a convergent formula, and $\not\alpha^{+,-}$ marks that formula as being given up through agreement, respectively.

[30] On this, see the discussion below in relation to linear logic, and see also the discussion of G3 sequent calculi in [23].

For every positive combining schema with focal insertion $\#_n^c$ of the form

$$\frac{;\Gamma_1^+ \vdash\!\!\dashv ;\beta_1^+ \quad \ldots \quad ;\Gamma_n^+ \vdash\!\!\dashv ;\beta_n^+}{;\Gamma_1^+ \ldots \Gamma_n^+ \vdash\!\!\dashv ;\#_n^c(\beta_1^+, \ldots, \beta_n^+)} \text{ (Combining-I}^+\text{) } \#_n^c$$

(where $n \geq 2$), there exist n non-focal insertion schema of the form:

$$\frac{;\Gamma_i^+, \beta_i^+ \vdash\!\!\dashv ;\sigma^+}{;\Gamma_i^+, \#_n^c(\beta_1^+, \ldots, \beta_n^+) \vdash\!\!\dashv ;\sigma^+} \text{ (Combining-N}^+\text{) } \#_n^c$$

For every negative combining schema with focal insertion $\#_n^c$ of the form

$$\frac{\beta_1^-; \vdash\!\!\dashv \Delta_1^-; \quad \ldots \quad \beta_n^-; \vdash\!\!\dashv \Delta_n^-;}{\#_n^c(\beta_1^-, \ldots, \beta_n^-); \vdash\!\!\dashv \Delta_1^- \ldots \Delta_n^-;} \text{ (Combining-I}^-\text{) } \#_n^c$$

(where $n \geq 2$), there exist n non-focal insertion schema of the form:

$$\frac{\sigma^-; \vdash\!\!\dashv \Delta_i^-, \beta_i^-;}{\sigma^-; \vdash\!\!\dashv \Delta_i^-, \#_n^c(\beta_1^-, \ldots, \beta_n^-);} \text{ (Combining-N}^-\text{) } \#_n^c$$

Where positive splitting operation $\#_n^s$ has n focal insertion schema of the form:

$$\frac{;\Gamma_i^+ \vdash\!\!\dashv ;\beta_i^+}{;\Gamma_i^+ \vdash\!\!\dashv ;\#_n^s(\beta_1^+, \ldots, \beta_n^+)} \text{ (Splitting-I}^+\text{) } \#_n^s$$

(where $1 \leq i \leq n$), there exists a non-focal insertion schema of the form:

$$\frac{;\Gamma_1^+, \beta_1^+ \vdash\!\!\dashv ;\beta^+ \quad \ldots \quad ;\Gamma_n^+, \beta_n^+ \vdash\!\!\dashv ;\beta^+}{;\Gamma_1^+ \ldots \Gamma_n^+, \#_n^s(\beta_1^+, \ldots, \beta_n^+) \vdash\!\!\dashv ;\sigma^+} \text{ (Splitting-N}^+\text{) } \#_n^s$$

Where a negative splitting operation $\#_n^s$ has n focal insertion schema of the form:

$$\frac{\beta_i^-; \vdash\!\!\dashv \Gamma_i^-;}{\#_n^s(\beta_1^-, \ldots, \beta_n^-); \vdash\!\!\dashv \Gamma_1^- \ldots \Gamma_n^-;} \text{ (Splitting-I}^-\text{) } \#_n^s$$

(where $1 \leq i \leq n$), there exists a non-focal insertion schema of the form:

$$\frac{\sigma^-; \vdash\!\!\dashv \Gamma_1^-, \beta_1^-; \quad \ldots \quad \sigma^-; \vdash\!\!\dashv \Gamma_n^-, \beta_n^-;}{\sigma^-; \vdash\!\!\dashv \Gamma_1^- \ldots \Gamma_n^-, \#_n^s(\beta_1^-, \ldots, \beta_n^-);} \text{ (Splitting-N}^-\text{) } \#_n^s$$

For every positive conditional operation $\#_n^h$ with $m \geq 1$ focal insertion schema of the form:

$$\frac{;\Gamma_1^+, \beta_1^{+\prime} \vdash\!\!\dashv ;\beta_1^+ \quad \ldots \quad ;\Gamma_n^+, \beta_n^{+\prime} \vdash\!\!\dashv ;\beta_n^+}{;\Gamma_1^+ \ldots \Gamma_n^+ \vdash\!\!\dashv ;\#_n^h(\beta_1^+, \ldots, \beta_n^+)} \text{ (Conditional-I}^+\text{) } \#_n^h$$

(where $n \geq 2$), there exists n non-focal insertion schema of the form:

$$\frac{;\Gamma_i^+ \vdash\!\dashv ;\beta_i^+ \qquad ;\Gamma_i^+, \beta_i^{+\prime} \vdash\!\dashv ;\alpha}{;\Gamma_i^+ \ldots \Gamma_n^+, \#_n^h(\beta_1^+, \ldots, \beta_n^+) \vdash\!\dashv ;\alpha^+} \text{(Conditional-N}^+\text{) } \#_n^h$$

(where $1 \leq i \leq n$).

For every negative conditional operation $\#_n^h$ with $m \geq 1$ focal insertion schema of the form:

$$\frac{\beta_1^-; \vdash\!\dashv \Gamma_1^-, \beta_1^{-\prime}; \qquad \ldots \qquad \beta_n^-; \vdash\!\dashv \Gamma_n^-, \beta_n^{-\prime};}{\#_n^h(\beta_1^-, \ldots, \beta_n^-); \vdash\!\dashv \Gamma_1^- \ldots \Gamma_n^-;} \text{(Conditional-I}^-\text{) } \#_n^h$$

(where $n \geq 2$), there exists n non-focal insertion schema of the form:

$$\frac{\beta_i^-; \vdash\!\dashv \Gamma_i^-; \qquad \alpha^-; \vdash\!\dashv \Gamma_i^-, \beta_i^{-\prime};}{\alpha^-; \vdash\!\dashv \Gamma_1^- \ldots \Gamma_n^-, \#_n^h(\beta_1^-, \ldots, \beta_n^-);} \text{(Conditional-N}^-\text{) } \#_n^h$$

(where $1 \leq i \leq n$).

For any positive combining parameterised operation $\#_{x\in\mathcal{D}}^{cp_{x_n}}$ with $m \geq 1$ focal insertion schema of the form:

$$\frac{;\Gamma_n^+ \vdash\!\dashv ;\alpha^+[x_i/x]}{;\Gamma_n^+ \vdash\!\dashv ;\#_{x\in\mathcal{D}}^{cp_{x_n}} x.\alpha^+(x)} \text{(Combining Paramaterised}-I^+\text{) } \#_{x\in\mathcal{D}}^{cp_{x_n}}$$

(where each variable x_i must be fresh such that it can not occur in Γ_n^+ or in $\#_{x\in\mathcal{D}}^{cp_{x_n}} x.\alpha^+(x)$; the box $[x_i/x]$ indicates that a fresh variable $x_i \in \mathcal{D}$ is substituted for x.), there exists n non-focal insertion schema of the form:

$$\frac{;\Gamma_i^+, \alpha^+[x_i/x] \vdash\!\dashv ;\sigma_i^+}{;\Gamma_i^+, \#_{x\in\mathcal{D}}^{cp_{x_n}} x.\alpha^+(x) \vdash\!\dashv ;\sigma_i^+} \text{(Combining Paramaterised}-N^+\text{) } \#_{x\in\mathcal{D}}^{cp_{x_n}}$$

(where $1 \leq i \leq n$, and the variable x is free in the conclusion, and there is no restriction upon freshness of variables substituted in $[x_i/x]$ since we require only one variable to be substituted in order to introduce the operation into the non-focus). For any negative combining parameterised operation $\#_{x\in\mathcal{D}}^{cp_{x_n}}$ with $m \geq 1$ focal insertion schema of the form:

$$\frac{\alpha^-[x_i/x]; \dashv \Gamma_n^-;}{\#_{x\in\mathcal{D}}^{cp_{x_n}} x.\alpha^-(x); \dashv \Gamma_n^-;} \text{(Combining Paramaterised}-I^-\text{) } \#_{x\in\mathcal{D}}^{cp_{x_n}}$$

(where x_i must be fresh such that it can not occur in Γ_n^- or in $\#_{x\in\mathcal{D}}^{cp_{x_n}} x.\alpha^-(x)$; the box $[x_i/x]$ indicates that a fresh variable $x_i \in \mathcal{D}$ is substituted for x.), there exists n non-focal insertion schema of the form:

$$\frac{\sigma_i^-; \vdash\!\dashv \Gamma_i^-, \alpha^-[x_i/x];}{\sigma_i^-; \vdash\!\dashv \Gamma_i^-, \#_{x\in\mathcal{D}}^{cp_{x_n}} x.\alpha^-(x);} \text{(Combining Paramaterised}-N^-\text{) } \#_{x\in\mathcal{D}}^{cp_{x_n}}$$

(where $1 \leq i \leq n$, and the variable x is free in the conclusion, and there is no restriction upon freshness of variables substituted in $[x_i/x]$ since we require only one variable to be substituted in order to introduce the operation into the non-focus).
For any positive splitting parameterised operation $\#^{sp_{x_n}}_{x \in \mathcal{D}}$ with $m \geq 1$ focal insertion schema of the form:

$$\frac{;\Gamma_n^+ \vdash ; \alpha^+[x_i/x]}{;\Gamma_n^+ \vdash ; \#^{sp_{x_n}}_{x \in \mathcal{D}} x.\alpha^+(x)} \text{(Splitting Paramaterised}-I^+\text{)}\ \#^{sp_{x_n}}_{x \in \mathcal{D}}$$

(where the variable x is free in the conclusion, and there is no restriction upon freshness of variables substituted in $[x_i/x]$ since we require only one variable to be substituted in order to introduce the operations), there exists n non-focal insertion schema of the form:

$$\frac{;\Gamma_i^+, \alpha^+[x_i/x] \vdash ; \sigma_i^+}{;\Gamma_i^+, \#^{sp_{x_n}}_{x \in \mathcal{D}} x.\alpha^+(x) \vdash ; \sigma_i^+} \text{(Splitting Paramaterised}-N^+\text{)}\ \#^{sp_{x_n}}_{x \in \mathcal{D}}$$

(where $1 \leq i \leq n$, and each variable x_i must be fresh such that it can not occur in Γ_i^+ or in $\#^{sp_{x_n}}_{x \in \mathcal{D}} x.\alpha^+(x)$; the box $[x_i/x]$ indicates that a fresh variable $x_i \in \mathcal{D}$ is substituted for x).

For any negative splitting parameterised operation $\#^{sp_{x_n}}_{x \in \mathcal{D}}$ with $m \geq 1$ focal insertion schema of the form:

$$\frac{\alpha^-[x_i/x]; \vdash \Gamma_n^-;}{\#^{sp_{x_n}}_{x \in \mathcal{D}} x.\alpha^-(x); \vdash \Gamma_n^-;} \text{(Splitting Paramaterised}-I^-\text{)}\ \#^{sp_{x_n}}_{x \in \mathcal{D}}$$

(where the variable x is free in the conclusion, and there is no restriction upon freshness of variables substituted in $[x_i/x]$ since we require only one variable to be substituted in order to introduce the operations), there exists n non-focal insertion schema of the form:

$$\frac{\sigma_i^-; \vdash \Gamma_n^-, \alpha^-[x_i/x];}{\sigma_i^-; \vdash \Gamma_i^-, \#^{sp_{x_n}}_{x \in \mathcal{D}} x.\alpha^-(x);} \text{(Splitting Paramaterised}-N^-\text{)}\ \#^{sp_{x_n}}_{x \in \mathcal{D}}$$

(where $1 \leq i \leq n$, and each variable x_i must be fresh such that it can not occur in Γ_i^- or in $\#^{sp_{x_n}}_{x \in \mathcal{D}} x.\alpha^-(x)$; the box $[x_i/x]$ indicates that a fresh variable $x_i \in \mathcal{D}$ is substituted for x).

3 Local Completeness and the Reconstruction of Logic

In standard form, the above schema conform to the intuitive interpretation of "tests" between positive and negative moves given above. Moreover, these schema are constructed from prior constraints on the coherence of interaction, so cohering with the general idea that this form of inferentialism does not require that the meaning of statements involving logical constants is determined by a fixed set of rules that are given in a formal language (as is the view of Dummett-Prawitz). Instead, this form of *interactive* inferentialism argues that the meaning of statements involving logical constants is determined by a coherent interaction between agents. As such, we can allow that meanings are not determined in advance of the process of an interaction, but rather, are "built-up" over the course of a coherent interaction, dependent only upon agents' understanding how to respond to each move as it is made, and in that context. So, the notion of interaction involving positive and negative moves *generalises* formal notions of proofs and refutations as long as they are understood (as by Dummett-Prawitz) as reducible to formal derivation. Indeed, there are *no* sequents, theorems, or counter-theorems that are valid *simply by dint of their formal derivability* in IPR. Nonetheless, it is possible to define a valid proof, or refutation, in terms of a terminating interaction as follows. First, we define the notion of a terminating coherent interaction, which requires that the justification of a final formula passes the relevant tests. This can be formalised as follows:

Definition 36 (*Terminating Coherent Interaction*) A terminating coherent interaction between players, $\mathcal{P}$, $\mathcal{R}$, (A^+, A^-) is a finite closed set of moves in schematic form, satisfying the following conditions:

- If (A_1^+, A_1^-) represents a stage of the interaction, then for each subsequent positive move α_n^+, this is followed by exactly one negative move α_n^-, such that α_n^+ and α_n^- are both in (A^+, A^-).[31]
- The final move in (A^+, A^-) is either a positive or negative resolution move by exactly one of $\mathcal{P}, \mathcal{R}$, which marks the counterpart of the convergent formula, where the convergent formula must be a result of a negative or positive insertion move.
- The interaction (A^+, A^-) is coherent such that terminal-cut is admissible on each move (bearing in mind that some moves may require transformation as per the definition of coherence) without removing the *convergent focal formula* $\alpha_m^{+,-}$. Then, taking a little liberty with formalism and allowing A^+, A^- to represent the focal insertion formulae only (that is, where $\alpha_1^+, \alpha_2^+ \ldots \alpha_n^+$ are formulae inserted into positive focus over the course of A^+, and $\alpha_1^-, \alpha_2^- \ldots \alpha_n^-$ over A^-) then either:

$$\frac{A^-; \vdash\!\dashv; \qquad ; \vdash\!\dashv; A^+}{\alpha_m^-; \vdash\!\dashv;}\ \text{(Terminal-cut)}$$

---

[31]This may be liberalised a little, along the line suggested in the following chapter, where it is suggested that, rather than require subsequent moves, it is enough to require that there are "matching" moves.

or;

$$\frac{A^{-}; \vdash; \qquad ; \vdash; A^{+}}{; \vdash; \alpha^{+}_{m}} \text{ (Terminal-cut)}$$

We will say of the convergent focal formula (α^{+}_{m}, or α^{-}_{m}), of a terminating coherent interaction, that the formula is JUSTIFIED by the interaction, since it is on this that the interaction converges with an agreement between players, so that the counterpart formula is given up. The above definition of a terminating coherent interaction ensures that, for every justified formula α, it has been tested by an attempt to justify the counterpart formula $\alpha_{\circ}$.

Following a standard sequent presentation, we may take the convergent formula to just be the base formula of a sequent-style proof, or refutation, from which we proceed upwards, considering positive and negative tests of each sub-formula, until each assumption has been tested. Whilst we do not require an interaction to *actually* proceed in this way, it is nevertheless possible to *reconstruct* terminating coherent interactions in this way. To see this, we use the following definition of the validity of a proof or refutation in terms of the schematic presentation of IPR as follows, bearing in mind that each element of the definition is already available in the definition of a terminating coherent interaction.

Definition 37 (*Validity*) A proof or refutation for a formula α is valid iff:

- Each sub-formula of α is tested by each player until some formula fails to pass a test (and a resolution move is made)
- Each move is coherent
- α is a result of making a positive or negative insertion move
- α° is marked with a positive or negative resolution move

As such, α in the above definition is identifiable with the convergent formula of a terminating coherent interaction. This may be clarified by considering the relationship between terminating coherent interactions and the Prawitz-Dummett approach (discussed in Chap. 2). Their definition requires that a proof is *canonical*, or can be reduced to a canonical proof. In this context, canonicity requires the terminal formula to be introduced into the focus of the move by a right-positive rule, or left-negative rule, which is just what we require in the definition of a terminating coherent interaction. Moreover, the requirement that a non-canonical proof can be reducible to a canonical proof has an analogy in the definition of coherence. The latter definition requires that any insertion move is either preceded, or followed, by a move (by the opposite agent) that is either an insertion move of dual form and inserting the counterpart formula, or that it can be transformed into such in a single move. Their definition also requires that a proof is *closed*, by which it is meant that there are no open assumptions (or free variables) in the proof. In this context, being "closed" corresponds to looking at each assumption, so that all assumptions are tested.

Let us consider a fairly simple example which more or less operates by starting with the convergent formula and decomposing it. Say we take an interaction where $\mathcal{P}$ begins with a move of the splitting schematic form: $(\alpha^{+}\#^{c}_{2}\sigma^{+})$ supports

$(\alpha^+\#^c_2\sigma^+)\#^s_2(\beta^+\#^c_2\sigma^+)$. This is responded to, by $\mathcal{R}$, with the combining move of form: $(\alpha^-\#^s_2\sigma^-)$, $(\beta^-\#^s_2\sigma^-)$ supports $(\alpha^-\#^s_2\sigma^-)\#^c_2(\beta^-\#^s_2\sigma^-)$. Then, $\mathcal{P}$ makes the move α^+, σ^+ in support of $(\alpha^+\#^c_2\sigma^+)$, whilst $\mathcal{R}$ makes the move α^- in support of $(\alpha^-\#^s_2\sigma^-)$. This is followed by a move in support of α^+, which passes the test by α^- made by $\mathcal{R}$. The following move made by $\mathcal{P}$ in support of σ^+, which also passes the test made by $\mathcal{R}$ by σ^-. The interaction finishes with $\mathcal{R}$ marking $(\alpha^-\#^s_2\sigma^-)\#^c_2(\beta^-\#^s_2\sigma^-)$ with the resolution move. Now, if we take each pair of moves, it is simple to see that each pair of moves is dual in schematic shape to one another, and also the focus of each contains the counterpart formula of its opposite *apart from the resolution move*, with which thc interaction terminates. This can be formalised, in sequent form as follows:

$$\dfrac{\dfrac{;\Gamma^+ \dashv\vdash ;\alpha^+ \qquad ;\Gamma^+ \dashv\vdash ;\sigma^+}{;\Gamma^+ \dashv\vdash ;\alpha^+\#^c_2\sigma^+}(\#^c_2) \qquad \dfrac{\alpha^-; \dashv\vdash \Gamma^-;}{(\alpha^-\#^s_2\sigma^-);\dashv\vdash\Gamma^-;}(\#^s_2) \qquad \dfrac{\sigma^-;\dashv\vdash\Gamma^-;}{(\alpha^-\#^s_2\sigma^-);\dashv\vdash\Gamma^-;}(\#^s_2)}{\cancel{(\alpha^-\#^s_2\sigma^-)\#^c_2(\beta^-\#^s_2\sigma^-)};\Gamma^+\dashv\vdash\Gamma^-;(\alpha^+\#^c_2\sigma^+)\#^s_2(\beta^+\#^c_2\sigma^+)}(\#^s_2)/(\#^c_2)$$

Of course, this is not a standard "proof" in sequent calculus, not only because we also have refutations running alongside it, but also because we have the redundant, attempted, refutation in the central tree. This highlights that the definition of validity *will not* correspond to any standard formal definition and it will contain many redundancies and wrong-turns.

The above definition of validity ensures that the semantic role of proofs and refutations are understood in a local context, and without reference to idealised or global constraints upon soundness and completeness.[32] It is not the case that rules of inference determine meaning by themselves, but rather that the correctness of these rules is already dependent upon interactive coherence. So, the meaning of a proposition is built-up through this process of interaction between positive and negative moves to the point where there exists adequate justification for the proof, or refutation, of that proposition. That is to say, the proposition is constructed whenever there exists *agreement* between players that adequate justification has been provided for the relevant statement's truth, or falsity. In this sense, far from requiring access to idealised Lindenbaum chains, or semantic models, a local completeness theorem arises naturally from the constraints upon the validity of proofs and refutations given above:

Theorem 38 (Completeness) *Let $|P|$ denote a terminating coherent interaction with a positive convergent formula, and $|R|$ a terminating coherent interaction with a negative convergent formula. Then, for any coherent interaction, with convergent formula α, which has terminated, there exists either a valid proof P of α in the schema of IPR such that $P = |P|$, or a valid refutation R of α in the schema of IPR such that $R = |R|$. Then, $|P|$ or $|R|$, may be said to interpret the formula α.*

[32] Interactions are not intended to be determinate, so in some cases the interaction will not terminate, and also, a player's move may fail (involving a negation) without determining whether the other player will be successful.

Proof RIGHT TO LEFT: We require a valid proof P of α in IPR, whilst there is no equivalent terminating coherent interaction, with α as convergent formula. By definition, α is a result of making a positive insertion move, with counterpart $\alpha_\circ$ marked with a negative resolution move, so α^+ is the convergent formula. By definition, each move must be coherent, so we can work backwards from the convergent formula, ensuring that each pair of moves is coherent through the proof. Similarly, by definition each sub-formula of α must tested by each player until some formula fails to pass a test (and a resolution move is made). In other words, we have a terminating coherent interaction with α^+ the convergent formula $|P|$, where $|P| = P$.

LEFT TO RIGHT: We require a terminating coherent interaction, $|P|$, with α^+ as convergent formula, whilst there is no valid proof P of α in IPR. By definition, a terminating coherent interaction ends with a resolution move, in this case negative, on α^-, identifying α^+ as the convergent formula. So, we have that α^+ is a result of making a positive insertion move, and α° marked with a negative resolution move. By the definition of $|P|$, each move must be coherent, with only α^+ "falling out" of terminal-cut on all focal formula. This is just another way of saying that each sub-formula of α is tested by each player until some formula fails to pass a test (and a resolution move is made). □

It is important to note that this theorem refers only to interactions that have terminated, that it does not require that every interaction will terminate, and is localised in that it makes no reference to formulae outside of the current context of the interaction in question. In fact, we can prove a stronger result:

Theorem 39 *Where we are limited to the standard forms of the schema for moves (i.e. where $n = 2$ for conjunction and disjunction, and so on), then, for any coherent interaction, with focus α, which has terminated, there exists either a valid proof P of α in the schema of IPR such that $P = |P|$, which is equivalent to a derivation in standard sequent calculus for L_I, typically known as LJ; or there exists a valid refutation R of α in the schema of IPR such that $R = |R|$, which is equivalent to a derivation in standard sequent calculus for L_C, typically known as LDJ (insofar as that calculus is interpreted as a refutation calculus).*[33]

Let us consider a few examples before we show this. Take, for example, take the case of a dialogue involving an initial assertion of $\alpha \wedge \beta$, and with an attempted refutation through α (the notation $[S_i]$ just indicates subproofs (or subrefutations)):

$$\dfrac{\dfrac{\begin{matrix}[S_1]\\ ;\Gamma^+ \vdash\!\dashv ;\alpha^+\end{matrix} \qquad \begin{matrix}[S_2]\\ ;\Gamma'^+ \vdash\!\dashv ;\beta^+\end{matrix}}{;\Gamma^+,\Gamma'^+ \vdash\!\dashv ;\alpha^+ \wedge \beta^+}(\wedge R^+) \qquad \dfrac{\begin{matrix}[S_3]\\ \alpha^-; \vdash\!\dashv \Delta^-;\end{matrix}}{\alpha^- \wedge \beta^-; \vdash\!\dashv \Delta^-;}(\wedge L_1^-)}{;\Gamma^+,\Gamma'^+ \vdash\!\dashv \Delta^-;}\text{(Terminal-cut)}$$

[33]On the latter, see [18].

Equally, "refuter" might attempt to refute via β, with the same result. As is obvious, this also means that the dominant formulas can be eliminated through the usual process of cut-elimination by pushing cuts upwards. For example, for α:

$$\dfrac{\overset{[S_4]}{;\Gamma^+ \vdash\!\dashv ; \alpha^+} \qquad \overset{[S_5]}{\alpha^-; \vdash\!\dashv \Delta^-;}}{;\Gamma^+ \vdash\!\dashv \Delta^-;}\ \text{(Terminal-cut)}$$

Let us also consider a case in which a refutation is valid. This requires that a sub-refutation $[S_i]$ of all $\sigma \in \Delta^+$ is valid, which we denote just by $[S_V]$. Then, the case where there exists a valid refutation of $\alpha \wedge \beta$ looks like this:

$$\dfrac{\dfrac{\overset{[S_6]}{;\Gamma^+ \vdash\!\dashv ; \alpha^+} \qquad \overset{[S_7]}{;\Gamma'^+ \vdash\!\dashv ; \beta^+}}{;\Gamma^+, \Gamma'^+ \vdash\!\dashv ; \cancel{\alpha^+ \wedge \beta^+}}\ (\wedge R^+) \qquad \dfrac{\overset{[S_V]}{\alpha^-; \vdash\!\dashv \Delta^-;}}{\alpha^- \wedge \beta^-; \vdash\!\dashv \Delta^-;}\ (\wedge L_1^-)}{\alpha^- \wedge \beta^-; \Gamma^+, \Gamma'^+ \vdash\!\dashv \Delta^-;}\ \text{(Terminal-cut)}$$

Where we use $\cancel{\alpha^+ \wedge \beta^+}$ to indicate that the formula $\alpha^+ \wedge \beta^+$ is not introduced. We could equally denote this by an empty stoup, but this notation better indicates resolution on the part of prover, which is, in this case, due to the existence of a valid refutation of α^- so the subproof attempt, $[S_6]$, of α^+ fails, and there can be no insertion of $\alpha^+ \wedge \beta^+$. With this in mind, it is fairly simple to give an informal proof of Theorem 39.

Proof First, consider that the schematic presentation for IPR contains both *LJ* and *LDJ* (as refutation calculus) as specific cases, whilst also ensuring that they are syntactically separated due to the parameterisation of formulae by their role in interactions. Subsequent to an interaction, it is possible to remove the trace of an interaction so that, for example, for some $P = |P|$, or $R = |R|$, P is equivalent to a derivation P_I carried out in *LJ*, and R is equivalent to a derivation R_C carried out in *LDJ*. For example, the above valid refutation R may, after the process of interaction has been carried out, be written as follows:

$$\dfrac{\overset{[S_V]}{\alpha^-; \vdash\!\dashv \Delta^-;}}{\alpha^- \wedge \beta^-; \vdash\!\dashv \Delta^-;}\ (\wedge L_1^-)$$

Formally, this valid refutation R_C is equivalent with the version that went via a terminating interaction, so $R_C = R = |R|$ holds. It is important to note two features of this equivalence:

- It does not, however, mean that we can reduce interactions to formally valid derivations, rather, it clarifies that it is possible to identify interactions with formal proofs (and refutations) of propositions only *after* an interaction has been carried out.
- The equivalence does not work in the opposite direction, since, there are derivations in *LJ* (and *LDJ*) that are not merely not immediately valid according to Definition 14, but that are never valid in the schema for IPR. For example, $\alpha \to \neg\neg\alpha$ is

derivable in *LJ* but we could never have a terminating coherent interaction that ends with a move introducing this formula. Bearing in mind that by definition of validity a proof of $\alpha \rightarrow \neg\neg\alpha$ must be closed, we also require all subproofs to be valid. Take an obvious subformula, $\neg\alpha$, defined as $\alpha \rightarrow 0$. According to the definitions given above, a positive move by $\mathcal{P}$ of form $\alpha^+ \rightarrow 0^+$ can never be terminating, since it would require $\mathcal{P}$ to provide a valid proof of α^+ such that it passes all tests of α^+ made by $\mathcal{R}$ in the form of moves involving α^-, which is impossible. If we limit ourselves to standard forms of the schema described above, then the *formal* derivations of IPR at the point of termination are equivalent with a combination of the positive and negative fragments of intuitionistic and co-intuitionistic logic, respectively. That is, we can define positive formulae as just those formulae formed with only conjunction, disjunction, conditional, and quantifiers, and used in positive schematic forms; negative formulae are just those formulae formed with only conjunction, disjunction, conditional, or quantifiers, and used in negative schematic forms. In essence, these are just sublogics of intuitionistic, and co-intuitionistic, logics, respectively, that are negationless.[34]

□

3.1 Logic and Transcendental Constraints

In [25], Tennant suggests that "harmony for logical operators would be a transcendental constraint on the very possibility of logically structured communication" (p. 629). In Chaps. 1 and 2, I pointed out that harmony fails to provide an adequate justification for logical rules that avoids Carrollian regress. Instead we have practically reversed the dictum, such that we may say that the very possibility of a coherent interaction places transcendental constraints on the internal structure of dialogical reasoning. It is logical rules that "fall out" from this prior constraint, whilst also providing normative reasons for the acceptance or rejection of some proposition or other. On this view, the meaning of logical constants is given determinately only at the end of a process of interactive reasoning, whilst internal to the interaction is a normative set of co-ordinating constraints ensuring coherence.

Let us briefly return to the spurious connective, *tonk*. The obvious issue here is that, if we attempt to reverse engineer "tonk-like" rules from schematic moves, then we would end up with an imbalance from the point of view of interaction. For example, if refuter is to test prover's assertion of a splitting move $\alpha Tonk \beta$, they must test both α and β. If refuter wants to test prover's assumption of $\alpha Tonk \beta$,

[34] We may draw attention here to arguments in favour of negationless logic by Griss, which is based on somewhat similar ideas to those discussed in Chap. 2 regarding negation and proofs as acts. As discussed in [24], Griss argues that, thinking of negation in terms of reasoning that is carried to an absurdity, is at odds with how we think of asserting, since it would require us to assert something true, which later becomes false. Moreover, similar to our own account, according to Griss, $\alpha \rightarrow \beta$ may be asserted only if β follows from α, and α is true. Griss' own suggestions regarding what to do about this situation differ significantly from those presented here.

they need only test either α or β, so violating the symmetry of schematic moves. Alternatively, we may attempt to replicate the problem presented by tonk by taking a splitting move to introduce a "tonkish" formula into positive focus, whilst also taking a splitting move to introduce a "tonkish" formula into negative focus. The problem with this suggestion is that *there are no rules to define prior to interaction*. Instead, we take schematic moves that are coherent with each other to *subsequently* define logical rules. Quite clearly, no such "tonkish" rule could ever be reconstructed from a terminating coherent interaction. For, if we tried to *force* an interaction wherein prover's assertion of $\alpha Tonk\beta$ using a splitting move was responded to, by refuter's test in the form of a splitting move *supposed* to introduce $\alpha Tonk\beta$, negatively, then the interaction would not be coherent: prover and refuter are now simply talking past each other. So, tonk falls by the wayside due to the prior normative conditions on the coherence of interactions.

4 Comparison with Other Systems

4.1 Refutability Systems

In standard logical structures, a valid sequent "$\alpha \vdash \beta$" is ordinarily interpreted as saying that, under the assumption that there exists a proof of α, there exists a proof of β also. But, this interpretation is not forced upon us, and we should notice this is an argument in the form of a conditional.[35] Say, instead that we are inclined to assume that β is refuted, then "$\alpha \vdash \beta$" may be interpreted as saying that there exists a refutation of α, under that assumption.[36] To clarify this, we have introduced a "negative" turnstile, $\dashv$, to indicate when the latter interpretation is being used, where $\beta \dashv \alpha$ is just $\alpha \vdash \beta$ under the refutation interpretation. In this context, just as $\vdash \alpha$ indicates that α is a theorem, $\dashv \alpha$ indicates that α is a counter-theorem. But now, note that, in many logics, this switching of interpretation will not coincide with negation, so that $\beta \dashv \alpha$ is inequivalent to $\neg\beta \vdash \neg\alpha$, as are $\dashv \alpha$ and $\vdash \neg\alpha$.

Broadly speaking, this follows the line of thought instigated by Restall's [27] account of symmetric sequents, but it also has much deeper roots in refutation based logics. For example, Łukasiewicz (e.g. [28]) constructed a logic operating as a refutation process for certain formulae, demonstrating through a refutation procedure that the invalid formulae of classical sentential logic can be specified. In this vein, Łukasiewicz follows Aristotle's suggestion that the systematic rejection of non-valid arguments is just as necessary as the acceptance of valid arguments. Perhaps, with some relationship to the suggestions above, this would allow us to manage without

35 As pointed out in [26].

36 Or if you prefer to think in terms of truth, then roughly, if $\alpha \vdash \beta$ is interpreted as saying that, if α is true, then β is true also, then $\alpha \dashv \beta$ is interpreted as saying that, if α is false, then β is false also.

a counter-model which provides a semantic analysis of relationships between false conclusions and true premises. Instead, this would allow us to infer directly, from other statements to the non-validity of the argument.[37] Aristotle, and Łukasiewicz, use a rule that operates like "modus tollens", allowing for deductive systems for inferring non-valid rather than valid statements. In particular, Łukasiewicz develops a system containing as axiom that "p is rejected" (where p is a fixed propositional variable), together with two rules called reverse modus ponens and inverse substitution. The former is essentially modus tollens, and the latter says just that if a substitution instance of a formula β is rejected, then β itself is rejected also. As such, Łukasiewicz does not identify refutation with logical negation, but takes rejection to be the complement of acceptance insofar as his work is intended to formalise Aristotle's approach to syllogistics, and to directly construct a class of unprovable statements.[38] Later, a student of Łukasiewicz's, Slupecki (e.g. [30, 31]), constructed a general Tarskian approach to rejected propositions by investigating dual consequence relations (understood in the standard Tarskian sense). There, a complete refutation system for the modal logic *S5* was given, and later extended by Goranko [29] to include general modal logics such as *K*. In a similar vein, [32] develops a general approach to dual consequence operations for refutation systems.

In all of the above systems, the refutable formulae constructed by a refutation system are proven to be equivalent to the set of unprovable formulae in the corresponding deductive system. So, if we think of these in the context of a Kripke-semantics, for example, then this set of unprovable (= refutable) formulae is just those formulae that are *non-valid*. Importantly, then, whilst it is not the case that, for some formula α, α is refuted whenever $\neg\alpha$ is proved (in the corresponding deductive system), this brings with it an asymmetry of the kind that was noted at the end of Chap. 2. For example, as Dummett ([33], p. 190) points out, this refutation process is 'parasitic upon the proof-procedure whereby the validity of formulas was demonstrated and hence rested upon quite heavy assumptions'. For instance, the reverse modus ponens rule is a hybrid rule, as discussed by Skura (e.g. [34, 35]), which involves both provability and refutability. This is typically of the form: "If $\alpha \to \beta$ is provable, and β is refutable, then α is refutable also". Skura uses this rule as the basis for developing dual proof and refutation structures to great effect, for example, showing that they are equivalent to multiple-conclusion consequence relations of a certain kind.[39] However, there remains the worrying issue that the generation of non-valid formulae is inequivalent to a symmetry between the process of proving and the process of refuting formulae, that would put them on an equal footing. Also, as Dummett [33] points out, 'there is something highly unnatural about a purely *deductive* refutation of a (purported) logical law'. In this context, it is of interest to note that Łukasiewicz's attempt to extend his system to intuitionistic logic was shown to be incomplete by

[37]This is also discussed in detail in [29].

[38]Though, in later work (as is well-known), he introduces a third semantic value intermediate between truth and falsity (or acceptance, and rejection) in order to deal with indeterminate propositions.

[39]See [36], and also my own discussion of this relationship in [37].

Kreisel and Putnam (the general idea was to add a disjunction rule to the classical refutation system that enforces a kind of disjunction property for refutations). As discussed by Dummett [33], whilst Dana Scott provided such an extension (also Skura provides an axiomatisation in [34]), this is quite opaque, and, in any case, any such attempt to provide standard completeness results that would carry over to predicate logic are surely doomed to fail. That is, the joint existence of complete effective proof and refutation procedures would require first-order logic to be decidable, which it is proven not to be.

These issues are suggestive of a more constructive approach to proofs and refutations. For example, in the work of Nelson (e.g. [8]), there is, added to intuitionistic logic, the suggestion that formulae may be constructively falsified, alongside formulae that may be constructively verified. In Nelson's system, we have a "strong" negation operator by which formulae may be constructively falsified, where constructive falsity is understood neither as reduction to absurdity (as in standard constructive logics), nor as merely "non-valid" (by comparison with intuitionistic logic). Rather, this approach to falsification is direct, requiring a refutation in the form of a construction mirroring the usual BHK-syle constructions for verifications. Nelson does, however, take negation to define an exclusive relationship between verifications and falsifications, and also requires an asymmettry for falsifying the conditional. There, again, we have a hybrid construction, in which a verification of the antecedent must be combined with a falsification of the consequent. This is similarly the case in the work of Lopez-Escobar [38], which suggests that we adopt constructions that directly refute formulae, in addition to constructions directly proving formulae: 'in addition of having the concept of *a construction c proves a formula B* there is at hand the concept of *a construction d refutes a formula C*' (p. 362). Again, negation is defined as the operation turning a proof of α into a refutation of α, and a refutation of α into a proof of α.

In the previous chapter, I pointed out that forms of co-intuitionistic logic (labelled L_C above) have been suggested by a number of writers to deal with refutation directly, without reference to either negations or verifications. Of particular interest is the work of Tranchini [39], where a natural-deduction framework for co-intutitionistic logic is constructed as a system of refutation. There, as with our own approach, refutation is no longer understood in the terms of Łukasiewicz, and those working on similar refutability systems, where refutability is a matter of constructing a set of non-theorems of a deductive system. Instead, Tranchini argues that we ought to construct a system capable of 'grasping a notion of refutation satisfying certain intuitive constraints', and in a *direct* manner, analogous to the way in which we ordinarily approach the notion of proof. To do so, Tranchini constructs a bottom to top natural deduction system for refutation-derivation, where refutations of the assumptions are made from refutations of the conclusions, which, in many ways echoes the negative turnstile used above (insofar as we may think of it as an ordinary turnstile being read from right to left). In this light, Tranchini draws attention to the notion that dual-intuitionistic logic provides a decent account of constructive falsificationsism, echoing Miller's [40] earlier suggestions that it may be used as a "Popperian" logic (later developed by Shramko [41] as discussed in the previous chapter). As such, and unlike the Łukasiewicz-style systems, the set of refutated statements can only be constructively accumulated as and when a construction refuting a statement is provided. In this vein, we might consider the two structures, L_I and L_C as concurrent systems that provide a general framework for constructive proofs and constructive refutations, respectively, and where the two structures construct disjoint classes of proved and refuted statements over time.[40]

4.2 Bi-Intuitionistic Logic

It is these latter attempts to produce constructive approaches to refutations, rather than Łukasiewicz-style systems, that are most similar to our own approach: We require that a terminating interaction which ends with a refutation is on symmetric grounds with an interaction terminating with a proof; that these represent disjoint classes; that these are not determined by means of a set of non-theorems; that logical negation is not involved in their definition. Moreover, the system that we are investigating is non-deterministic in the sense that validity is not pre-determined to hold in any logical system, and so, rather than take the negative turnstile to be interpreted by means of a notion of valid refutation, we are, rather, determining what a refutation is by prior notions involving negative moves as tests of assertions. Nonetheless, as I said above, in the simplified case where schematic moves are limited to standard forms, it is possible to understand interactions to be constructed from two concurrent processes with L_I for the positive structure, L_C, for the negative structure (though both of these are generalised by the context of an interaction). As such, this builds upon analysis of the relationship between intuitionistic and co-intuitionistic logic as discussed in [6, 17, 22, 41–46].

The typical relationship between intuitionistic and co-intuitionistic logic in the literature is not without significant issues, and often the relationship between the two is left without intuitive interpretation.[41] There have been a number of attempts to allow the two logics to "cohabit" in the same structure that have been explored by means of bi-intuitionistic calculi. For example, [46] develops the algebraic analysis of co-intuitionism given in [42], by means of bi-Heyting algebra, which combines a Heyting algebra with the dual, co-Heyting (or Brouwerian), algebra. This is equivalent, on Rauszer's [46] approach, to an extension of intuitionistic logic with a co-conditional operator, which is sometimes called subtraction. Just as negation can

[40]For example, in Goodman's [42] account of what he calls "anti-intuitionistic" logic (our L_C), he argues that if we think of the logic by means of a Kripke structure, then: "[i]n the intuitionistic case, the Kripke structures have the property that any formula, once true remains true. The anti-intuitionistic Kripke structures, on the other hand, are Popperian. That is, they have the property that any formula, once false, remains false. Instead of proving new theorems as we go along, we are refuting new conjectures" (p. 124).

[41]With exception to the latter: [41] explores co-intuitionistic logic as Popperian logic as I say, though he does not consider co-conditional, nor a combined system; [22] develops a polarized proof-theory for the logic of pragmatics as extension of classical logic.

be defined by means of conditional and 0 in a Heyting algebra, so negation can be defined by means of co-conditional and 1 in a co-Heyting algebra: $\neg_C\alpha =_{df} 1 \leftarrow \alpha$.

However, this construction of bi-intuitionism runs into significant difficulties regarding the construction of an adequate proof-theory, and proof-theoretic semantics. Broadly speaking, it is difficult to ensure that bi-intuitionism does not simply collapse to a non-constructive single logic due to the simulation of an involution within the combined structure.[42] For example, the difficulties in constructing a bi-intuitionsitic logic include: (a) a collapse to a single pre-order when co-exponential is added to a cartesian-closed category (e.g. [17])[43]; (b) an inability for many systems to internalise deduction and dual deduction theorems without collapsing to classical logic (e.g. [17]); (c) a collapse result to a Boolean topos for cases for a standard topos extended with a complement classifier and vice-versa (e.g. [19]); (d) a lack of cut-elimination results for many bi-intuitionistic inference structures (e.g. [22]). As a simple example of thes difficulties, take the standard sequent calculus *LJ* to define derivability in the logic L_I, and its dual *DLJ* to define derivability in the logic L_C (see (e.g. [6]) for the calculus), and the symmetric calclus *LK* to define derivability in classical logic L_{CL}, then we have the following theorem:

Theorem 40 *There is no connective $\leftarrow$ definable in L_I such that $\alpha \vdash \beta$ is derivable iff $\alpha \leftarrow \beta \vdash$ is derivable. Similarly, there is no connective $\rightarrow$ definable in L_C such that $\alpha \vdash \beta$ is derivable iff $\vdash \alpha \rightarrow \beta$ is derivable.*

Proof We broadly follow the proof given in [6], and consider consider L_I only (the proof for L_C is analogous). Take as a hypothesis that a connective $\leftarrow$ exists, so that in L_I, $\alpha \vdash \beta \leftrightarrow \beta \Leftarrow \alpha \vdash$. If $\alpha \vdash \beta$ is derivable in L_I then it is derivable in classical logic, L_{CL}. So, by hypothesis, $\alpha \leftarrow \beta \vdash$ would be derivable in L_I and L_{CL}, and so $\leftarrow$ would be definable in L_{CL}. Since $\alpha \vdash \alpha$ is derivable in L_I, it would then be the case that $\alpha \leftarrow \alpha \vdash$ would be derivable in L_I, and so L_{CL}. Classically, we can rewrite any α with its double negation, so it is possible to replace any α by $\neg\neg\alpha$ for any counter-theorem of L_{CL}. It follows that $\neg\neg\alpha \leftarrow \alpha \vdash$ would be derivable in L_{CL}. However, it is well-known that the counter-theorems of L_I are equivalent to the counter-theorems of L_{CL}, so $\neg\neg\alpha \leftarrow \alpha \vdash$ would be derivable in L_I also. This would allow us to derive $\neg\neg\alpha \vdash \alpha$ in L_I which is false. □

[42] Note that this is not the case for bi-intuitionistic linear logic, which does not collapse when merging both intuitionistic and co-intuitionistic logic [47].

[43] The issue is clarified on consideration of the possibility of a categorical interpretation of bi-intuitionism (I shall return to this in the following chapter). It is well-known that constructive logic suspends the principle of excluded middle as theorem. Of course, this does not mean that principle of excluded middle is always rejected, rather its use is controlled by context. This can be studied structurally by means of a cartesian closed category (CCC), which can adequately characterize the notion of constructive proof (e.g. [48]). However, it is shown by Crolard [17], that, for any CCC (with final and initial objects), if it is extended by co-exponentials (which represents the co-conditional operator as defined above), then it collapses to a single partial order, and so fails to provide adequate proof-theoretic semantics.

There are, of course, ways around this latter problem, but they are often rather *ad hoc*. For example, the most obvious, if ad hoc, approach to bi-intuitionistic logic would seem to be to formalise it using a multiple-conclusion sequent calculus, whilst restricting the right conditional insertion rule so that one one formula is allowed to occur in the succedent of the premise:

$$\frac{\Gamma, \alpha \mathrel{\vdash\!\!\dashv} \beta}{\Gamma \mathrel{\vdash\!\!\dashv} \alpha \to \beta, \Delta}\ (\to\text{-R})$$

Similarly, for the co-conditional left-rule, only one formula is allowed to occur in the antecedent of the premise:

$$\frac{\beta \mathrel{\vdash\!\!\dashv} \alpha, \Delta}{\Gamma, \beta \leftarrow \alpha \mathrel{\vdash\!\!\dashv} \Delta}\ (\leftarrow\text{-L})$$

These are ad hoc restrictions with little intuitive support, but they do ensure that it is not possible, for example, to pass from a proof of $\beta \leftarrow \alpha \mathrel{\vdash\!\!\dashv} \alpha \to \beta$ to a proof of $1 \leftarrow (\alpha \to \beta) \mathrel{\vdash\!\!\dashv} (\beta \leftarrow \alpha) \to 0$, unless the restrictions to singularity in these rules were violated.[44] It is possible, in this way, to deal with the issues regarding a system containing operators obeying deduction and dual-deduction theorems. For example, in Crolard's work [17], he uses a deduction theorem in the form: for every proof of $1 \vdash \beta$ from the assumption $1 \vdash \alpha$, there is an associated proof of $\alpha \vdash \beta$ that does not depend upon $1 \vdash \alpha$.[45] He then constructs the dual-deduction theorem as: for every refutation of $0 \dashv \beta$ from the assumption $0 \dashv \alpha$, there is an associated refutation of $\alpha \dashv \beta$ that does not depend upon $0 \dashv \alpha$.[46] The deduction theorem holds in the intuitionistic calculus as long as co-conditional is not introduced into that system, whilst dual deduction theorem holds in the co-intuitionistic calculus as long as conditional is not introduced into that system. As Crolard goes on to show, if both are available in a system without restrictions to singularity for conditional and co-conditional, then it is simple to derive $\neg\neg\alpha \vdash \alpha$, for example by application of dual-deduction theorem followed by deduction theorem.

Even with such restrictions, however, as pointed out by Tarmo Uustalu (discussed in [49]), cut is not eliminable in the combined calculus. The counter-example given by Uustalu is the sequent: $\alpha \vdash \beta, \sigma \to ((\alpha \leftarrow \beta) \wedge \sigma)$. This is provable with cut as follows:

$$\dfrac{\dfrac{\alpha \vdash \alpha \qquad \beta \vdash \beta}{\alpha \vdash \alpha \leftarrow \beta, \beta} \qquad \dfrac{\dfrac{\alpha \leftarrow\vdash \alpha \leftarrow \beta \qquad \sigma \vdash \sigma}{\alpha \leftarrow \beta, \sigma \vdash (\alpha \leftarrow \beta) \wedge \sigma}}{\alpha \leftarrow \beta \vdash \sigma \to ((\alpha \leftarrow \beta) \wedge \sigma)}}{\alpha \vdash \beta, \sigma \to ((\alpha \leftarrow \beta) \wedge \sigma)}\ (\text{Cut})$$

44 My thanks to Gianluigi Bellin for instructive discussion of these issues.

45 This is drawn from categorical logic (e.g. [48]).

46 I adapt his formulation to our own use of the negative turnstile.

However, it is not possible to eliminate cut in the usual way by permuting it above the cut on the right, since the restriction to a single formula in the succedent of the premise of the conditional-right rule would then be violated. As such, it is clear that a distinct approach to proof theory for bi-intuitionistic logic would be required. In this vein, Goré [49] constructs a combined bi-intuitionistic logic using a display calculus, which is capable of dealing with proofs and refutations on a par. In order for cut-elimination, however, it must be possible to trace the direction (positive or negative) of entailment, and so there is a metalogical decision that, for any sequent, its direction is determined to go in one of the two ways. In sum, there is no cut-free standard sequent calculus for bi-intuitionistic logic, with formulae in antecedent and succedent positions related by a turnstile (even if rules are restricted as above). There is also, as Tranchini [39] points out, no way to combine constructive conditional and co-conditional in natural deduction calculi, which is capable of dealing with verifications and falsifications (and without collapsing together in some way).[47] Whether or not it is possible to construct such a proof theory, I leave aside for present purposes. For our purposes, it is enough to note that the embedding of the abstract structure (i.e. the structure as abstraction from the duals L_I and L_C) inside an account of interaction, is enought to ensure that it is syntactically separated, and so no possible collapse is possible. In fact, as I shall show in the following chapter, this allows us to formulate conditions under which a controlled collapse might be possible for considering domains where certain classical laws may be considered to hold.

4.3 Linear Logic and Ludics

There are clear similarities between aspects of the overarching approach to logic discussed here, and some elements of the research projects initiated by Jean-Yves Girard, particularly regarding geometry of interaction, and ludics (e.g. [51]). There are, for example, philosophical similarities in terms of the deflation of model-theoretic approaches to logic in favour of approaches that deal, foremost, with syntax and certain constraints thereon. The motivations for this project are distinct from that of inferentialism, however, such that meaning is often eradicated from syntax altogether in Girard's work. There are, however, attempts to bring the two together, more recently, in the work of Porello [52], which reconstructs aspects of Brandom's modal semantics in terms of linear logic, and in the work of Alain Lecomte (e.g. [53]), which places ludics in the context of Brandom's work to some degree. Girard also attempts to construct logic from prior geometrical relationships between syntactical elements, which to an extent has an analagous relationship with the symmetry and coherence constraints upon interactions discussed above. There is also an abstraction away from

[47] Though Wansing [50] attempts to construct a logic, called *2Int*, that dualises natural deduction rules for intuitionistic propositional logic, and that may be used to support construct truth and falsity relations. However, it is unclear that system is capable of dealing with Crolard's analysis of the collapse of categorical versions of these logics, and there are a number of isssues arising regarding whether or not it would be possible to extend *2Int* to predicate logic.

rules in Girard's work, particularly in ludics, where the "logical" structure we are left with is just a skeleton in which no syntax for formulae is given, and only the location of sub-formulae matters for the construction of "designs" (similar to strategies in game semantics).[48] Moreover, as I said above, the notion of "tests" is something that I have borrowed from Girard's work (though arguably this also is to be found in Lorenzen [54]), where a structure involving proofs and tests is also required to allow for "incorrect proofs".

However, whilst there are overarching similarities, these are somewhat superficial, particularly given the way in which certain structural rules are implicitly admissible in the above system, whilst in linear logic (from which ludics is born), the structural rules of weakening and contraction are both dropped so that they may be later controlled. Unlike standard logical rules, which concern the conditions under which connectives may be inserted, or used as assumptions, structural rules concern the properties of the turnstile itself. In the above system, we have adopted a presentation in which structural rules are *implicit* in each of the individual structures (positive, or negative), because the formulae used as reasons in support of a move come in "sets", rather than sequences, or lists. This allows us to absorb structural rules such as weakening, contraction, and permutation, since these make no difference in the context of a set. We could explicitly write these rules into the system above, and adopt sequences, or lists, instead, but this would make no difference other than to clutter the presentation unnecessarily. However, let us do just that now, to make transparent the properties of the turnstiles as they are operating in the above. So, for example, weakening and contraction are admissible in the above structure in the following form:

$$\frac{;\dashv\vdash \Delta^-;}{\alpha^-;\dashv\vdash \Delta^-;}\ (\text{Weak-}L^-) \qquad \frac{;\Gamma^+ \dashv\vdash ;\beta^+}{;\Gamma^+,\alpha^+ \dashv\vdash ;\beta^+}\ (\text{Weak-}L^+)$$

$$\frac{\beta^-;\dashv\vdash \Delta^-;}{\beta^-;\dashv\vdash \Delta^-,\alpha^-;}\ (\text{Weak-}R^-) \qquad \frac{;\Gamma^+ \dashv\vdash ;}{;\Gamma^+ \dashv\vdash ;\alpha^+}\ (\text{Weak-}R^+)$$

$$\frac{\beta^-;\dashv\vdash \Delta^-,\alpha^-,\alpha^-;}{\beta^-;\dashv\vdash \Delta^-,\alpha^-;}\ (\text{Cont}^-) \qquad \frac{;\Gamma^+,\alpha^+,\alpha^+ \dashv\vdash ;\beta^+}{;\Gamma^+,\alpha^+, \dashv\vdash ;\beta^+}\ (\text{Cont}^+)$$

In a similar context to the argument that was made in Chap. 2 concerning the distinction between proofs as objects and proofs as acts, Dubucs and Marion [55] argue that adopting the latter view also provides reason to adopt a logic, linear logic, where weakening and contraction are not admissible. As a result of dropping weakening and contraction, perhaps the most significant alteration to logic is that we have to deal with two forms of logical connectives, which are dependent upon the way in which certain forms of weakening and contraction are admissible in controlled ways. This split into *additive* and *multiplicative* connectives has counter-intuitive consequences, as pointed out by Bonnay and Cozic [56]. For example, the multiplicative

[48]For a full account, see [21].

form of conjunction does not allow for a move from "α and β" to "α", whilst additive conjunction does not allow for a move from "α, β" to "α and β". Bonnay and Cozic make an argument that attempts to suggest that conjunction is rooted in natural language usage, and because there is only one, non-ambiguous, word for "and" in natural languages, this provides reason to think that splitting the connective in this way is unwarranted. I am not overly convinced by appeals to natural language to settle arguments about formal language in this way (as discussed in Chap. 1). Nonetheless, a similar sort of argument may be made in relation to the schematic forms of the moves in the terms of interactions outlined above. For example, in order for the kind of symmetry between moves to operate as we have shown above, a combining positive move is typically met with a response involving a negative splitting move. But, this would no longer make sense, insofar as either the combining move itself would no longer be permitted (additive), or the splitting move does not adequately test that move insofar as it must be possible to extract information in order to see where to test the move (multiplicative). As such, this looks like an unnecessary complication that does not map well onto the way in which schematic moves may be made in our system.[49]

Let us consider reasons for dropping structural rules a little further. Originally, the case for dropping contraction in linear logic is that the logic takes each formula as a *resource* that is no longer available for that derivation once it has been used. As such, the number of formulae available, even where identical, makes a difference to which derivations are valid because of this resource sensitivity. For example, as in Girard's standard example, two ten pound notes do not buy the same amount of cigarettes as one ten pound note. As such, insofar as linear logic is a logic that is set up to keep track of resources, the number of identical formulae will make a difference to derivations. So, this is why contraction must be inadmissible (over some formulae at least). Similarly, against the backdrop to this approach to logic in terms of resources, weakening looks less reasonable for similar reasons to those motivating relevant logics (discussed briefly above). For example, the turnstile now takes on an interpretation in the vein of a causal link between premises and conclusion. So, if some premise causally effectuates some conclusion, we may have reason to think that adding a premise (an additional) cause may indeed have a distinct effect.

The issue with these arguments is that they rely upon an interpretation of the turnstile in terms of resources, and a certain kind of causal relation between them. However, there is little reason to think that the turnstile operates in this way if we adopt the "proofs as acts" view, or indeed the interactive interpretation suggested above. The proofs as acts view requires that an agent is required to have a proof to hand, or, on the

[49]In the arguments in favour of proofs as acts, which I have endorsed in Chap. 2 (though with the complication that these acts are social interactions), the major reason that is given for the revision of logical rules is that the domain of decidable formulae is altered. Bonnay and Cozic [56] highlight this, arguing that the major change effected by adopting the "proofs as acts" view is the shrinkage of the domain of decidable formulae from decidability in principle to decidability in practice. Then, just as in the argument for adopting intuitionistic logic, the scope of the validity of excluded middle (which holds only for decidable formulae) is shrunk accordingly. But, as they go on to say, this in itself provides no additional reason to adopt linear logic, and thereby abandoning structural rules.

view espoused above, that an interaction gives rise to a justification providing reasons for those agents to have entitlement to some statement or other. Once entitlement to a statement is in place, however, then it seems that both agents should be able to use that statement as many times as they would like in any ensuing interaction. During the process of an interaction, of course, agents may well revise their commitments under pressure of the testing of those commitments. However, we now have a rather different motivation for thinking that something similar to weakening and contraction may fail, which is due to the interrelations between assumptions, tests, and the process of justification (and is handled in this way). This is quite different to the motivation underlying the argument to adopt a logic dropping contraction such as linear logic. Moreover, this kind of dynamic can be handled, not by dropping structural rules, but by considering the relationship between agents over the course of an interaction. The aim of this, is to terminate in agreement, which to some degree may always be thought to involve the revision of one or both agents' commitments. A similar point may be made about the structural rule of exchange, or permutation, which has not yet been discussed. This rule concerns the order in which formulae are listed as assumptions. In a causal reading of the turnstile, the order of formulae may (fairly obviously) make a difference to what is entailed. But, unlike weakening and contraction, this also carries over to interactions to some degree, since the order in which an interaction goes on may make a difference to the way in which statements are justified, or otherwise. However, as with weakening and contraction, this dynamic is also better explained in terms of the interaction between negative and positive moves, rather than the structural rules governing the moves themselves.

It does certainly seem to be the case, as Bonnay and Cozic point out, that 'justification does not disappear when I use it to build another justification in the same way that buying a pack of cigarettes makes a dollar or two disappear out of my pocket', so there is little reason to think that the "proofs as acts" view should also bring with it the rejection of structural rules.[50] Consider the following example mentioned by Bonnay and Cozic, and taken from Blass' work [57] on game semantics for linear logic. This concerns an infinite game concerning an atom α where $\mathcal{R}$ (in our terminology) has a winning strategy for $\alpha \wedge \alpha$ (where $\wedge$ is multiplicative) whilst not having a winning strategy for α alone. According to Blass, this is supposed to provide some intuitive reason for the fact that "α and α" (where "and" is multiplicative) may be more difficult to justify than the justification of α alone. That is, $\mathcal{R}$ may be able to refute an assertion made by $\mathcal{P}$ that "α and α" without being able to refute the assertion that "α". I find this example to be quite enlightening, since it clarifies the extent to which an asymmetry is built-in to the framework of linear logic, and also that of ludics insofar as it may be understood to operate over a pre-constituted space of questions and answers (e.g. [21]).[51] The symmetry of the approach to interactions that I have

[50] There are additional reasons for thinking that structura rules are important for our purposes, since they are often necessary to ensure that the schematic transformations of moves to ensure maintenance of focus runs smoothly (that is, the generalisation of harmony discussed above).

[51] We should also mention the work of Rahman (e.g. [58]), which analyses extensions of Lorenzen-style dialogues to forms involving dropping certain structural rules.

adopted as central to the approach outlined above ensures that to test an assertion of "α and α" requires there to be a test of "α" (or at least a test that can be transformed into such). As such, there is no possibility for it to be stronger to test the conjunction than it is the formula, and so no argument in favour of dropping contraction from a prior notion of testing in interactions is forthcoming.[52] Moreover, the flexibility that is gained by means of dropping structural rules is maintained in the above system by means of the requirement that no single move is counted as "valid", together with a structure of interactions organised by the dynamics of testing. As such, whilst ludics certainly shares many philosophical, and some formal, similarities with the system described here, its grounding in a logic of resources suggests that it is a separate, yet related project, and one that ought to be of interest as such.[53]

4.4 Non-monotonic Logics and Negation as Failure

In general, non-monotonic logics have been developed to deal with defeasible reasoning, in which an entailment may be defeated by the addition of information. This may also be seen as a way of attempting to integrate the issues raised in Harman's [60] arguments against the normativity of logic, into logic itself. For Harman, reasoning is understood as "change in view", which is precisely what is captured in defeasible reasoning contexts. Take the following example: Sophie, and her colleague, Rose, who at 3 pm, usually returns from the refectory with a Camomile tea, before the refectory closes. On this particular day, Rose returns to the office without her tea, and Sophie infers that the refectory has closed early that day. But, then, Sophie notices her other colleague, Matt, walking into the office with a coffee from the refectory, from which she infers that the refectory is not closed after all. Schematically, in the first case, we have $\alpha \vdash \beta$, whilst later, we have $\alpha, \sigma \nvdash \beta$. Of course, this second inference violates the standard structural rule of left weakening (discussed in the previous subsection). Non-monotonic logics, in the main, attempt to retain as much as possible of classical logic, whilst also allowing for updating processes in the context of such overriding information.[54] As such, they may be understood, philosophically, as attempts to inte-

[52]This is carried over to the construction of designs in ludics, which are much more asymmetric than the above construal, in terms of providing a structure for assertions and requests, rather than a symmetric account of assertions, denials, and tests between the two.

[53]In this vein, perhaps the best text to provide an overview that is relevant to the some of the discussions of the present text is in Lecomte [59]. It is highly likely that additional connections between the interactive structure outlined above, and Girard's ludics may be made, particularly from a formal point of view. However, I leave such formal translations for further work, particularly since they inevitably come with significant difficulties due to the fact that in neither system are atomics justified, nor are formulae valid by dint of derivability alone. As such, we can not, for example, simply start by listing the set of theorems, or key valid derivations.

[54]See [61, 62] for an overview.

grate ordinary reasoning into logic, bypassing Harman's objections discussed in the introductory chapter.[55]

These systems are intended to clarify the way in which retractions are handled in logic. The standard examples used to motivate non-monotonicity take some form of normality assumptions, and show how they operate in ordinary (defeasible reasoning). So, we may infer from "Tweety is a bird" to "Tweety can fly", on the basis of heuristics, normality, context, and so on. However, if we then learn that Tweety is a penguin, then we have good reason to think that our earlier inference is unjustified, and should be retracted. Similar examples are typically used to motivate the addition of abductive reasoning to formal systems.[56] A different, though standard, example used to motivate abductive reasoning is this: one morning, upon leaving her flat, Lizi notices that the streets are wet, and infers that it has rained during the night; later on, she remembers, however, that this is the day that the streets are cleaned, and, noticing that the grass next to the road is dry, she retracts her earlier inference.

One mechanism by which non-monotonicity is typically formalised is by a default reasoning tool in the form of NEGATION AS FAILURE: that, if α cannot be proved, then $\sim \alpha$. The idea is that, if each possible proof of α fails, then we may infer $\sim \alpha$. This is developed by Reiter's [63] requirements that $\sim \alpha$ may be assumed for any atomic α that is not provable, which is equivalent to taking a minimal model in which α is false (e.g. [67]). However, negation as failure also typically relies on the CLOSED WORLD ASSUMPTION, which requires that it is possible to list each potential proof search, such that they are exhausted prior to inferring the negation of the focus of the search.[57] In other words, we must have complete information about the facts over which the search ranges in order to discern that the search fails. The closed world assumption thus ensures that negation as failure operates much like classical negation. But, to do so, it also requires the assumption that agents have complete knowledge of the truths of the "world" in which they are making inferences. This allows the negation of all atoms to be added to the knowledge base that are not entailed by it. In other words, we must circumscribe the context of inference, and say that, for some logical theory T, T expresses a knowledge set that is assumed to be complete, containing complete knowledge regarding which atomic formulas are true.[58] That this is non-monotonic arises because, for two theories, T and T', and where $T \subseteq T'$, it may not hold that $(CWA)T \models \alpha$ implies that $(CWA)T' \models \alpha$, for all α (where (CWA) indicates that the theory T is subject to the closed world assumption, i.e. that $(CWA)T \models\sim \alpha$ iff $T \not\models \alpha$).

There are a number of issues facing the closed world assumption, including, for example, the requirements placed upon agents' knowledge. As such, a number of restrictions on closed world assumption are suggested in the literature (e.g. [70]), where certain elements of the system may be allowed to vary with context. In addition,

[55]There are many other approaches such as default logic [63]; adaptive logics [64]; default assumptions logics [65], to name just three.

[56]See, for example, the work of Magnani (e.g. [66]).

[57]See also [68, 69].

[58]Note that there is an obvious analogy here with Hintikka's approach to game semantics.

an alternative approach, termed the OPEN WORLD ASSUMPTION has been proposed (e.g. [71–73]). This allows for reasoning involving incomplete information (in fact, this is assumed by default). As such, it is arguable that it is better positioned to deal with areas of inquiry such as, scientific reasoning, in which complete information is impossible. So, unlike closed world assumption, the "default" rule is that any information that is not currently specified about the domain of reasoning is considered unknown. As such, these systems do not require non-monotonicity in the same way as those with the closed world assumption. However, they are clearly non-monotonic from the point of view of a refutation relation. It is only because the standard implication relations privilege truth that this is non-obvious. For example, the open world assumption, in this context, just says that, for two negative theories, T and T', and where $T \subseteq T'$, it may not hold that $(OWA)T \models^{-} \alpha$ implies that $(OWA)T' \models^{-} \alpha$, for all α, where $\models^{-}$ is just a refutability form of implication (falsity preserving) (where (OWA) indicates that the theory T is subject to the open world assumption, that $(OWA)T \models^{-} \sim \alpha$ iff $T \not\models \alpha$).[59]

As such, from the interactive point of view, both CWA and OWA are problematically asymmetric, and bring with them dual problems, but most obviously with regard to their semantics-first approach to logic.[60] Whilst it is clearly the case that any approach to logic by means of interaction and dialogue will require a kind of non-monotonicity, the way in which this is dealt with by the system suggested here is quite distinct from these approaches. It is, rather, the interaction between agents that gives rise to a kind of non-monotonicity *in process*, whilst also allowing for monotonicity to hold for those statements which are DETERMINED in the sense that they are justified via terminating interactions. As such, non-monotonicity is not handled by making semantic assumptions (or requiring the requisite completeness results), nor by altering structural rules, but rather by the dynamics of interactions. Consider, for example, that Sophie's reasoning may be construed as an (internalised) interaction that has a *dynamics* of reasoning along the following lines. In the first case, we have a move of form $\alpha^{+} \vdash \beta^{+}$, quickly followed, in the second case, by $\sigma^{-} \dashv \beta^{-}$, with overriding justification awarded to the refutation with focus β^{-}. Similarly, take the case of Tweety, which can be handled as followed. Let T_B indicate the utterance that "Tweety is a bird", T_F that "Tweety can fly"; P_F that "Penguins can fly". Then: $\mathcal{P}$ makes a positive move of the form $T_B^{+} \vdash T_F^{+}$. This is followed by $\mathcal{R}$ who makes a move of the form $\dashv P_F^{-}$, which is transformed into $P_F^{-} \dashv T_F^{-}$, where the interaction concludes with T_F^{-} justified.

There is, to some extent, an analogy with the role played by $\neg_{+}$ and $\neg_{-}$ in the context of interactions, since these denote something like "failure to prove", and "failure to refute", respectively. As such, these are WEAK negations, that correspond

[59]In other words, we have exactly the kind of duality that we discussed in Chap. 3 with regard to a purely refutative version of co-intuitionism as accounting for a Popperian epistemology of science. See also the discussion in [74].

[60]In a sense, we might consider the interactive approach as a radicalised, and symmetric, version of the open world assumption, insofar as we allow all formulae to be undecided outside of the context of interactions.

to a form of negation as failure (though we do not require either CWA or OWA to hold). As such, for any α, for which $\neg_{+,-}$ holds, we have only that α is retracted *by that agent* in the context of that specific interaction. These are, therefore, "weak" negations, of the form discussed towards the end of in Chap. 2, since they do not rule α out altogether (either positively or negatively). That is to say, whilst we allow that $\neg_{+}\alpha^{+}$ retracts α at that point in the interaction (from the perspective of $\mathcal{P}$, this does not prevent interactions concerning α from taking place at a later stage.

This way of treating the defined negations is not entirely distinct from Neil Tennant's [75] argument that negation should not be understood as a proof of a false sentence (0). Instead, Tennant takes a view similar to Brouwer's, that negation can be understood in terms of a "construction that no longer goes", suggesting that negation indicates a "dead-end" of a construction. The context of interaction clarifies this further still, by allowing the introduction of 0 and 1 to perform the function of "dead-ends", bringing with them the retraction of anything that was intended to map into them (positively or negatively). So, rather than treat 0 and 1 as falsity and truth, these should be understood much more like forms of "punctuation marks" for interactions. That is to say, a move of the form $\alpha^{+} \rightarrow 0^{+}$ indicates that $\mathcal{P}$, in attempt to prove α, has come to a dead end, and this allows them to retract α from the interaction. This does not, however, bring with it that α is refuted, since no refutation of α has been given, and, in fact, no negative move involving α may have been made whatsoever. It is possible that a negative move intending to provide a refutation of α may yet come up later in the interaction. This goes hand-in-hand with the idea that the negation of a formula can *never be valid* according to the view of proofs-as-acts, since there can not (by definition) be a proof that the antecedent is inhabited in order to map into 0 (or 1, if a negative move). As such, we can express, in the object language, that a move is unsuccessful such that it leads to a "dead-end", without also ensuring that the opposite move is successful.[61]

References

1. Jean-Yves Girard. *The Blind Spot: lectures on logic*. European Mathematical Society, 2011.
2. Jaroslav Peregrin. What is the logic of inference? *Studia Logica*, 88(2):263–294, 2008.
3. Nissim Francez. Bilateralism in proof-theoretic semantics. *Journal of Philosophical Logic*, (2-3):1–21, 2013.
4. Luca Tranchini. Proof-theoretic semantics, paradoxes and the distinction between sense and denotation. *Journal of Logic and Computation*, 2014.
5. Nissim Francez and Roy Dyckhoff. A note on harmony. *Journal of Philosophical Logic*, 41(3):613–628, 2012.

[61] This way of thinking about 0 and 1 also alleviates some of the issues with defining 0 as falsity, which, whilst fairly simple mathematically (e.g. as an arithmetic absurdity), is much trickier in other contexts. Though, as pointed out in [76], even in the domain of arithmetic, there appears to be an ineliminable (and vicious) circularity in the typical definition of negation as $0 = 1$. One way around this (that I will not follow up here) might be to weaken the conditions on derivation of a contradiction, so that, for example, following [45], we think of 0 °C as expressing something that is antithetical to the inquiry that we are interested in.

6. Igor Urbas. Dual-intuitionistic logic. *Notre Dame Journal of Formal Logic*, 37(3):440–451, 1996.
7. Paul-André Mellies. A micrological study of negation. *manuscript, available at the author's web page*, 2014.
8. David Nelson. Constructible falsity. *Journal of Symbolic Logic*, 14(1):16–26, 1949.
9. Andreas Kapsner. *Logics and Falsifications: A New Perspective on Constructivist Semantics*, volume 40. Springer, 2015.
10. Enrico Martino and Gabriele Usberti. Temporal and atemporal truth in intuitionistic mathematics. *Topoi*, 13(2):83–92, 1994.
11. Dorothy Edgington. On conditionals. *Mind*, 104(414):235–329, 1995.
12. Karl R. Popper. *The Logic of Scientific Discovery*. Routledge, 1959
13. Andrew Tudor. *Beyond Empiricism: Philosophy of Science in Sociology*. Routledge & K. Paul, 1982.
14. Stephen Read. *Relevant Logic: A Philosophical Examination of Inference*. B. Blackwell, 1988.
15. S. Awodey. Structure in mathematics and logic: A categorical perspective. *Philosophia Mathematica*, 4(3):209–237, 1996.
16. Saunders Mac Lane. *Categories for the working mathematician*, volume 5. springer, 1998.
17. Tristan Crolard. Subtractive logic. *Theoretical computer science*, 254(1):151–185, 2001.
18. James Trafford. Co-constructive logic for proofs and refutations. *Studia Humana*, 3(4):22–40, 2015.
19. James Trafford. Structuring co-constructive logic for proofs and refutations. *Logica Universalis*, 2016.
20. Michael De. Empirical negation. *Acta Analytica*, 28(1):49–69, 2013.
21. Jean-Yves Girard. Locus solum: From the rules of logic to the logic of rules. *Mathematical Structures in Computer Science*, 11(03):301–506, 2001.
22. Gianluigi Bellin, Massimiliano Carrara, Daniele Chiffi, and Alessandro Menti. Pragmatic and dialogic interpretations of bi-intuitionism. part i. *Logic and Logical Philosophy*, 2014.
23. Morten Heine Sørensen and Pawel Urzyczyn. *Lectures on the Curry-Howard isomorphism*, volume 149. Elsevier, 2006.
24. Miriam Franchella et al. Brouwer and Griss on intuitionistic negation. *Modern Logic*, 4(3):256–265, 1994.
25. Neil Tennant. Rule-circularity and the justification of deduction. *Philosophical Quarterly*, 55(221):625–648, 2005.
26. Greg Restall. Truth values and proof theory. *Studia Logica*, 92(2):241–264, 2009.
27. G. Restall. Multiple conclusions. In Valdes-Villanueva L. Hajek P. and Westerstpahl D., editors, *Logic, Methodology, and Philosophy of Science: Proceedings of the Twelfth International Congress*, pages 189–205. Kings College Publications, 2005.
28. Jan Lükasiewicz. *Aristotle's Syllogistic From the Standpoint of Modern Formal Logic*. Garland Pub., 1957.
29. Valentin Goranko. Refutation systems in modal logic. *Studia Logica*, 53(2):299–324, 1994.
30. Jerzy Słupecki, Grzegorz Bryll, and Urszula Wybraniec-Skardowska. Theory of rejected propositions. i. *Studia Logica*, 29(1):75–123, 1971.
31. Jerzy Słupecki, Grzegorz Bryll, and Urszula Wybraniec-Skardowska. The theory of rejected propositions. ii. *Studia Logica*, 30(1):97–145, 1972.
32. Ryszard Wójcicki. Dual counterparts of consequence operations. *Bulletin of the Section of Logic*, 2(1):54–57, 1973.
33. Michael A. E. Dummett. *The Logical Basis of Metaphysics*. Harvard University Press, 1991.
34. Tomasz Skura. On pure refutation formulations of sentential logics. *Bulletin of the Section of Logic*, 19(3):102–107, 1990.
35. Tomasz Skura. On refutation rules. *Logica Universalis*, 5(2):249–254, 2011.
36. Tomasz Skura. Refutability and post-completeness. http://www.uni-log.org/contest2013/skura1.pdf, 2013.
37. James Trafford. Duality and inferential semantics. *Axiomathes*, pages 1–19, 2015.

38. E. G. K. López-Escobar. Constructions and negationless logic. *Studia Logica*, 30(1):7–22, 1972.
39. Luca Tranchini. Natural deduction for dual-intuitionistic logic. *Studia Logica*, 100(3):631–648, 2012.
40. David Miller. Paraconsistent logic for falsificationists. In *Proceedings of the First Workshop on Logic and Language (Universidad de Sevilla)*, pages 197–204, 2000.
41. Yaroslav Shramko. Dual intuitionistic logic and a variety of negations: The logic of scientific research. *Studia Logica*, 80(2-3):347–367, 2005.
42. Nicholas Goodman. The logic of contradictions. *Zeitschrift fur Mathematische Logic und Grundlagen der Arithmetik*, 27:119–126, 1981.
43. Andreas B. M. Brunner and Walter A. Carnielli. Anti-intuitionism and paraconsistency. In *URL* =http://www.cle.unicamp.br/e-prints/vol3,n1,2003.html, 2003.
44. Luis Estrada-González. Complement-topoi and dual intuitionistic logic. *Australasian Journal of Logic*, 9:26–44, 2010.
45. Graham Priest. Dualising intuitionictic negation. *Principia*, 13(2):165–184, 2009.
46. Cecylia Rauszer. Applications of Kripke models to Heyting-Brouwer logic. *Studia Logica*, 36(1-2):61–71, 1977.
47. Gianluigi Bellin. Categorical proof theory of co-intuitionistic linear logic. arXiv preprint arXiv:1407.3416, 2014.
48. Joachim Lambek and Philip J Scott. *Introduction to higher-order categorical logic*, volume 7. Cambridge University Press, 1988.
49. Rajeev Goré and Linda Postniece. Combining derivations and refutations for cut-free completeness in bi-intuitionistic logic. *Journal of Logic and Computation*, 20(1):233–260, 2010.
50. Heinrich Wansing. Proofs, disproofs, and their duals. In Lev D. Beklemishev, Valentin Goranko, and Valentin Shehtman, editors, *Advances in Modal Logic*, pages 483–505. College Publications, 2010.
51. Jean-Yves Girard. On the meaning of logical rules i: syntax versus semantics. In *Computational logic*, pages 215–272. Springer, 1999.
52. Daniele Porello. Incompatibility semantics from agreement. *Philosophia*, 40(1):99–119, 2012.
53. Alain Lecomte. Ludics, dialogue and inferentialism. *Baltic International Yearbook of Cognition, Logic and Communication*, 8(1), 2013.
54. Paul Lorenzen and Kuno Lorenz. *Dialogische logik*. Wissenschaftliche Buchgesellschaft, 1978.
55. Jacques Dubucs and Mathieu Marion. Radical anti-realism and substructural logics. In A. Rojszczak, J. Cachro, and G. Kurczewski, editors, *Philosophical Dimensions of Logic and Science*, pages 235–249. Kluwer Academic Publishers, 2003.
56. Denis Bonnay and Mikaël Cozic. Which logic for the radical anti-realist? In *The Realism-Antirealism Debate in the Age of Alternative Logics*, pages 47–67. Springer, 2012.
57. Andreas Blass. A game semantics for linear logic. *Annals of Pure and Applied logic*, 56(1):183–220, 1992.
58. Shahid Rahman. Non-normal dialogics for a wonderful world and more. In *The age of alternative logics*, pages 311–334. Springer, 2006.
59. Alain Lecomte. *Meaning, logic and ludics*. World Scientific, 2011.
60. Gilbert Harman. *Change in View*. MIT Press, 1986.
61. Christian Strasser and G. Aldo Antonelli. Non-monotonic logic. In Edward N. Zalta, editor, *The Stanford Encyclopedia of Philosophy*. Fall 2015 edition, 2015.
62. Dov M Gabbay and John Alan Robinson. *Handbook of Logic in Artificial Intelligence and Logic Programming: Volume 5: Logic Programming*. Clarendon Press, 1998.
63. Raymond Reiter. A logic for default reasoning. *Artificial intelligence*, 13(1):81–132, 1980.
64. Diderik Batens. A universal logic approach to adaptive logics. *Logica universalis*, 1(1):221–242, 2007.
65. David Makinson. General patterns in nonmonotonic reasoning. In *Handbook of Logic in Artificial Intelligence Nad Logic Programming, Vol. Iii*. Clarendon Press, 1994.
66. Lorenzo Magnani. The eco-cognitive model of abduction. *Journal of Applied Logic*, 13(3):285–315, 2015.

67. Keith L Clark. Negation as failure. In *Logic and data bases*, pages 293–322. Springer, 1978.
68. David Pearce. *What is Negation?*, chapter From Here to There: Stable Negation in Logic Programming, pages 161–181. Springer Netherlands, Dordrecht, 1999.
69. Raymond Reiter. *Logic and Data Bases*, chapter On Closed World Data Bases, pages 55–76. Springer US, Boston, MA, 1978.
70. Michael Gelfond and Halina Przymusinska. Negation as failure: Careful closure procedure. *Artificial Intelligence*, 30(3):273–287, 1986.
71. Ronald Fagin, Phokion G Kolaitis, and Lucian Popa. Data exchange: getting to the core. *ACM Transactions on Database Systems (TODS)*, 30(1):174–210, 2005.
72. Timothy L Hinrichs, Eric Jui-Yi Kao, and Michael R Genesereth. Inconsistency-tolerant reasoning with classical logic and large databases. In *SARA*, 2009.
73. Leonid Libkin and Cristina Sirangelo. Open and closed world assumptions in data exchange. *Description Logics*, 477, 2009.
74. José Júlio Alferes, Luís Moniz Pereira, and Teodor C Przymusinski. 'classical' negation in nonmonotonic reasoning and logic programming. *Journal of Automated Reasoning*, 20(1–2):107–142, 1998.
75. Neil Tennant. Negation, absurdity and contrariety. In *What is Negation?*, pages 199–222. Springer, 1999.
76. Roy T. Cook and Jon Cogburn. What negation is not: Intuitionism and '0=1'. *Analysis*, 60(265):5–12, 2000.

Chapter 6
Structuring Interactions

In this chapter, I first consider mode standard constructive semantic approaches to the formal propositional logics for proofs and refutations. Whilst these illuminate aspects of the structure of interactive logic, they also highlight the shortcomings of standard semantics, even in dualised form. Nonetheless, the way in which these structures "objectify" propositions is useful for considering propositions as stable objects *subsequent to their construction in interaction*. These ideas are made more precise through the notion of bisimulation equivalence, and closure under bisimulation. To finish, I provide a number of examples, using these to highlight a slightly more liberal conception of the coherence of interactions.

1 Proof Objects and More Standard Semantical Approaches

As I said at the end of the previous chapter, for the subset of interactions that are terminating and coherent, an interaction may be subsequently identified with a proof-object. This allows us to define a proposition by means of these proof-objects, which itself is constructed over the course of interaction. In addition, we also saw there, that, if we limit ourselves to standard forms for move-schema, then, for any coherent interaction, with focus α, which has terminated, there exists either a valid proof P of α in the schema of IPR such that $P = |P|$, which is equivalent to a derivation in standard sequent calculus for L_I, typically known as LJ; or there exists a valid refutation R of α in the schema of IPR such that $R = |R|$, which is equivalent to a derivation in standard sequent calculus for L_C, typically known as LDJ (insofar as that calculus is interpreted as a refutation calculus). In this setting, it is interesting to consider more standard approaches to proof-theoretic semantics in terms of these *proof-objects*. For the sake of ease, I shall consider the standard formulation in what

J. Trafford, *Meaning in Dialogue*, Studies in Applied Philosophy, Epistemology and Rational Ethics 33, DOI 10.1007/978-3-319-47205-8_6

follows only, and just show some features of standard semantics that are dualised to include refutation-objects. In so doing, several limitations of these approaches will also be made transparent.

1.1 Dualising Types, Algebras, and Forcing

Let us (very briefly) show how this interacts with the theory of proof-objects in type theoretic formulations.[1] What is important about type-theoretic formulations is that types are always indexed to a specific range, so they involve a kind of context, or domain-restriction for all variables. In constructive type theory (e.g. [2]), according to the Curry-Howard isomorphism (e.g. [3]), propositions may be considered as types. Moreover, elements of propositions, as types, are just proof-objects. The type-theoretic framework builds upon the BHK-interpretation of intuitionistic logic, and provides a generalised way of considering the construction of a proof-object by means of non-formal explanation of justification.[2] Meaning, on this approach, is explained in terms of evidence making something true, so that, we may say that whenever a proposition is true, it is "inhabited" by a proof, which is just a function computing a specific term in the syntax of the type theory.[3]

In this light, according to Martin-Löf [2], type-theory concerns two elementary forms of judgment. These judgments are, in brief, that a mathematical statement is true, and that an object has the form of a mathematical object. There, $A : \mathsf{prop}$ is the judgment that A expresses a proposition, and $A : \mathsf{true}$ that the proposition is true. In particular, the former is used to show that A is well-formed (and so the kind of thing that may be combined by means of logical operations), and the latter is used to show the judgment that A is true, which requires that it is inhabited by a proof. We are interested in the way that a proposition is given meaning by means of its being inhabited by a proof. This is constructed on the basis of the distinction between types and terms, where $\alpha : A$ is a typing declaration that says that α is a term of type A, and propositions are types, which bear formal relationship to proofs, as terms. So, for example, following Kolmogorov's [5] interpretation of constructive logic, we can think of a proposition $A : \mathsf{prop}$ as expressing a problem, whose solution is given by a proof, and where a proposition may be judged true when solved (i.e. given the existence of a proof). There is an obvious relationship between this and the formal notion of a problem and a program meeting its specification, which is formalised by

[1] To be brief, I leave details aside, and assume some familiarity with constructive type theory. For a simple introduction, see (e.g. [1]). Even in the absence of familiarity, the discussion should be fairly self-evident.

[2] This is in distinction with formal construction of mathematical models inside a set-theory, for discussion see (e.g. [1, 4]).

[3] The use of the term "computing" is not accidental. Types are foundational in many areas of programming, particularly functional approaches to programming, so types are often understood to specify a task carried out by a program, and as such, a proof is also identifiable with the running of such a programme (at least in some cases).

the Curry-Howard isomorphism between propositions and types and between proofs and programs. So, whenever α is a proof-term, it is indexed to a type A, so that $M : A$ will say that α is a proof-term of type A; that is, α is a proof of the proposition A. Since $\alpha : A$ says that A is inhabited, this may be rewritten as $A : \mathsf{true}$.

Constructive type theory suffers from exactly the same issues regarding negation and falsity as intuitionistic logic (discussed in Chap. 2). As such, we may consider also that a proposition may be given meaning by means of its being inhabited by a refutation. Then, an element of a proposition may equally be a refutation-object, defined by means of co-intuitionistic logic (translated into the syntax of types and terms). For example, appending $(-)^-$ to indicate that we are working with refutations, we would still maintain a basic distinction between types and terms, and write $\alpha^- : A^-$ as a typing declaration that says that α^- is a term of type A^-. Since $\alpha^- : A^-$ says that A^- is inhabited, this may be rewritten as $A^- : \mathsf{false}$. In order to properly grasp the machinery of refutation here, we need to have some grasp on what it would take to construct the judgment $A^- : \mathsf{false}$. Just as previously the truth of a proposition requires some sort of justification for that proposition, so too we may say that the falsity of a proposition requires justification that it is false, which is something like a refutation (using formation rules derived from LDJ).[4] In this context (unlike the interactive account discussed here, where a testing relationship is natural), this is somewhat tricky to spell out. However, we might follow Estrada-Gonzalez [7], who introduces the notion of the *anti-extension* of a propositional function, where a type A^- is considered to make the propositional function false. Then, the basic proposition that we are concerned with is false, and the basic judgment, $A^- : \mathsf{false}$. Here, the terms that we are concerned with are not proofs, but refutations, which are considered direct as in previous chapters (rather than by means of the proof of negation).

Leaving details aside, recall the brief discussion of category-theoretic duality in the previous chapter. In the terms of the current discussion, a falsification of $A \times B$ in this dual type-theory would be isomorphic with a verification of $A \times B$ in constructive type theory. The key distinction (as suggested there) is that, where $\times$ corresponds to the logical operation of conjunction in L_I, in the dual theory, $\times$ corresponds to the logical operation of disjunction (in L_C). As such, given that we have an idea of the relationship between the falsification of some proposition A, and its inhabitation by a refutation, it is simple to sketch BHK-style falsification clauses as follows:

- c is a falsification of $A^- \wedge B^-$ iff c has the form (inl^{α^-}) such that α^- is are futation ofA^- or c has the form (inr^{β^-}) such that β^- is a refutation of B^-.
- c is a falsification of $A^- \vee B^-$ iff c is a pair (α^-, β^-) such that α^- is a refutation of A^- and β^- is a refutation of B^-.
- c is a falsification of $B^- \Leftarrow A^-$ iff c is a function that converts each refutation α^- of A^- into a refutation $c(\alpha^-)$ of B^-.
- c is a falsification of $\neg A^-$ iff c is a function that converts each refutation α^- of A^- into a refutation $c(\alpha^-)$ of 1^-.
- $1^\perp$ (the Unit type) has no refutations.

---

[4] See [6] for details of this system in natural deduction form.

Whilst we would need to develop the type theory in full, the isomorphism corresponding to the Curry-Howard isomorphism would then be this: a proposition is identified as a type of a refutation-object, where these refutations are formally valid in co-intuitionistic logic L_C. This relationship may be clarified further still by means of a simple dualising operation on algebraic semantics for intuitionistic propositional logic.

It is well known that the posetal structure of any intuitionist propositional logic is equivalent to a Heyting algebra.

Definition 41 (*Heyting algebra*) Let $(H, \leq)$ be a partial order with the following properties[5]:

Minimal element 0, such that, for all α, $0 \leq \alpha$;
Maximal element 1, such that, for all α, $\alpha \leq 1$;
$\alpha \wedge \beta \leq \alpha$;
$\alpha \leq \alpha \vee \beta$;
$(\alpha \wedge \beta \leq \gamma)$ iff $(\alpha \leq \beta \Rightarrow \gamma)$
$\neg\alpha = (\alpha \Rightarrow 0)$.

The relationship between a Heyting algebra and intuitionist propositional logic, L_I (with derivability relation denoted $\vdash_I$), is as follows.[6] Elements of $(H, \leq)$ are formulas of the syntax of L_I; $\leq$ interprets $\vdash_I$; an intuitionistic sequent $\Gamma \vdash_I \alpha$ is $\beta_1 \wedge \cdots \wedge \beta_n \leq \alpha$ for all $\beta_i \in \Gamma$; and $\emptyset \vdash_I \alpha$ is interpreted as $1 \leq \alpha$; $\alpha \vee \beta = sup(\alpha, \beta)$; $\alpha \wedge \beta = inf(\alpha, \beta)$. By the definition of implication, we can rewrite the definition of negation as $\alpha \leq \neg\beta$ iff $\alpha \wedge \beta = 0$. It follows that $\alpha \wedge \neg\alpha = 0$, though it may not be the case that $\alpha \vee \neg\alpha = 1$, or $\alpha = \neg\neg\alpha$. It is a further property of implication that we have the familiar rule modus ponens so that $(\beta \Rightarrow \gamma) \wedge \beta \leq \gamma$. Then, the posetal structure of a Heyting algebra is equivalent to the logical poset for intuitionist logic, as defined by its typical sequent calculus LJ.

For any poset L, it is simple to define its dual.

Theorem 42 *If $L = (S, \leq)$ is a poset, then so is $L^d = (S, \leq^d)$, where $\leq^d$ is the converse of $\leq$ such that, for all $\alpha, \beta \in S$, $\beta \leq^d \alpha$ iff $\alpha \leq \beta$.*

Together with the definition of a logic, this gives us that: whenever $\alpha \leq_L \beta$, then $(\alpha, \beta) \in \vdash_L$; it follows that $(\beta, \alpha) \in \vdash_{L^d}$, so $\beta \leq_{L^d} \alpha$. This has particular significance in the context of constructing a dual algebra to Heyting algebra, which is commonly called co-Heyting algebra.[7]

Definition 43 (*Co-Heyting algebra*) Let $(C, \leq)$ be a partial order with the following properties [10]:

Minimal element 0, such that, for all α, $0 \leq \alpha$;

[5]The following is folklore, but further details can be found in [8, p. 50ff].
[6]What follows is well-known, for details see [9].
[7]Though sometimes it is called a Brouwerian algebra, for example, [10].

Maximal element 1, such that, for all α, $\alpha \leq 1$;
$\alpha \leq \alpha \wedge \beta$;
$\alpha \vee \beta \leq \beta$;
$(\alpha \leq \gamma \vee \beta)$ iff $(\alpha \Leftarrow \beta \leq \gamma)$
$\neg\alpha = 1 \Leftarrow \alpha$.

We take co-intuitionistic logic L_C, and hence co-Heyting algebra, as a structure of refutation-objects. In this setting, the relationship between a co-Heyting algebra and co-intuitionistic propositional logic is as follows. Elements of $(C, \leq)$ are formulas of the syntax of L_I; $\leq$ interprets $\dashv_C$; a sequent $\Gamma \dashv_C \alpha$ is $\alpha \leq \beta_1 \vee \cdots \vee \beta_n$ for all $\beta_i \in \Gamma$; and $\emptyset \dashv_I \alpha$ is interpreted as $\alpha \leq 0$; $\alpha \vee \beta = inf(\alpha, \beta)$; $\alpha \wedge \beta = sup(\alpha, \beta)$. By the definition of co-implication, we can rewrite negation as $\neg\beta \leq \alpha$ iff $(\alpha \vee \beta) = 1$. Here, $\alpha \vee \neg\alpha = 1$, but $\alpha \wedge \neg\alpha = 0$ need not hold, nor $\neg\neg\alpha = \alpha$. So, the posetal structure of a co-Heyting algebra is equivalent to the logical poset for co-intuitionist logic, L_C, as defined by the usual sequent calculus LDJ taken as a calculus of refutation.

Both algebras may be taken to provide a "semantics" for L_I and L_C, respectively. A formula of propositional intuitionistic logic is just a term in the syntax of Heyting algebras; a formula of propositional co-intuitionistic logic is just a term in the syntax of co-heyting algebras. More formally, for H, take a function $V : \mathcal{P} \Rightarrow H$ as a valuation, where $\mathcal{P}$ is a set of propositional variables, and V is extended to all formulas by recursive definition:

$$\begin{aligned} V(1) &= 1 \\ V(0) &= 0 \\ V(\alpha \wedge \beta) &= V(\alpha) \wedge V(\beta) \\ V(\alpha \vee \beta) &= V(\alpha) \vee V(\beta) \\ V(\alpha \rightarrow \beta) &= V(\alpha) \rightarrow V(\beta) \end{aligned}$$

Then, say that a formula α is H-valid whenever $V(\alpha) = 1$ for all H-valuations V. Soundness and completeness results are well-known (e.g. [11]). For co-Heyting algebras, define a C-valuation as a function $V : \mathcal{P} \Rightarrow C$, extended to formulae by the same definitions above, apart from $\rightarrow$, which is replaced by:

$$V(B \leftarrow \alpha) = V(\beta) \leftarrow V(\alpha)$$

Then, say that a formula α is C-valid whenever $V(\alpha) = 0$ for all C-valuations V. Soundness and completeness results are easily constructed given the duality relationship between H and C, and may be found in (e.g. [10, 12]). It should already be obvious that if 1 and 0 are identified in C and H, then we will soon collapse if we allow to the two algebras to "cohabit". For example, any refutation valid in C will be identical with a valid negation in H. We will return to this issue in the context of category- and topos-theoretic approaches below.

The relationship between Heyting semantics defined as above, and standard Kripke-semantics for propositional intuitionistic logic is also well-known (e.g. [1]), though a Kripke-semantics for L_C is less obvious (though again, this is discussed albeit only sketchily in [10]). A standard Kripke-semantics for L_I may be given as follows.

Definition 44 Define a Kripke-model for L_I as the structure $\mathcal{M} =_{df} \langle \mathcal{S}, \leq, \models^+ \rangle$. Here, $\mathcal{S}$ is a collection of "states", or "stages" of reasoning; $\leq$ is transitive and reflexive on $\mathcal{S}$, $\models^+$ denotes a constructive forcing relation between formulae of the syntax of L_I and elements of $\mathcal{S}$. We let s_i indicate a stage of the construction where i, $1 \leq i \leq n$. We also say that $\models^+$ is monotonic, in other words it satisfies the following persistence property:

If $s_1 \leq s_2$ then $\forall \alpha \in (\mathcal{S}, s_1), \alpha \in (\mathcal{S}, s_2)$; and If $s_1 \leq s_2$ then $\forall \alpha \in (\mathcal{S}, s_1), \alpha \in (\mathcal{S}, s_2)$.

This ensures that, whenever a formula is forced at a stage, that "state" remains at every stage upstream, so, if $s_1 \models^+ \alpha$, then $s_2 \models^+ \alpha$. The clauses defining compound formulae are defined recursively as follows:

1. $[\wedge] s_i \models^+ (\alpha \wedge \beta)$ iff $s_i \models^+ \alpha$ and $s_i \models^+ \beta$
2. $[\vee] s_i \models^+ (\alpha \vee \beta)$ iff $s_i \models^+ \alpha$ or $s_i \models^+ \beta$
3. $[\rightarrow] s_i \models^+ (\alpha \rightarrow \beta)$ iff $s_i \models^+ \alpha$ then $s_i \models^+ \beta$
4. $[\neg_I] s_i \models^+ (\neg_I \alpha)$ iff $s_i \not\models^+ \alpha$

Definition 45 (*Satisfaction*) We say that a model $\mathcal{M}$ satisfies α, $\mathcal{M} \models^+ \alpha$ iff $s_i \models^+ \alpha$ for every $s_i \in \mathcal{S}$. A formula is valid if it is satisfied in all models.

Again, soundness and completeness are well-known, and may be found in (e.g. [11]).

As may be expected from the ease by which dual algebraic semantics for L_C were constructed, a dual Kripke-semantics is also easily constructed. In [10], Goodman suggests that such a semantics will treat formulae with persistence in the "opposite" direction to usual. That is, whilst in standard Kripke-semantics, once a formula is *true*, it remains true across further stages of reasoning. In the dual Kripke-semantics, once a formula is *false*, it remains false across further stages of reasoning. Whilst Goodman does not develop the semantics as such, this has a clear analogy with the refutation interpretation of co-intuitionistic logic taken here.

Definition 46 Define a Kripke-model for L_C as the structure $\mathcal{M} =_{df} \langle \mathcal{S}, \leq, \models^- \rangle$. Here, $\mathcal{S}$ is a collection of "states", or "stages" of reasoning; $\leq$ is transitive and reflexive on $\mathcal{S}$, $\models^-$ denotes a constructive, negative, forcing relation between formulas of the syntax of L_C and elements of $\mathcal{S}$. We let s_i indicate a stage of the construction where i, $1 \leq i \leq n$. We also say that $\models^-$ is monotonic, in other words it satisfies the following persistence property:

If $s_1 \leq s_2$ then $\forall \alpha \in (\mathcal{S}, s_1), \alpha \in (\mathcal{S}, s_2)$; and If $s_1 \leq s_2$ then $\forall \alpha \in (\mathcal{S}, s_1), \alpha \in (\mathcal{S}, s_2)$.

This ensures that, whenever a formula is forced at a stage, that "state" remains at every stage upstream, so, if $s_1 \models^- \alpha$, then $s_2 \models^- \alpha$. The clauses defining compound formulas are defined recursively as follows:

1. $[\wedge]s_i \models^- (\alpha \wedge \beta)$ iff $s_i \models^- \alpha$ or $s_i \models^- \beta$
2. $[\vee]s_i \models^- (\alpha \vee \beta)$ iff $s_i \models^- \alpha$ and $s_i \models^- \beta$
3. $[\Leftarrow]s_i \models^- (\beta \Leftarrow \alpha)$ iff $s_i \models^- \alpha$ then $s_i \models^- \beta$
4. $[\neg_C]s_i \models^- (\neg_C\alpha)$ iff $s_i \not\models^- \alpha$

Definition 47 (*Satisfaction*) We say that a model $\mathcal{M}$ satisfies α, $\mathcal{M} \models^- \alpha$ iff $s_i \models^- \alpha$ for every $s_i \in \mathcal{S}$. A formula is valid if it is satisfied in all models.

The relevant completeness theorem here is this:

Theorem 48 (Completeness) *For any α, if $\dashv_C \alpha$, then for all models $\mathcal{M}$, $\mathcal{M} \models^- \alpha$*

Proof The proof is given in [12, 13], and is easily constructed given the duality between the underlying algebras.

For intuitionistic and co-intuitionistic propositional logics, then, it is possible to extend the equivalence between validity in L_I with constructive type theory, validity in H, and satisfaction in $\mathcal{M}$, to an equivalence between validity in L_C with dualised constructive type theory, validity in C, and satisfaction in $\mathcal{M}$.

1.2 Categorial Collapse

An issue which immediately becomes apparent with the above concerns how we are to construe any sort of relationship between the two structures, aside from noting their duality. It is simplest to see where this is problematic from the point of view of adding an operator equivalent to $\leftarrow$ to a Cartesian closed category. This is also useful as preliminary structure for understanding the construction of topoi in the following subsection. In the previous chapter, we have already defined a $\mathbb{C}_{\mathbb{H}}$, together with initial and terminal objects, product, exponential. If we allow that, in $\mathbb{C}_{\mathbb{H}}$ we also have finite co-products (as defined in the previous chapter) and initial object, then we have a Heyting algebra as a category. As such, it is possible to interpret intuitionistic calculus in $\mathbb{C}_{\mathbb{H}}$ much as above. Objects of $\mathbb{C}_{\mathbb{H}}$ interpret formulae α, β; morphisms $f : A \to B$ model proof attempts $\alpha \vdash_I \beta$; composition models cut; products model conjunction; co-products model disjunction; initial and terminal objects are 0, 1 respectively; exponentiation models implication. On the latter, for example, if we allow that B^A interprets $(\alpha \Rightarrow \beta)$ in an intuitionistic calculus, then it is obvious that the evaluation morphism is equivalent to modus ponens. The general idea is to construe morphisms in $\mathbb{C}_{\mathbb{H}}$ as proofs in L_I, and with 1 understood as "locally true". This ensures that the structure can be understood syntactically, so that morphisms such as $f : 1 \to A$ interpret theorems of L_I.

Consider now what happens when we attempt to add an operator such as $\leftarrow$ to C_H, which is just co-exponential (as defined in the previous chapter) such that every $- \oplus A : \mathbb{C}_{\mathbb{H}} \to \mathbb{C}_{\mathbb{H}}$ has left adjoint. This is equivalent to the definition of co-conditional given for co-Heyting algebras. Of course, this also allows us to construct two negations by means of exponentials and co-exponentials within C_H. In this vein, attempts to allow L_I and L_C to "cohabit" in the same structure have been explored by means of bi-intuitionistic calculi, which include both intuitionistic and co-intuitionistic negation. Algebraically, these are expressed by bi-Heyting structures [14] combining Heyting and co-Heyting algebras, which are equivalent to C_H with co-exponential. Importantly, Crolard [15] shows that, for any CCC, if we add co-exponential, it will collapse to a single partial order. This is related to Joyal's lemma, which proves any CCC that is self-dual is a pre-order. Importantly for later, this can be strengthened since, whenever the self-duality is due to a dualizing object (such as a negation operator), then the posetal reflection of the preorder will be a Boolean algebra. This is due to the fact that, whenever a CCC contains exponentials, initial object is strict (so $0 \times 0 \cong 0$). Then, by presence of co-exponentials, we know that $A \oplus - : \mathbb{C}_{\mathbb{H}} \rightarrow \mathbb{C}_{\mathbb{H}}$ has left adjoint, which distributes over product as follows: $A \cong A \oplus 0 \cong A \oplus (0 \times 0) \cong (A \oplus 0) \times (A \oplus 0) \cong A \times A$, and this entails that, any two maps into A (providing they have the same domain) are equivalent.[8]

1.3 Dualising Topoi

To pinpoint the exact mechanisms of collapse both algebraically and logically, we now construct separate topos-theoretic structures for L_I and L_C. In doing so, it will become possible to control collapse to the extent it can be utilized.

A topos can be understood as an extension of a CCC with a subobject classifier, which means that it is particularly useful for providing a structural analysis of propositional logics. Following the motivation given in [16], we may think of a topos as a kind of category consisting of '(i) objects with some arbitrary structure, i.e., with morphisms between them, and (ii) "everything that can be constructed from these by logical means"' [16, p. 223]. In [17] (and developed in [7]), it is shown that it is possible to construct topoi whose internal logical structure is paraconsistent. This is in sharp distinction with the commonly-held [8, 11] assumption that the internal logic of any elementary topos is intuitionistic.[9] Further still, it is argued in [7] that the internal logic of a topos is not fully determined by its mathematical structure, but rather, considered as an abstract structure, a topos can support a variety of internal logics. We make use of this here, by constructing dual topoi for proofs and refutations, before identifying the manner in which they collapse to a Boolean topos so that it may be controlled.

[8]This follows an informal proof given in correspondence by Peter Johnstone available at: http://permalink.gmane.org/gmane.science.mathematics.categories/7045, and [15] provides a formal proof with discussion.

[9]Estrada-González [7] also points out that this has also been taken to provide support for the far stronger suggestion that the 'universal, invariant laws of mathematics are those of intuitionistic logic', in [18].

Definition 49 (*Subobject classifier* [8]) In a category C with finite limits, a subobject classifier is a monic: $t : 1 \rightarrow \Omega$, where Ω is an object of C, such that, for every object A and any monic $m : S \rightarrow A$ in C, there is a unique arrow $\theta : A \rightarrow \Omega$ which forms the pullback square:

$$\begin{array}{ccc} S & \longrightarrow & 1 \\ {\scriptstyle m}\downarrow & & \downarrow{\scriptstyle t} \\ A & \underset{\theta}{\longrightarrow} & \Omega \end{array}$$

Then, we call Ω the subobject classifier of C, and θ is the classifying map of m.

We are now in a position to define a topos.

Definition 50 (*Topos*) A topos T is a category with finite limits and colimits; exponentials; a subobject classifier. This is equivalent to the simpler definition that T is a topos whenever T is a category with terminal object, pullbacks, exponentials, and a subobject classifier. In a topos, a proposition is simply a morphism $1 \rightarrow \Omega$.

Theorem 51 [8, p. 198ff] *For any topos T, and for any object A in T, the subobjects of A, Ω^A, form a Heyting algebra.*[10]

Proof We first provide categorial definitions of the standard connectives of a Heyting algebra in the context of T by means of morphisms of the subobject classifier. Conjunction, is defined as the characteristic morphism of $\langle t, t \rangle : 1 \rightarrow \Omega \times \Omega$, so that $\Omega \times \Omega \overset{\wedge}{\rightarrow} \Omega$, makes the following a pullback diagram:

$$\begin{array}{ccc} 1 & \longrightarrow & 1 \\ {\scriptstyle \langle t,t \rangle}\downarrow & & \downarrow{\scriptstyle t} \\ \Omega \times \Omega & \underset{\wedge}{\longrightarrow} & \Omega \end{array}$$

Disjunction, is defined as the characteristic morphism of $[\langle t, id \rangle, \langle id, t \rangle]$, so that $\Omega \times \Omega \overset{\vee}{\rightarrow} \Omega$, makes the following a pullback diagram:

$$\begin{array}{ccc} \Omega \oplus \Omega & \longrightarrow & 1 \\ {\scriptstyle [\langle t,id \rangle, \langle id,t \rangle]}\downarrow & & \downarrow{\scriptstyle t} \\ \Omega \times \Omega & \underset{\vee}{\longrightarrow} & \Omega \end{array}$$

Conditional is defined as the characteristic morphism of $e :\leq \overset{e}{\rightarrow} \Omega \times \Omega$ (where $\leq$ is the partial order formed by the elements of Ω), which is the equalizer of $\wedge$ and the

[10]As is typical, we may slightly abuse notation and say that a subobject may be a member of Ω, since this makes little difference in practice (in the context of propositional logic), though to be more precise we should say that the equivalence class for a subobject is a member of Ω.

canonical projection, π_1, on the first component of the product, i.e. $\leq \xrightarrow{e} \Omega \times \Omega \underset{\pi_1}{\overset{\wedge}{\rightrightarrows}} \Omega$, so that $\Omega \times \Omega \overset{\Rightarrow}{\rightarrow} \Omega$, makes the following a pullback diagram:

$$\begin{array}{ccc} \leq & \longrightarrow & 1 \\ {\scriptstyle e}\downarrow & & \downarrow{\scriptstyle t} \\ \Omega \times \Omega & \underset{\Rightarrow}{\longrightarrow} & \Omega \end{array}$$

Note that this mirrors the fact that in a Heyting algebra, $\alpha \leq \beta \leftrightarrow \alpha \wedge \beta = \alpha$. Let us define the morphism $f_T : 1 \rightarrow \Omega$ by means of the initial object 0 by the following diagram:

$$\begin{array}{ccc} 0 & \longrightarrow & 1 \\ \downarrow & & \downarrow{\scriptstyle t} \\ 1 & \underset{f_T}{\longrightarrow} & \Omega \end{array}$$

Then, negation is defined as the characteristic morphism of f_T, so that $\Omega \overset{\neg}{\rightarrow} \Omega$ makes the following a pullback diagram:

$$\begin{array}{ccc} 1 & \longrightarrow & 1 \\ {\scriptstyle f_T}\downarrow & & \downarrow{\scriptstyle t} \\ \Omega & \underset{\neg}{\longrightarrow} & \Omega \end{array}$$

Then, since $\Omega \times \Omega \Rightarrow \Omega$ constructs an exponential object Ω^{Ω}, so we can define a Heyting algebra by the poset structure of Ω. □

Since propositional intuitionistic logic is equivalent to any Heyting algebra, it is unsurprising that the same is true in a topos T. We can extend the translation into Heyting algebra given above, and interpret a formula α of a formal language, whose set of formulae is S, by means of a topos T, to be denoted as $T \models_T \alpha$, which says that α is equivalent to the morphism t in T. Similarly, $\alpha \models_T \beta$ says that whenever α is equivalent to the morphism t in T, so too is β.[11]

Theorem 52 *For any topos T and proposition* α*,* $T \models \alpha$ *whenever* $\vdash_I \alpha$*.*

Proof By the fact that Ω is a Heyting algebra, and given in [8]. □

Of course, every Boolean algebra is also a Heyting algebra when each element, α of a Heyting algebra is complemented so that $\neg\neg\alpha = \alpha$. In the topos-theoretic setting, we will say that a topos T_B is Boolean whenever Ω forms a Boolean algebra

[11] This follows the account given in [16].

[8, p. 270]. This is the case whenever f is complement to t, so that the negation operator $\neg : \Omega \to \Omega$ satisfies $\neg\neg = id$, and, by Diaconescu's theorem, this is also equivalent to a topos having axiom of choice. For example, the topos SET has a subobject classifier which forms a Boolean algebra, where the objects of SET are sets, and morphisms are functions.

Now, consider a topos adequate to a co-Heyting algebra. By analogy with the relationship between T and intuitionistic logic, we require a topos with Ω forming a co-Heyting algebra, and by which it is possible to construct an interpretation of co-intuitionistic logic as a logic of refutation.

As pointed out in [7], there is little reason to interpret the morphism t as indicating "truth", or "proof" as is ordinarily the case, and such an interpretation is a matter of external decision, which is to say that it is not forced upon us by the internal language of a standard topos. So, whilst the construction of a paraconsistent topos may be seen by some as a simple matter of relabelling, this will have a significant impact upon the structure of the propositional logic that we interpret in it. For example, the process of constructing a complement topos begins with the subobject classifier, which is renamed "complement classifier", which, as such, is a different generalized element than $t : 1 \to \Omega$ in T.

Definition 53 (*Complement classifier* [7, 17, 19]) In a category T_C with terminal object 1, a complement classifier is a monic: $f : 1 \to \Omega$, where Ω is an object of T_C, such that, for every object A and any monic $m : S \to A$ in T_C, there is a unique arrow $\sigma : A \to \Omega$ which forms the pullback square:

$$\begin{array}{ccc} S & \longrightarrow & 1 \\ {\scriptstyle m}\downarrow & & \downarrow{\scriptstyle f} \\ A & \xrightarrow[\sigma]{} & \Omega \end{array}$$

Then, we call Ω the complement classifier of C, and σ is the classifying map of m.

Definition 54 (*Complement topos* [7, 17, 19]) A complement topos T_C is a category with finite limits and colimits, exponentials, and a complement classifier. This is equivalent to the simpler definition that T_C is a complement topos whenever T_C is a category with terminal object, pullbacks, exponentials, and a complement classifier. In a complement topos, a (refutation) proposition is a morphism $1 \to \Omega$.

Theorem 55 [7, 17, 19] *For any complement topos T_C, and for any object A in T_C, the subobjects of A, Ω^A, form a co-Heyting algebra.*

Proof We first provide categorial definitions of the standard connectives of a co-Heyting algebra in the context of T_C by means of morphisms of the subobject classifier. Conjunction, is defined as the characteristic morphism of $[\langle f, id\rangle, \langle id, f\rangle]$, so that $\Omega \times \Omega \xrightarrow{\wedge} \Omega$, makes the following a pullback diagram:

$$\begin{array}{ccc} \Omega \oplus \Omega & \longrightarrow & 1 \\ {\scriptstyle [\langle f, id\rangle, \langle id, f\rangle]}\downarrow & & \downarrow{\scriptstyle f} \\ \Omega \times \Omega & \xrightarrow[\wedge]{} & \Omega \end{array}$$

Disjunction, is defined as the characteristic morphism of $\langle f, f\rangle : 1 \rightarrow \Omega \times \Omega$, so that $\Omega \times \Omega \overset{\vee}{\rightarrow} \Omega$, makes the following a pullback diagram:

$$\begin{array}{ccc} 1 & \longrightarrow & 1 \\ {\scriptstyle \langle f, f\rangle}\downarrow & & \downarrow{\scriptstyle f} \\ \Omega \times \Omega & \xrightarrow[\vee]{} & \Omega \end{array}$$

Co-implication is defined as the characteristic morphism of $\overline{e} :\geq \overset{\overline{e}}{\rightarrow} \Omega \times \Omega$, which is the equalizer of $\vee$ and the canonical projection, π_1, on the first component, i.e. $\geq \overset{\overline{e}}{\rightarrowtail} \Omega \times \Omega \underset{\pi_1}{\overset{\vee}{\rightrightarrows}} \Omega$, so that $\Omega \times \Omega \overset{\Leftarrow}{\rightarrow} \Omega$, makes the following a pullback diagram:

$$\begin{array}{ccc} \geq & \longrightarrow & 1 \\ {\scriptstyle \overline{e}}\downarrow & & \downarrow{\scriptstyle f} \\ \Omega \times \Omega & \xrightarrow[\Leftarrow]{} & \Omega \end{array}$$

Note that this mirrors the fact that in a co-Heyting algebra, $\alpha \geq \beta \leftrightarrow \beta \vee \alpha = \beta$.

Let us define the morphism $t_{T_C} : 1 \rightarrow \Omega$ by means of the initial object 0 by the following diagram:

$$\begin{array}{ccc} 0 & \longrightarrow & 1 \\ \downarrow & & \downarrow{\scriptstyle f} \\ 1 & \xrightarrow[t_{T_C}]{} & \Omega \end{array}$$

Then, negation is defined as the characteristic morphism of t_{T_C}, so that $\Omega \overset{\neg}{\rightarrow} \Omega$ makes the following a pullback diagram:

$$\begin{array}{ccc} 1 & \longrightarrow & 1 \\ {\scriptstyle t_{T_C}}\downarrow & & \downarrow{\scriptstyle f} \\ \Omega & \xrightarrow[\neg]{} & \Omega \end{array}$$

Then, since $\Omega \times \Omega \Rightarrow \Omega$ constructs an exponential object Ω^{Ω}, we can define a co-Heyting algebra by the poset structure of Ω. □

Since propositional co-intuitionistic logic is equivalent to a co-Heyting algebra, it is unsurprising that the same is true in T_C. We can extend the translation into co-Heyting algebra given above, and interpret a formula α of a formal language, whose set of formulae is S, by means of T_C, to be denoted as $T_C \models_{T_C} \alpha$, which says that α is equivalent to the morphism f in T_C. Similarly, $\alpha \models_{T_C} \beta$ says that whenever α is equivalent to the morphism f in T_C, so too is β.[12]

Theorem 56 *For any complement topos T_C and proposition α, $T_C \models_{T_C} \alpha$ whenever $\dashv_C \alpha$.*

Proof By the fact that Ω is a co-Heyting algebra, and by duality with the proof for Heyting algebra and ordinary topos [19]. □

Interpreting existential and universal quantification in a topos-theoretic setting requires more complicated machinery than we have introduced here. For a full discussion, the interested reader should consult [11, p. 245ff]. We can sketch the general direction of how this works by considering a universal operator that assigns truth values to properties. Take, for example a statement, $\forall x \phi(x) = true$, which we can think of as saying that $\forall$ assigns the property ϕ the value 1, bearing in mind that properties of any A are just arrow $A \to \Omega$. So, for any properties ϕ_A of A, assigning a truth-value to $\forall_A x \phi(x)$ and also for $\exists_A x \phi(x)$ will send an element of Ω^A to an element of Ω. Without further detail, and using the fact that the subobject classifier of a topos T is a Heyting algebra, we can define equalisers with the following truth-conditions:

$$\text{Universal quantifier: } \forall_A x \phi(x) = inf(\phi(x)) \text{Existential quantifier:}$$
$$\exists_A x \phi(x) = sup(\phi(x))$$

And, in complement-topoi, T_C, since the subobject classifier is a co-heyting algebra, we can define equalisers with the following falsity conditions[13]:

$$\text{Existential quantifier: } \exists_A x \phi(x) = inf(\phi(x)) \text{Universal quantifier:}$$
$$\forall_A x \phi(x) = sup(\phi(x))$$

As is obvious, a complement topos is a natural structure for articulating co-intuitionistic logic as a logic of refutation, just as a topos is a natural structure for articulating intuitionistic logic as a logic of proof. For example, the following two properties hold in the respective topoi.

Corollary 57 *The disjunction property holds in T [11]; the conjunction property holds in T_C.*

Proof In intuitionistic logic, we have the disjunction property: $\vdash_I \alpha \vee \beta$ iff $\vdash_I \alpha$ or $\vdash_I \beta$; in co-intuitionistic logic (as refutation) we have the conjunction property: $\dashv_C$

[12] This follows the account given in [16].

[13] See [7] for a similar connection between the algebraic reading, and see [19] for a full account of quantifiers in topoi and complement-topoi.

$\alpha \wedge \beta$ iff $\dashv_C \alpha$ or $\dashv_C \beta$. In the topoi as defined above, these hold by definition of disjunction in T and conjunction in T_C, respectively. Whenever $\alpha \vee \beta$ is equivalent to the morphism t in T, then so is the morphism α or the morphism β. Whenever $\alpha \wedge \beta$ is equivalent to the morphism f in T_C, then so is the morphism α or the morphism β. □

As is typically the case, the intuitionistic disjunction property is connected to its constructive nature, since the law of excluded middle is not a theorem. This is mirrored in co-intuitionistic logic, as constructive logic of refutation, where law of non-contradiction is not a theorem. Moreover, just as the simultaneous proof of α and $\neg\alpha$ in L_I would trivialize the logic such that it is possible to prove any formula, so the simultaneous refutation of α and $\neg\alpha$ in L_C would trivialize the logic such that it is possible to refute any formula. In fact, these follow from the typical conditions on initial objects, which, if interpreted as "non-proved" in T, says that from "non-proved" everything follows, or if interpreted as "non-refuted" in T_C, says that from "non-refuted" everything follows.

Corollary 58 *Intuitionistically, conditional is an operator for which deduction theorem holds, such that* $\alpha \vdash_I \beta \leftrightarrow \vdash_I \alpha \rightarrow \beta$. *In co-intuitionistic logic, co-conditional is an operator for which dual-deduction theorem holds, such that* $\alpha \dashv_C \beta \leftrightarrow \dashv_C \beta \leftarrow \alpha$. *Both of these are the case in* T *and* T_C *(respectively), by definition.*

As shown in the previous chapter, it is not possible to define an operator in L_C by means of $\wedge, \vee, \neg, \leftarrow$ that obeys deduction theorem. Furthermore, whilst $\leftarrow$ will not satisfy modus ponens, it is simple to derive the dual, refutation, form of modus ponens:

$$\dfrac{\overset{[\text{ass.}]}{\dashv_C \alpha \leftarrow \beta} \qquad \dfrac{\overset{[\text{ass.}]}{\dashv_C \beta} \qquad \overset{[\text{Identity}]}{\alpha \dashv_C \alpha}}{\alpha \leftarrow \beta \dashv_C \alpha}\,(\leftarrow L)}{\dashv_C \alpha}\,(\text{Cut})$$

If we were interested in L_C (and T_C) as a construction ranging over proofs, then the lack of a definable operator for which deduction theorem holds would be problematic, since it would suggest that such a logic is incapable of defining a decent conditional.[14] The reason that it is impossible to define the respective operators inside the opposite structure is obviously connected with the collapse to a preorder when co-exponential is added to CCC. In this context, we can go further still, and show that, any attempt to construct a topos whose internal logic supports both intuitionistic and co-intuitionistic logic will collapse to a topos with Boolean algebra such as SET. Consider this in terms of the logical structure first. As shown in the previous chapter, any attempt to

[14]The point is arguable. For example, [19] discusses the suggestion in [17] that, in fact, this is not a problematic deficiency since mathematics depends less on an object-language implication operator, than it does a deducibility relation, which can be constructed in terms of the ordering on the complement classifier. I will not take a stance on this here, since the current construction does not require such an operator.

extend L_I with an operator for which dual-deduction theorem holds, or L_C with an operator for which deduction theorem holds is impossible. Moreover, again discussed in the previous chapter, any attempt to bring L_I and L_C together faces significant difficulties, particularly if they are to avoid *ad-hoc* alterations. Nonetheless, whilst it is certainly *possible*, if tricky, for intuitionistic and co-intuitionistic logic to be supported within the same structure, both logically, and algebraically, this is not the case for topos-theoretic interpretations.[15] For example, it is not possible for a standard topos T with Ω forming a Heyting algebra to also have Ω form a co-Heyting algebra unless the algebra formed is Boolean.

Theorem 59 *In a topos T, the characteristic morphism of $0 \to 1$ is $f_T : 1 \to \Omega$; in a complement topos T_C, the characteristic morphism of $0 \to 1$ is $t_{T_C} : 1 \to \Omega$. Whenever a topos T also has a complement classifier such that both $t : 1 \to \Omega$ and $f : 1 \to \Omega$ classify, then T is Boolean. Whenever a complement topos T_C also has a classifier such that both $f : 1 \to \Omega$ and $t : 1 \to \Omega$ classify, then T_C is Boolean.*

Above, we defined a Boolean topos, T_B by means of Ω forming a Boolean algebra, which is the case whenever $\neg\neg = id$; equally, this is the case whenever $\neg f = t$ and $\neg t = f$, so, for any α $\neg\neg\alpha = \alpha$. Equivalently, in any topos T, $1 \oplus 1$ is definable by presence of co-product, and, by definition $t, f : 1 \to \Omega$, so we have $1 \oplus 1 \to \Omega$. Whenever the latter morphism is an isomorphism, the topos T is Boolean since every proposition α in T is equivalent with either t or f. The above lemma is equivalent to saying that the topos forms subobjects with Heyting algebra, and co-Heyting algebra. Letting $f_t = f$ (in T_C), and $t_{T_C} = t$ (in T), we know by the respective definitions of negation that both $\alpha \leq \neg\neg\alpha$ and $\neg\neg\alpha \leq \alpha$, by which it follows that $\alpha = \neg\neg\alpha$ for every α. In other words, whenever $\neg\alpha$ is equivalent to the morphism t, then α is equivalent to the morphism f, and vice-versa, so we have $\neg\neg = id$.[16]

Since the internal logic of a topos is defined by the algebra that is formed by Ω, the internal logic of a Boolean topos is classical [11, Chap. 7]. This gives us the following corollary:

Corollary 60 *In a topos-theoretic setting, L_I and L_C can not co-habit, since any attempt to combine L_I and L_C (or T and T_C) will collapse to classical logic (or Boolean topos).*

1.4 Harnessing "Collapse" Between Topoi

From the point of view of the construction of proof and refutation *objects* as subsequent to terminating coherent interactions we are in an interesting position. I will discuss this in detail in the following chapter, but consider the following. Limited to the schematic moves as here, we may identify positive terminating coherent interactions with valid proofs in L_I, and negative terminating coherent interactions with

[15] Thanks to Dave Ripley for pushing me on this issue.

[16] For example, [19] points out that in effect, SET, therefore always has both classifiers.

refutations in L_C. As part of the definition of a terminating coherent interactions is that it is impossible to terminate with the introduction of both a proof- and refutation-object (of the same formula). Call the latter local truth and local falsity, to foreground the discussion in the following chapter. Then, in simple terms, a coherent interaction is defined such that it may not terminate with a formula as locally true and locally false (at the same time). In other words, for proof- and refutation-objects, a sort of *bivalence* holds.

One way to mimic this situation in this, topos-theoretic, setting, albeit in an incredibly coarse manner, would be to syntactically separate L_I and L_C (and so T and T_C), allowing the individual topoi to represent concurrent processes by which a proof-search (in T) and refutation-search (in T_C) may be carried out.[17] To do so, we simply employ the positive and negative polarity superscripts adopted in the previous chapter. In this system, there is no way to account for a structure of dialogue between the two processes, but, nonetheless, it is interesting to consider their relationship in terms of their being "in-process", where bivalence does not hold, and "subsequently" where local truth and local falsity hold, as does bivalence. In the former state, there is obviously not any restriction on the simultaneous proof and refutation searches since they are separate structures. As such, there is nothing in the above topoi T and T_C, that requires the following to hold in general:

- $T \models_T \alpha^+$ or $T_C \models_{T_C} \alpha^-$
- $T \not\models_T \alpha^+$ or $T_C \not\models_{T_C} \alpha^-$

Let us say that, whenever a proof-search comes to an end with the conclusive justification for α^+, then α^- can not be refuted, so any refutation-search is terminated. Dually, let us say that whenever a refutation-search comes to an end with the conclusive justification for α^-, then α^+ can not be proved, so any proof-search is terminated. I will leave open how the notion of "conclusive justification" is supposed to be cashed out in these situations, noting only that, since constructive type theory also requires this, any account that fills in for the type-theorist may be employed here.[18]

In a typical constructive logic such as L_I, we say that a formula α is decidable iff $\alpha \vee \neg\alpha$ holds for α; we say that a formula α is stable iff $\neg\neg\alpha$ implies α. Neither of these is a theorem of L_I, though they hold in certain contexts. By analogy with this, in the current structure we shall say that a formula α is *determined* whenever there exists a *conclusive* proof of α^+, or a *conclusive* refutation of α^-. What is important about the set of formulae for which a *conclusive proof or refutation* exists is that

[17] As such, the key distinction between the approach suggested here, and the construction in [7, 17, 19] is that, in the latter, a paraconsistent topos is understood as a complete mathematical universe, which is, at the least, capable of expressing a "full" propositional logic. Here, we take an ordinary topos and a paraconsistent topos to express two "halves" of a logic, with the former dealing proof-objects, and the latter, refutation-objects. This allows us to deal with complications that can they can not, such as the fact that a paraconsistent topos (and logic in general) is incapable of expressing a detachable conditional operator for proofs.

[18] Of course, on the interactive approach, we *do* have such an account, and without requiring any external approach to justification, since justification is achieved through the testing relationship.

we are then in a position to *know*: the proof of α^+ *rules out* any further refutation attempts of α^-; or the refutation of α^- *rules out* any further proof attempts of α^+. As such, over these (and only these formulae), it becomes possible to define an *external* relationship between formulae of L_I and L_C by means of a kind of metalinguistic negation operation, since, if, for some $\alpha^{+,-}$, α^+ is conclusively proved, then α^- can *not* be conclusively refuted; if α^- is conclusively refuted, then α^+ can *not* be conclusively proved. As such, we can define the following *external* negation between L_I and L_C:

Definition 61 For all and only those formulae which are conclusively proved (in L_I) or conclusively refuted (in L_C), the following metalinguistic negation operation, $\sim$ holds between formulae of the logical structures:

- $\sim \alpha^+ := \alpha^-$
- $\sim \alpha^- := \alpha^+$
- $\sim\sim \alpha^{+,-} = \alpha^{-,+}$

Since $\sim$ is an external negation, it negates a formula as a whole, rather than some component of that formula. So, here, it carries the simple meaning that, for example, $\sim (\alpha^+ \wedge \beta^+)$ is equivalent to $(\alpha^- \wedge \beta^-)$. In other words, this says that, whenever there exists a conclusive refutation of "α and β", we also know that there can *not* exist a conclusive proof of "α and β".

We can then use $\sim$ to simulate a metalinguistic (i.e. external) relationship between the topos-theoretic structures introduced above by using them to map those formulae into a Boolean topos. Let us call this relationship EXCLUSION:

Definition 62 Say that T and T_C are exclusive iff it is the case that whenever $T \models_T \alpha$ then $\sim (T_C \models_{T_C} \alpha)$, and whenever $T_C \models_{T_C} \alpha$ then $\sim (T \models_T \alpha)$, and, for every α, either $T \models_T \alpha$ or $T_C \models_{T_C} \alpha$, and either $\sim (T \models_T \alpha)$ or $\sim (T_C \models_{T_C} \alpha)$. So, $\sim \alpha^+$ is equivalent to saying that whenever $T_C \models_{T_C} \alpha$ then $\sim (T \models_T \alpha)$; $\sim \alpha^-$ is equivalent to saying that whenever $T \models_T \alpha$ then $\sim (T_C \models_{T_C} \alpha)$, so the above clauses defining $\sim$ hold for coherent elements of T and T_C.

It can be seen from the above, and by the fact that $\sim\sim = id$, that it is possible to construct a topos (T_B) containing only *determined* formulas, with $\sim f = t$ and $\sim t = f$, so, for any α, $\sim\sim \alpha = \alpha$. In other words, the following holds:

Corollary 63 *Whenever T and T_C are exclusive, they can be represented by a Boolean topos T_B.*

This also means that the internal logic of a Boolean topos (i.e. classical logic), can be simulated by considering T alongside T_C, but *only* when they are in a relationship of exclusion. In addition, we also have the result that, for all and only those formulae which are conclusively proved (in L_I) or conclusively refuted (in L_C), they may be interpreted in a Boolean topos such that $T_B \models \alpha$ or $T_B \models \sim \alpha$ for each α.

This is analogous to the well-known result that a double-negation morphism, $\sim\sim\colon \Omega \to \Omega$, on subobjects of a topos T, defines a closure operation called a Lawvere-Tierney topology (or double negation topology) on T. Then, the associated topos of sheaves, $T_{\sim\sim} \hookrightarrow T$ that corresponds to the double negation topology is a Boolean topos (i.e. $T_{\sim\sim} := T_B$).[19] As such, the topos of sheaves $T_{\sim\sim}$ is a model of classical logic. Here, we are simulating this construction by taking only those formulae that exist in the two topoi, T and T_C that are conclusively proved (refuted). By the above definition of exclusion over these formulae, we can then define a double-negation morphism, $j = \sim\sim = \Omega \to \Omega = id$, on the respective subobjects of T and T_C such that the determined subobject $J \to \Omega$ classified by j includes only the set of conclusively proved and refuted formulae. The subtopos $T_{\sim\sim}$ of T and T_C formed in this way is obviously Boolean by the fact that $\sim\sim = id$.

The machinery of this interpretation may be clarified by means of a *hybrid* Kripke-Joyal semantics for presheaf topos, complement topos, and Boolean topos. For sake of simplicity, let us use the standard Kripke-style construction of a model, with the distinction being that we have a hybrid structure combining objects of a Boolean topos with objects of "dialogue" topoi above them.[20] Only objects of a Boolean topos will be counted as satisfying formulae in the sense that they are given a definite truth-value, which is retained by persistence (at each stage of the construction of the relevant subobject). Objects of dialogue topoi, on the other hand, are self-contained and do not "reach-into" other stages at all, so they represent a space of dialogue existing prior to the forcing of a definitive truth-value.[21] So, we shall consider a proposition as a function from stages of reasoning as represented by an object A_i of a Boolean topos T_B.

Definition 64 Take the syntax of L_I and L_C (call these S and S^d respectively), with the three topoi, T, T_C, T_B as defined in the previous section.[22] Then, since we are primarily interested in constructing an interpretation $\mathcal{M} : S, S^d \to T_B$, we take an object A_i of T_B, α a formula of S, S^d and $A_i \xrightarrow{a_i} X_i$, $1 \leq i \leq n$, then we call the morphism a_i the generalised element of X_i at stage A_i. We also have, in addition, interpretations $S \to T$, and $S^d \to T_C$, which we will define over objects A_i of T and T_C.

We can then define the forcing relations $A_i \models^+ \alpha$ and $A_i \models^- \alpha$, and say that α positively or negatively holds at stage A_i, respectively, for T_B. The relation $A_i \models^+ \alpha$

---

[19]Proofs and further discussion may be found in detail in [20], and are also given by Todd Trimble here: https://ncatlab.org/nlab/show/Heyting+algebra#ToBooleanAlgebras.

[20]For an alternative presentation, and for further details of the semantics for an ordinary topos, T, and Boolean topos, T_B, see [20, p. 302ff] and also [21, p. 783ff], the latter of which the following definition draws upon.

[21]Note that, whilst we are calling these dialogue topoi, there is no real relationship between them in terms of the structure of interaction, proofs (refutations) carried out within them are much more like concurrent processes.

[22]We should say something here about the syntax of the Boolean topos T_B, which we are loosely specifying in the following as simply S, S^d. This looseness is harmless since it is possible to define all of the syntax of both within a topos whose internal language is Boolean, and this follows from the collapse result in the previous section. For example, $\neg_I$ is definable by means of $\to$ and 0 as usual, and $\neg_C$ by $\leftarrow$ and 1, as they were for L_I and L_C, respectively. The difference here is that they collapse into each other by the fact that $\neg_I \neg_C = id$ in T_B. As such, and due to the (non-formal) interpretation of conclusive proofs and refutations, it is safe to replace $\neg_I$ and $\neg_C$ in T_B by the single external negation $\sim$ defined above.

also holds for T, whilst the relation $A_i \models^- \alpha$ holds for T_C, and these are taken to denote proof and refutation *attempts*, respectively.

We can also define a poset, $\leq$, as per usual, over stages $\mathcal{S}$ such that $\leq$ is transitive and reflexive over $\mathcal{S}$. The idea is that we have "Boolean" stages, $\mathcal{B} \subseteq \mathcal{S}$ which are just those for which a conclusive proof or refutation has been established. At these (and only these) stages, $\models^+$ and $\models^-$ are monotonic for each $A_i \in \mathcal{B}$. In other words, they satisfy the following properties:

If $A_1 \leq A_2$ then $\forall\alpha \in (\mathcal{B}, A_1), \alpha \in (\mathcal{B}, A_2)$; and If $A_1 \leq A_2$ then $\forall\alpha \in (\mathcal{B}, A_1), \alpha \in (\mathcal{B}, A_2)$.

This ensures that, whenever a formula is conclusively proved or refuted it can be interpreted at a Boolean stage, where that "state" remains at every stage upstream, so, if $A_1 \models^+ \alpha$, then $A_2 \models^+ \alpha$; if $A_1 \models^- \alpha$, then $A_2 \models^- \alpha$. This is not, however, the case for stages outside of $\mathcal{B}$, i.e. at stages of dialogue which are not yet terminated such that a formula is conclusively proved, or conclusively refuted.

For any generalised element $A_i \overset{a_i}{\to} X_i$ of an object X_i in a topos, there are rules specifying when this generalised element belongs to a subobject of X defined in terms of the forcing relation at stage A_i. This allows us to consider proof and refutation attempts at stage A_i in T, T_C, whilst also interpreting conclusive proofs and refutations in T_B. The clauses defining compound formulae are as follows, with Boolean stages reaching above themselves by the monotonicity property. Also, note that Boolean stages do not include the internal negations of L_I and L_C, since these are only ever involved in proof and refutation *attempts*, but they do include the external negation, $\sim$, as defined above, whilst the stages that are non-Boolean do not.

Definition 65 (*Compound formulae (dropping superscripts throughout)*): $S \to T$:

- $[\wedge] A_i \models^+ (\alpha \wedge \beta)$ iff $A_i \models^+ \alpha$ and $A_i \models^+ \beta$
- $[\vee] A_i \models^+ (\alpha \vee \beta)$ iff $A_i \models^+ \alpha$ or $A_i \models^+ \beta$
- $[\to] A_i \models^+ (\alpha \to \beta)$ iff $A_i \models^+ \alpha$ then $A_i \models^+ \beta$
- $[\neg_I] A_i \models^+ (\neg_I \alpha)$ iff $A_i \not\models^+ \alpha$
- $[\exists] s_i \models^+ \exists\alpha\phi(\alpha)$ iff $s_i \models^+ \phi(s_i/\alpha)$
- $[\forall] s_i \models^+ \forall\alpha\phi(\alpha)$ iff $s_i \models^+ \phi(s_i/\alpha)$ for each term $t_i \in$ the domain $\mathcal{D}$ of s_i

$S^d \to T_C$:

- $[\wedge] A_i \models^- (\alpha \wedge \beta)$ iff $A_i \models^- \alpha$ or $A_i \models^-$
- $[\vee] A_i \models^- (\alpha \vee \beta)$ iff $A_i \models^- \alpha$ and $A_i \models^- \beta$
- $[\leftarrow] A_i \models^- (\beta \leftarrow \alpha)$ iff $A_i \models^- \alpha$ then $A_i \models^- \beta$
- $[\neg_C] A_i \models^- (\neg_C \alpha)$ iff $A_i \not\models^- \alpha$
- $[\exists] s_i \models^- \exists\alpha\phi(\alpha)$ iff $s_i \models^- \phi(s_i/\alpha)$ for each term $t_i \in$ the domain $\mathcal{D}$ of s_i
- $[\forall] s_i \models^- \forall\alpha\phi(\alpha)$ iff $s_i \models^- \phi(s_i/\alpha)$

$S, S^d \to T_B$

- $[\wedge] A_i \models^+ (\alpha \wedge \beta)$ iff $A_i \models^+ \alpha$ and $A_i \models^+ \beta$
- $[\wedge] A_i \models^- (\alpha \wedge \beta)$ iff $A_i \models^- \alpha$ or $A_i \models^-$
- $[\vee] A_i \models^+ (\alpha \vee \beta)$ iff $A_i \models^+ \alpha$ or $A_i \models^+ \beta$

- $[\vee]A_i \models^- (\alpha \vee \beta)$ iff $A_i \models^- \alpha$ and $A_i \models^- \beta$
- $[\rightarrow]A_i \models^+ (\alpha \rightarrow \beta)$ iff $\forall A_i'$ and $A_i \leq A_i'$, if $A_i' \models^+ \alpha$ then $A_i' \models^+ \beta$
- $[\leftarrow]A_i \models^- (\beta \leftarrow \alpha)$ iff $\forall A_i'$ and $A_i \leq A_i'$, if $A_i' \models^- \alpha$ then $A_i' \models^- \beta$
- $[\sim]$For all $A_i \in \mathcal{B}$, $A_i \models^+ \alpha$ iff $A_i \models^- \sim \alpha$
- $[\sim]$For all $A_i \in \mathcal{B}$, $A_i \models^- \alpha$ iff $A_i \models^+ \sim \alpha$
- $[\exists]b_i \models^+ \exists\alpha\phi(\alpha)$ iff, for some $b_n \in \mathcal{B}$, $b_n \models^+ \phi(b_n/\alpha)$
- $[\exists]b_i \models^- \exists\alpha\phi(\alpha)$ iff, for each b_i' and $b_i \leq b_i'$, and $b_i' \models^- \phi(b_i'/\alpha)$
- $[\forall]b_i \models^+ \forall\alpha\phi(\alpha)$ iff, for each b_i' and $b_i \leq b_i'$, and $b_i' \models^+ \phi(b_i'/\alpha)$
- $[\forall]b_i \models^- \forall\alpha\phi(\alpha)$ iff, for some $b_n \in \mathcal{B}$, $b_n \models^- \phi(b_n/\alpha)$

Definition 66 (*Satisfaction*) We say that a model $\mathcal{M}$ satisfies α positively or negatively, ($\mathcal{M} \models^+ \alpha$ or $\mathcal{M} \models^- \alpha$ respectively) iff $B_i \models^+ \alpha$ for every $B \in \mathcal{B}$ or $B_i \models^- \alpha$ for every $B \in \mathcal{B}$, respectively. Whenever a formula is satisfied positively or negatively, it is called TRUE or FALSE, respectively. A formula is called conclusively proved or refuted iff it is satisfied in all models.

In other words, satisfaction at a single *Boolean* stage implies satisfaction at all stages occurring before it, and so conclusive proof and refutation is equivalent to being forced at all Boolean stages. Moreover, whenever a topos T is Boolean, it can be easily be shown that the semantics defined by the above is just the standard classical set-theoretic semantics [20]. The distinction between the above construction and the latter is that this is a *constructive* form of classical semantics, and, as is the case with any sheaf-construction, this is always a local truth or falsity. It is a requirement, for example, that for $\alpha \vee \sim \alpha$ to be satisfied by T_B, that we have produced a conclusive proof of α, or a conclusive refutation of α. Once a proposition is established true or false is, therefore, dependent upon the state of reasoning at a given time, but once it is so established, by the property of monotonicity, it remains (locally) true or false at subsequent stages. In other words, by syntactically separating T and T_C we are capable of controlling collapse, allowing for collapse to Boolean topos only in case there exists a conclusive proof or refutation for some formula.

2 Processes and Types

Without doubt, situating proof- and refutation-theoretic semantics in a topos-theoretic context as above, provides some interesting structure that we may characterise a little more intuitively here. First, whilst we have used standard schematic shapes in the above (that is, to L_I and L_C), the duality between topoi suggests that it is not possible to collapse the dialogical structure into a single logical structure, without that also bringing with it a collapse to classicality. In other words, this would bring about the eradication of any constructive dynamics, and certainly, any kind of simultaneous interaction concerning the same formula. Second, we might also think of what we called, above, *conclusive* justification, by means of terminating coherent interactions. As such, we know that, for proof- and refutation-objects, considered subsequent to interaction, these may be represented by a Boolean topos, and hence, translated by external negation to a kind of constructive classical logic.

Having said all of this, the limitations of the approaches outlined in the previous section in relation to IPR are clear. In the above setting, we have nothing that tracks the construction of proof- and refutation-objects by means of interactions, to the extent that there is no real structure attributed to the processual dynamics of interaction at all. Moreover, the above approaches are, for the most part, useful only insofar as they mirror the *formal* validity of proofs (in LJ), and refutations (in LDJ), and so come with all of the issues that we discussed in detail in Chaps. 1 and 2 when defining the validity of a proof with formal derivability over a set of rules. The slightly more flexible approach, discussed in the previous subsection, involves a notion of *conclusive* validity, which also requires an account of justification (similar to constructive type theory). Again, this sort of account runs into problems that were discussed in relation to the approach of Prawitz and Dummett, in Chap. 2.

In distinction, the internal dynamics of interaction involves a *testing* relationship that brings with it a means by which justification is reached between agents, and which is internal to interactions. This problem is brought to light most clearly in the context of types. For type-theory, just as for intuitionistic logic more generally, formulae may be identified with propositions *from the outset*, since in order to be counted as a type requires only that a formula is "well-formed" in the syntax being worked with. This is just the problem, discussed in Chap. 2, where the meaning of formulae in intuitionistic logic is, problematically, supposed to be given in advance of the process by which a proof is carried out (as in constructive type theory). A term is always given with its type in type-theory, whereas, for the interactional approach given in the previous chapter, terms must be allowed to be given prior to any typing. So, in IPR, *interactions* can not be typed, nor types defined simply by syntactic well-formedness. As such, interactions must be untyped systems in which any move formed as part of an interaction is allowed to interact with any other, so that no type restrictions can apply.

Whilst standard approaches (including proof-theoretic semantics) foreground proof-objects over proof-acts, I have argued that the *process* of constructing a proof or refutation should be somehow accounted for by any formal semantic structure that is supposed to represent them. That is to say, whatever the proof- or refutation-semantics is to be, it should be fine-grained enough to deal with process, in addition to object, particularly since proof-objects, when typed, lose this specificity. In the language of type theory, this would mean that types (just as proof- and refutation-objects) are also constructed over the course of terminating coherent interactions. So, whilst in constructive type theory a proposition may be identified by its set of proofs, in IPR, a proposition may be identified by the coherent interactions that construct it.

We need to be more careful with retaining the processual nature of interactions, since we want to say that only specific processes (coherent interactions) construct propositions. Perhaps, then, we can take the set of coherent interactions terminating with a specific formula to construct its type. Even the most basic formula in type theory, $\alpha : A$, which is just an elementary assertion, requires that each term α be taken to instantiate a type, A. Instead, schematic moves involve formulae which are thought of, not as instantiating types, but as representing moves, and which have no well-formedness constraints beyond their role in moves. As I said in the previous chapter, and in concord with the discussion of meaning in Chap. 2, these may be better understood as "placeholders" for propositions subject to the termination of an interaction. As such, if we are to think of typing proofs, then a type must also be considered to be constructed from untyped schematic moves. The process of proving (or refuting) a proposition is just the process of *building* a proposition, where the specificity of that proposition is not known at the start of that process.

2.1 Testing and Type-Construction

The construction of a proposition is dependent upon an elementary testing relationship between moves of opposite polarity. We start with the elementary testing relationship between α^+ and α^-, and recall that, whenever these are involved in an interaction together, they are called coherent. From this, we can build up a set of interactions that are coherent. That is, an interaction may be considered as made up of two, dual, processes, which sometimes stand in a specific relationship with each other, called coherence, and these sometimes terminate with one labelled in resolution. Recall that an interaction, (A^+, A^-) is coherent such that terminal-cut is admissible on all *focal* formulae $\alpha_1^+, \alpha_2^+ \ldots \alpha_n^+ \in A^+$ and $\alpha_1^-, \alpha_2^- \ldots \alpha_n^- \in A^-$ without removing the convergent focal formula $\alpha_m^{+,-}$.

Let us say that an interaction, (A^+, A^-), terminates with α^+ justified. Then, this means that, for every $\alpha^{+,-}$ in (A^+, A^-), each α is tested by its counterpart formula, or is reducible to such a test.[23] This relationship is symmetric, so, where A^+ contains a "pf" of α^+, so A^- contains a "refutation" of α^- (though this eventually fails). Moreover, elements of A^+ are also tests for elements of A^-, so the testing relationship operates in both directions. This is required, in fact, to ensure that each possible test is "passed" in order to terminate. As such, for each $\alpha^{+,-}$ in (A^+, A^-), the symmetry defined using the counterpart relationship is involutive: in passing from prover to refuter, we move from a positive formula to its negative counterpart; in passing from refuter to prover, we move from a negative formula to its positive formula. In other words, $\alpha_{\circ\circ} = \alpha$, for any $\alpha^{+,-}$. This involutive relationship holds *subsequent* to the process of interaction, and is the mathematical symmetry which is prior to, and determinative of local truth and falsity.[24] Indeed, as described in the previous chapter, it is the dynamism that is created by this symmetry that gives rise to the possibility of formal rules subsequent to the interaction itself.

---

[23]This latter point is important, since, for example, it allows that $\alpha^+ \#_2^c \beta^+$ and $\alpha^- \#_2^s \beta^-$ are coherent, whilst the move in A^+ requires support from *both* α^+ and β^+, and the move in A^- requires support from α^- *or* β^-.

[24]Further discussion of these, apparently contradictory "foundations" of local truth and falsity can be found in the following chapter. This testing relationship is heavily inspired by the constructions of geometry of interaction discussed in the previous chapter. As I say there, the details, and aims, substantially differ.

Interestingly, subsequent to interaction, the involution of the counterpart relationship gives rise to a kind of CLOSURE operation over elements of an interaction, which may be understood as a way of ensuring the stability of formulae after the process has ended. Closure operations are well-known in logic, particularly after Tarski's suggestion that a *consequence* operator may be defined in terms of a finitary closure operator on the set of sentences that follow from a logical theory. This is distinct from the case here, since the closure operator is not to be defined by means of a set of formulae derivable from a well-formed set of formulae. Rather, the operation is constructed by a relationship of symmetry between formulae of opposite polarities, understood as moves between agents. Setting details aside (for the moment), the idea that such a closure operation over elements of (some) interactions emerges subsequent to termination may provide an internal-to-interaction construction of a type. That is, we can define a type as just that which is stable under the closure operations induced by $\alpha_{\circ\circ} = \alpha$ at the termination of interaction. Then, say that a proposition, or type, is locally true or false whenever it is constructed by means of this symmetric closure operation over elements of an interaction. In order for that to be the case, each element must be "measured up" against its counterpart, and it is the symmetry of this "measuring up" that constructs a proposition.

Consider an interaction that begins with a positive utterance of the form "(α and σ) or (β and γ)". Then, an interaction may proceed by refuter testing the utterance by choosing to first test the first half of the utterance. Say they test α, which, by a splitting move allows refuter to use a test of α to insert into the negative focus $(\alpha^{-}\#_2^s\sigma^{-})$ which is the counterpart formula to $(\alpha^{+}\#_2^c\sigma^{+})$. Prover, then responds by making a move in support of α, which passes that test, so refuter moves on to test σ, where prover supports σ as a move used to insert into positive focus $(\alpha^{+}\#_2^c\sigma^{+})$. It is agreed that σ fails to pass that test, so $(\alpha^{+}\#_2^c\sigma^{+})$ is also successfully tested. Refuter then turns to the second half of the utterance, first testing β, which, by a splitting move allows refuter to insert into negative focus $(\beta^{-}\#_2^s\gamma^{-})$. Prover responds by making a move in support of β, which passes the test, then going on to offer a move in support of γ to insert into positive focus $(\beta^{+}\#_2^c\gamma^{+})$, whilst refuter tests γ, this is unsuccessful, so prover uses $(\beta^{+}\#_2^c\gamma^{+})$ in a splitting move to introduce into positive focus $(\alpha^{+}\#_2^c\sigma^{+})\#_2^s(\beta^{+}\#_2^c\gamma^{+})$. So, the interaction terminates with agreement that $(\alpha^{+}\#_2^c\sigma^{+})\#_2^s(\beta^{+}\#_2^c\gamma^{+})$ is justified.

Now, think of the construction of A^{+} and A^{-} in terms of terminal-cut. If we attempted to apply terminal-cut at any stage of the interaction prior to termination, then both negative and positive foci would be derived empty. For example, first we have $(\alpha^{-}\#_2^s\sigma^{-})$ inserted negatively, alongside $(\alpha^{+}\#_2^c\sigma^{+})$, inserted positively, so they are also "cut away". To consider support for $(\alpha^{+}\#_2^c\sigma^{+})$, we first consider α^{+} alongside α^{-}, which, again, are cut away. Then, σ^{+} is considered alongside σ^{-}, which are cut away. Subsequently, we have $(\beta^{-}\#_2^s\gamma^{-})$ inserted negatively, alongside $(\beta^{+}\#_2^c\gamma^{+})$, inserted positively, so they are "cut away". Support for both is considered, first by β^{+} alongside β^{-}, and then γ^{+} alongside γ^{-}, so these are cut away also. Consider, more formally, A^{+} and A^{-} separately to see their structure. First, A^{+} proceeds as follows:

$$\cfrac{;\vdash\!\dashv;\alpha^{+} \qquad ;\vdash\!\dashv;\sigma^{+}}{;\vdash\!\dashv;\alpha^{+}\#_2^c\sigma^{+}} \qquad \cfrac{\cfrac{;\vdash\!\dashv;\beta^{+} \qquad ;\vdash\!\dashv;\gamma^{+}}{;\vdash\!\dashv;\beta^{+}\#_2^c\gamma^{+}}}{;\vdash\!\dashv;(\alpha^{+}\#_2^c\sigma^{+})\#_2^s(\beta^{+}\#_2^c\gamma^{+})}$$

And, A^{-} as follows:

$$\cfrac{\alpha^{-};\vdash\!\dashv;}{\alpha^{-}\#_2^s\sigma^{-};\vdash\!\dashv;} \quad \cfrac{\sigma^{-};\vdash\!\dashv;}{\alpha^{-}\#_2^s\sigma^{-};\vdash\!\dashv;} \quad \cfrac{\beta^{-};\vdash\!\dashv;}{\beta^{-}\#_2^s\gamma^{-};\vdash\!\dashv;}$$

$$\cfrac{\cfrac{\gamma^{-};\vdash\!\dashv;}{\beta^{-}\#_2^s\gamma^{-};\vdash\!\dashv;}}{\cancel{(\alpha^{-}\#_2^s\sigma^{-})\#_2^c(\beta^{-}\#_2^s\sigma^{-})};\vdash\!\dashv;}$$

We should immediately recognise that these are not formal proofs, or refutations, for example, in A^{+} we have an attempt to support $(\alpha^{+}\#_2^c\sigma^{+})$, which fails, whilst the attempt to support $(\alpha^{+}\#_2^c\sigma^{+})\#_2^s(\beta^{+}\#_2^c\gamma^{+})$ via support of $\beta^{+}\#_2^c\gamma^{+}$ is successful. Moreover, the interaction is clearly non-deterministic, both in terms of choices that can be taken with regard to which elements to test, and in which order. That is, there are a number of ways in which the interaction might have proceeded, with a set of distinct processes by which the utterance might be tested. What is important is that a purported proof of the formula is met with a purported refutation of it, and so the two are coherent with each other.

2.2 *Operations over Processes*

Let us consider in a little more detail, a structure by which to consider the process of interactions themselves. Whilst the definition of coherence in the above (and the previous chapter) suggests that interactions proceed stepwise with each move coherent with the previous move, this may be too restrictive for some interactions.[25] As such, we may want to consider A^{+} and A^{-} as two processes, abstracting over their internal structure to some degree, so that we can look at them, subsequent to an interaction, in terms of their *behaviour*.

Bisimulation is a mathematical way of defining two processes by means of their "behaving the same", where that behaviour can be considered in abstraction from their internal structure. In the domain of logic, bisimulation equivalence, or bisimilarity was defined in the context of modal logics by Van Benthem (e.g. [22, 23]). An equivalence between two processes can be constructed by means of bisimulation equivalence, which is not as strong an equivalence as isomorphism between processes. The general idea is to think of *behavioural* equivalence as something that

[25] Though, note that the reducibility condition allows some maneuverability over the course of an interaction.

can be considered in a more objective way, though still from the inside of an interaction. For example, consider interactions that begin with different agents making the first move, that differ in the order of moves, and differ in terms of choice of tests. We may also want to abstract away from the specificities of these actions, whilst allowing interactions to be equivalent under certain circumstances. Now consider that we would like to ask what A^+ "looks like" from the point of view of A^-, and vice-versa, on termination of the interaction. We still require that coherence holds between them, but we may want to allow that this is retroactively specifiable rather than requiring stepwise coherence of each move. As such, equivalence between A^+ and A^- does not need to be a strict isomorphism, but also, it must be less coarse-grained than looking simply at the *outcome* of interactions. The latter would just eliminate the process of interaction altogether, just like the semantic structures discussed above.

So, in order that we can retain some sense of the internal structure of interactions, we turn to a structural comparison between them in terms of bisimilarity.[26] We need to define a few things first. One way of thinking about one "side" of an interaction is as a labelled transition system:

Definition 67 (*Labelled transition system* (*e.g.* [25])) A labelled transition system (LTS) is a triple, $(S^{+,-}, \mathcal{A}, \Rightarrow)$, where $S^{+,-}$ is a set of "stages" of an interaction, $\mathcal{A}$ of "moves", and $\Rightarrow$ is a relation over $S^{+,-} \times \mathcal{A} \times S^{+,-}$ providing labelled transitions $\Rightarrow^a$ between stages for each move.

The only major difference from the standard definition of a LTS is that we label the set of moves, and say that there are two LTS's per interaction, one positive and one negative. We are also thinking of the labelling of a move in terms of the action itself, where $\Rightarrow^a$ should be understood to incorporate both a schematic move *and its dual*, depending upon which LTS we are operating within. Note, importantly, that to maintain the symmetry between LTS's requires that any resolve move should not be thought of as an additional stage of an LTS, but rather a "rewriting" of a stage already undertaken.[27] In general, we may write $s_1 \Rightarrow^a s_2$ for $(s_1, a, s_2) \in \Rightarrow$, where s_i are stages in S. First, we define a notion of AGREEMENT on stages, which is somewhat similar to what van Bentham calls "atomic harmony" [22].

Definition 68 Take a stage, s_i^+, of the positive part of an interaction, and a stage, s_i^-, of the negative part of an interaction. Then, say that s_i^+ and s_i^- "agree" whenever, at s_i^+, a formula α^+ is inserted into the positive focus, and, at s_i^-, a formula α^- is inserted in to the negative focus, and where α^+ and α^- are counterparts, so $\alpha^+_\circ = \alpha^-$ and $\alpha^-_\circ = \alpha^+$.

With this, we can define a bisimulation as follows.

Definition 69 (*Bisimulation*) A binary relation $\mathcal{R}$ on stages of an interaction between two LTS's is a bisimulation if, whenever $s_i^+ \mathcal{R} s_i^-$ and $a \in \mathcal{A}$,

[26] It should be noted that we are using bisimulation in a manner that is quite different to its use in other mathematical structures, including in the context of logical games, where they are taken to represent relationships between bisimilar strategies. Further discussion of the latter may be found in [24].

[27] See the example in the following section.

- s_i^+ and s_i^- "agree"
- for all s_2^+ with $s_1^+ \Rightarrow^a s_2^+$, there is s_2^- with $s_1^- \Rightarrow^a s_2^-$ and $s_2^+ \mathcal{R} s_2^-$
- for all s_2^- with $s_1^- \Rightarrow^a s_2^-$ and $s_2^+ \mathcal{R} s_2^-$ s_2^+, there is with $s_1^+ \Rightarrow^a s_2^+$

Then, s_i^+ is bisimilar to s_i^-, $s_i^+ \sim s_i^-$, if there is a bisimulation $\mathcal{R}$ with $s_i^+ \mathcal{R} s_i^-$.[28]

The latter two clauses ensure that each action has a dual action with a counterpart focus, allowing us to trace the process of interaction, and ensuring that each action has a dual action with the counterpart focus. We note two important features of bisimulation relations in the current context.[29]

Theorem 70 *$\sim$ is an equivalence relation, which is to say it is reflexive, symmetric, and transitive.*

Theorem 71 *$s^+ \sim s^-$ iff for each $a \in \mathcal{A}$:*

- *if $s_1^+ \Rightarrow^a s_2^+$ then $s_1^- \Rightarrow^a s_2^-$ for some s_2^- such that $s_2^+ \sim s_2^-$*
- *if $s_1^- \Rightarrow^a s_2^-$ then $s_1^+ \Rightarrow^a s_2^+$ such that $s_2^+ \sim s_2^-$*

When each stage in one half of an interaction has a corresponding stage in the other half of the interaction (understood as LTS's), we say that the interaction is closed under bisimulation equivalence:

Definition 72 (*Closure under bisimulation equivalence*): An interaction, understood as a pair of LTS's, $(S^+, \mathcal{A}, \Rightarrow)$ and $(S^-, \mathcal{A}, \Rightarrow)$ is closed under bisimulation equivalence whenever, for each $s_i^+ \in S^+$ and $s_i^- \in S^-$, $s_i^+ \sim s_i^-$.

This gives us the following theorem:

Theorem 73 *For any terminating coherent interaction understood as a pair of LTS's, $(S^+, \mathcal{A}, \Rightarrow)$ and $(S^-, \mathcal{A}, \Rightarrow)$, the two LTS's are closed under bisimulation.*

Proof (Informal) First, note that two states are bisimilar whenever they agree, and their transitions "match-up". As such, they provide a notion of equivalence of processes, which are being considered here in terms of the processes internal to an interaction. So, if two states are not bisimilar, then, for some move in the positive LTS, no matching move is made in the negative LTS, and so the moves do not match-up to their tests.[30] Recall that, by definition, the stages of an interaction are defined as a sequence of formulae that characterise moves. So, the closure of stages of an interaction under bisimulation equivalence is equivalent to the closure of formulae of interactions under the counterpart relationship. In other words, if an interaction is not closed under bisimulation equivalence, then, A^+ and A^- would not be coherent, and, so, no justification would have been produced by the interaction. □

[28] I keep to the standard use of $\sim$ to identify bisimilarity, since, whilst used briefly above and in the following chapter to indicate an external negation, I trust that the context of the formalism makes clear the usage of the symbol.

[29] See also [23, 24].

[30] Interestingly, determining when a relationship between the two processes is not a bisimilarity requires us to look for transitions that do not match-up "from the inside", by enumerating all actions and looking to see if there are any non-bisimilar actions.

That bisimilarity is a symmetric relation between stages of a positive LTS and a negative LTS (unlike a mere simulation), ensures that the relationship, $\sim$, operates in both directions equivalently. As such, we can use this relation to structure the behaviour of the counterpart agent from the other agent's perspective, ensuring that no information is added or lost in the shift in perspective. So, another way of thinking about an interaction being closed under bisimulation equivalence is that it is invariant under bisimulation, which gives us a way of considering, formally, the *stability* of a proof- or refutation-object at the termination of interaction. We can, therefore, use the notion of closure under bisimulation to characterise the way in which propositions are constructed over the course of interactions: Propositions are just the stable, justified, result of interactions that are closed under bisimulation.

The above ensures that, for any proof (or refutation) that justifies some formula, that formula may be considered a proposition since it is constructed through the process of interaction, and it is stable in that it has been tested by an attempted refutation (or proof) of the same proposition. The general idea is that we are able to reconstruct *types*, or *propositions*, from untyped interactions, by means of closure under bisimulation equivalence. A proposition is that which is constructed by an interaction closed under bisimulation, such that the sequence of moves (or stages) of that interaction may be said to both *interpret* and *justify* that proposition. So, at this point, we can point to a type that is inhabited by a proof (or refutation), such that it has been tested *fully*, by an attempted refutation (or proof).

Recall the issue of categoricity that we discussed in Chaps. 1 and 2, where, for example, it was shown that there was a fissure in which a model-theory supposed to be sound and complete for derivability over a set of rules was not unique, in forms of both classical and intuitionistic semantics. This was also connected with the problem of compositionality in similar settings, where it is possible for $\alpha \vee \alpha$ to be valid whilst neither α is not valid. As is obvious, from the point of view of propositions, problems of compositionality and categoricity can not arise in this setting. That is, it is no longer possible for both α^+ and α^- to be valid (defined by a TCI), and neither is it possible for $(\alpha^+ \vee \alpha^-)$ to be valid whilst neither α^+ nor α^- is (for any formula, α). This is because, whilst we require no appeal to any sort of model-theory, the semantics constructing propositions here are *stronger* than any semantics for intuitionistic logic. This follows because we do not have that α is refuted whenever α can not be proved, but that we have a witness that α is unprovable whenever there exists a refutation of α.

2.3 Liberalising Coherence

Rather than require stepwise coherence between moves mapping between positive and negative "sides" of an interaction, bisimulation provides a way of abstracting away from the specificities of interaction to liberalise that relationship a little. As such, we can restate the definition of a coherent interaction as follows:

Definition 74 An interaction is coherent iff it is the case that for each positive move that inserts a certain schema into the positive focus, there exists *some* negative move inserting the counterpart schema into the negative focus, so that, if terminal-cut were applied to *that pair of moves*: both negative and positive *foci* would be derived to be empty in the course of a interaction; or a focal insertion of a formula is derived in either the positive or negative foci, which terminates the interaction (in both cases we should add the clause the formula derived may have to be transformed into a focal insertion if it is not already).

This is just a slight liberalisation of the definition in the previous chapter to enable us to deal with interactions where positive moves can be mapped onto negative moves (and vice-versa), whilst these moves need not be consecutive. An additional way in which coherent interactions may be liberalised, which is brought to the foreground by the discussion of bisimulation equivalence, is that we can identify elements of interactions that are coherent, whilst the interaction as a whole may not be.

Definition 75 (*Partial closure under bisimulation equivalence*): Take an interaction, understood as a pair of LTS's, $(S^+, \mathcal{A}, \Rightarrow)$ and $(S^-, \mathcal{A}, \Rightarrow)$. Then, say that a subset $S_I^+ \in S^+$ and $S_I^- \in S^-$ of stages of an interaction are closed under bisimulation whenever, for each $s_I^+ \in S_I^+$ and $s_i^- \in S_I^-$, $s_i^+ \sim s_i^-$.

This latter definition is already implicit in the definition of coherence given in the previous chapter. It is interesting, nonetheless, to spell out the notion of partial closure, since it amounts to something like "partial testing" of the moves made over the course of an interaction. Of course, it is often the case that ordinary interactions yield partial opportunities for testing, whilst, as a whole they fail to come to any agreement, and involve agents "talking past each other", and so on. Nevertheless, these points of coherence indicate stages at which an interaction is conducted on the same ground, and, so, identifying them is important from the point of view of thinking about how we account for "pockets" of reasoning within, otherwise asymmetric, interactions.

Let us finish this chapter by considering a couple of examples that connect sequent calculus derivations, terminating coherent interactions, and bisimilar LST's. Take an interaction in which, first, $\mathcal{P}$ makes a move in support of α^+, which is tested by $\mathcal{R}$ with α^-. Then, $\mathcal{P}$ goes onto use α^+ in support of $\alpha^+\#_2^s\beta^+$, which is answered, by $\mathcal{R}$ by a test of form $\alpha^-\#_2^c\beta^-$ with β^- as assumption, alongside the previous test α^-. The interaction continues with $\mathcal{P}$ using $\alpha^+\#_2^s\beta^+$ together with σ^+ in a combining move to support $(\alpha^+\#_2^s\beta^+)\#_2^c\sigma^+$. $\mathcal{R}$ responds with a move in support of σ^-, which $\mathcal{P}$ tests with a move in support of σ^+, which fails. This is subsequently used to support a splitting move of form $(\alpha^-\#_2^c\beta^-)\#_2^s\sigma^-$ by $\mathcal{R}$, where the interaction terminates, with $\mathcal{P}$ making a resolution move of the form $\cancel{(\alpha^+\#_2^s\beta^+)\#_2^c\sigma^+}$. First, consider the sequent-form reconstruction of the *refutation-only* of the second convergent formula is just this:

$$\frac{\sigma^-; \mathrel{\vert\!\!-\!\!\vert} \Gamma^-;}{(\alpha^-\#_2^c\beta^-)\#_2^s\sigma^-; \mathrel{\vert\!\!-\!\!\vert} \Gamma^-;}\ (\#_2^s)$$

The "fuller" sequent reconstruction is this:

$$\dfrac{\dfrac{;\Gamma^+ \dashv\vdash ;\alpha^+}{;\Gamma^+ \dashv\vdash ;\alpha^+\#^s_2\beta^+}(\#^s_2) \qquad \dfrac{\overset{[\beta^-_{ass}]}{\beta^-;\dashv\vdash\Gamma^-;} \qquad \alpha^-;\dashv\vdash\Gamma^-;}{\alpha^-\#^c_2\beta^-;\dashv\vdash\Gamma^-;}(\#^c_2 \qquad ;\Gamma^+\dashv\vdash;\sigma^+ \qquad \sigma^-;\dashv\vdash\Gamma^-;}{(\alpha^-\#^c_2\beta^-)\#^s_2\sigma^-;\Gamma^+\dashv\vdash\Gamma^-;\underline{\cancel{(\alpha^+\#^s_2\beta^+)\#^c_2\sigma^+}}}(\#^c_2 \text{ / Resolve})/(\#^s_2)$$

Think of this in terms of the two LTS's S^+ and S^-, taking the positive system first[31]:

$s_1^+ : \Gamma^+ \Rightarrow^{at^+_\alpha} \alpha^+$
$s_2^+ : \Gamma^+, \alpha^+ \Rightarrow^{\#^s_2} \alpha^+\#^s_2\beta^+$
$s_3^+ : \Gamma^+, (\alpha^+\#^s_2\beta^+), \sigma^+ \Rightarrow^{\#^c_2} (\alpha^+\#^s_2\beta^+)\#^c_2\sigma^+$
$s_4^+ : \Gamma^+ \Rightarrow^{at^+_\sigma} \sigma^+$
Terminates: $\underline{\cancel{(\alpha^+\#^s_2\beta^+)\#^c_2\sigma^+}}$

And, the negative system:

$s_1^- : \Gamma^- \Rightarrow^{at^-_\alpha} \alpha^-$
$s_2^- : \Gamma^-, \alpha^-, \beta^- \Rightarrow^{\#^c_2} \alpha^-\#^c_2\beta^-$
$s_3^- : \Gamma^- \Rightarrow^{at^-_\sigma} \sigma^-$
$s_4^- : \Gamma^- \Rightarrow^{\#^s_2} (\alpha^-\#^c_2\beta^-)\#^s_2\sigma^-$

For LTS's we do not take the resolve move to move the LTS to a different stage, but rather it simply "marks" one of the two LTS's as terminating in failure to insert the counterpart of the convergent formula. Let us mechanically check that S^+ and S^- are bisimilar. First, s_1^+ and s_1^- are bisimilar by definition. s_2^+ and s_2^- are moves to counterpart formulae, and dual moves are used, so this is in keeping with the definition above. s_3^+ and s_4^- are bisimilar, since counterpart formulas are inserted with dual moves, and s_4^+ and s_3^- are bisimilar by definition. So, for each $s_i^{+,-} \in S^{+,-}$, there exists some s_i^+ and some s_i^- such that $s_i^+ \sim s_i^-$. As such, $S^+ \sim S^-$.[32]

To finish, consider the following interaction involving conditionals. This time, let us begin by think of the interaction in terms of the two LTS's S^+ and S^-, taking the positive system first:

$s_1^+ : \Gamma^+, \alpha^+ \Rightarrow^{at^+_\beta} \beta^+$
$s_2^+ : \Gamma^+ \Rightarrow^{\#^h_2} \alpha^+\#^h_2\beta^+$
$s_3^+ : \Gamma^+ \Rightarrow^{at^+_\alpha} \alpha^+$
$s_4^+ : \Gamma^+ \Rightarrow^{at^+_\beta} \beta^+$
$s_5^+ : \Gamma^+, \beta^+ \Rightarrow^{at^+_\sigma} \sigma^+$
$s_6^+ : \Gamma^+ \Rightarrow^{\#^h_2} \beta^+\#^h_2\sigma^+$
$s_7^+ : \Gamma^+, \beta^+, \beta^+\#^h_2\sigma^+ \Rightarrow^{\#^h_2/Cut^+} \sigma^+$

---

[31] We let the superscript $\Rightarrow^{at^{+/-}_\alpha}$ indicate a move with reasons in support of an atomic formula. Of course, this may be further broken down by testing reasons for it, but in many cases this may not be required.

[32] Note that, whilst stages s_3 and s_4 are not pair-wise coherent, they do conform to the definition of coherence by means of the "transformable in a single move" clause.

And, the negative system:

$s_1^- : \Gamma^-, \alpha^- \Rightarrow^{at_\beta^-} \beta^-$
$s_2^- : \Gamma^- \Rightarrow^{\#_2^h} \beta^-\#_2^h\alpha^-$
$s_3^- : \Gamma^- \Rightarrow^{at_\alpha^-} \alpha^-$
$s_4^- : \Gamma^- \Rightarrow^{at_\beta^-} \beta^-$
$s_5^- : \Gamma^-, \beta^- \Rightarrow^{at_\sigma^-} \sigma^-$
$s_6^- : \Gamma^- \Rightarrow^{\#_2^h} \sigma^-\#_2^h\beta^-$
$s_7^- : \Gamma^+, \beta^-, \sigma^-\#_2^h\beta^- \Rightarrow^{\#_2^h/Cut^-} \sigma^-$
Terminates: $\cancel{\sigma}$

Let us begin by mechanically checking that S^+ and S^- are bisimilar. First, s_1^+ and s_1^- are bisimilar by definition. s_2^+ and s_2^- are moves to counterpart formulae, and dual moves are used, so this is in keeping with the definition above. s_3^+ and s_3^- are bisimilar by definition, as are s_4^+ and s_4^-, and s_5^+ and s_5^-. s_6^+ and s_6^- are moves to counterpart formulae, $(\beta^+\#_2^h\sigma^+)$ and $(\sigma^-\#_2^h\beta^-)$, respectively (both using dual conditional moves). s_7^+ and s_7^- are similar, both inserting $\sigma^{+,-}$. So, for each $s_i^{+,-} \in S^{+,-}$, there exists some s_i^+ and some s_i^- such that $s_i^+ \sim s_i^-$. As such, $S^+ \sim S^-$. In this case, the convergent formula is σ. Let us reconstruct the interaction by means of the two separate (positive and negative) sequent trees:

$$\dfrac{\dfrac{\dfrac{;\Gamma^+, \alpha^+ \vdash\!\dashv ; \beta^+}{;\Gamma^+ \vdash\!\dashv ; \alpha^+\#_2^h\beta^+} \qquad ;\Gamma^+ \vdash\!\dashv ; \alpha^+}{;\Gamma^+ \vdash\!\dashv ; \beta^+} \qquad \dfrac{;\Gamma^+, \beta^+ \vdash\!\dashv ; \sigma^+}{;\Gamma^+ \vdash\!\dashv ; \beta^+\#_2^h\sigma^+}}{;\Gamma^+ \vdash\!\dashv ; \sigma^+}$$

$$\dfrac{\dfrac{\dfrac{\beta^-; \vdash\!\dashv \Gamma^-, \alpha^-;}{\beta^-\#_2^h\alpha^-; \vdash\!\dashv \Gamma^-;} \qquad \alpha^-; \vdash\!\dashv \Gamma^-;}{\beta^-; \vdash\!\dashv \Gamma^-;} \qquad \dfrac{\sigma^-; \vdash\!\dashv \Gamma^-, \beta^-;}{\sigma^-\#_2^h\beta^-; \vdash\!\dashv \Gamma^-;}}{\cancel{\sigma}; \vdash\!\dashv \Gamma^-;}$$

Let us ensure that the sequent trees are equivalent with a terminating coherent interaction. First, the interaction terminates with a resolution move on σ^-, identifying it as the counterpart of the convergent formula, σ^+. Second, each move is immediately followed by a move made by the opposite agent. Third, check coherence with terminal cut: take the focus of each step: $A^+ : (\beta^+; \alpha^+\#_2^h\beta^+; \alpha^+; \beta^+; \sigma^+; \beta^+\#_2^h\sigma^+; \sigma^+)$, and $A^- : (\beta^-; \beta^-\#_2^h\alpha^-; \alpha^-; \beta^-; \sigma^-; \sigma^-\#_2^h\beta^-; \cancel{\sigma})$. Then, use terminal cut as follows:

$$\dfrac{A^-; \vdash\!\dashv; \qquad ; \vdash\!\dashv; A^+}{; \vdash\!\dashv; \sigma^+} \text{ (Terminal-cut)}$$

So, the interaction is terminating and coherent. Now consider the reduction process given in the previous chapter, which is simple given that we are working with standard forms of schematic moves, so we should be able to reduce the TCI to a proof in intuitionistic sequent calculus LJ. It is simple, in fact, to see that the positive sequent tree above may be identified (since there are no detours in this example) with this:

$$\dfrac{\dfrac{\dfrac{;\Gamma^+,\alpha^+ \vdash\!\dashv ;\beta^+}{;\Gamma^+ \vdash\!\dashv ;\alpha^+\to\beta^+} \qquad ;\Gamma^+ \vdash\!\dashv ;\alpha^+}{;\Gamma^+\vdash\!\dashv;\beta^+} \qquad \dfrac{;\Gamma^+,\beta^+\vdash\!\dashv;\sigma^+}{;\Gamma^+\vdash\!\dashv;\beta^+\to\sigma^+}}{;\Gamma^+\vdash\!\dashv;\sigma^+}$$

As such, we have that $|P|$ inserting σ^+ as convergent formula equivalent with a valid proof P of σ^+ in sequent schema IPR such that $|P| = P$, and, additionally in this case, that P is equivalent with a derivation of σ in LJ.

References

1. Erik Palmgren. Constructive logic and type theory, 2014.
2. Per Martin-Löf. On the meanings of the logical constants and the justifications of the logical laws. *Nordic Journal of Philosophical Logic*, 1(1):11–60, 1996.
3. W. A. Howard. The formulæ-as-types notion of construction. In Philippe De Groote, editor, *The Curry-Howard Isomorphism*. Academia, 1995.
4. Frank Pfenning and Rowan Davies. A judgmental reconstruction of modal logic. *Mathematical structures in computer science*, 11(04):511–540, 2001.
5. A. N Kolmogorov. Zur deutung der intuitionistischen logik. *Mathematische Zeitschrift*, 35:58–65, (English translation in Mancosu 1998, pp. 328–334) 1932.
6. Luca Tranchini. Natural deduction for dual-intuitionistic logic. *Studia Logica*, 100(3):631–648, 2012.
7. Luis Estrada-González. From (paraconsistent) topos logic to universal (topos) logic. In Buchsbaum Arthur (Eds.) Koslow, Arnold, editor, *The Road to Universal Logic*, chapter 12. Birkhäuser Basel, 2015.
8. Saunders Mac Lane and Ieke Moerdijk. *Sheaves in geometry and logic*. Springer Science & Business Media, 1992.
9. A. S. Troelstra. *Constructivism in Mathematics: An Introduction*. Sole Distributors for the U.S.A. And Canada, Elsevier Science Pub. Co., 1988.
10. Nicholas Goodman. The logic of contradictions. *Zeitschrift fur Mathematische Logic und Grundlagen der Arithmetik*, 27:119–126, 1981.
11. Robert Goldblatt. *Topoi: the categorial analysis of logic*. Elsevier, 2014.
12. Andreas B. M. Brunner and Walter A. Carnielli. Anti-intuitionism and paraconsistency. In *URL* = http://www.cle.unicamp.br/e-prints/vol3,n1,2003.html, 2003.
13. James Trafford. Co-constructive logic for proofs and refutations. *Studia Humana*, 3(4):22–40, 2015.
14. Cecylia Rauszer. Applications of Kripke models to Heyting-Brouwer logic. *Studia Logica*, 36(1-2):61–71, 1977.
15. Tristan Crolard. Subtractive logic. *Theoretical computer science*, 254(1):151–185, 2001.
16. S. Awodey. Structure in mathematics and logic: A categorical perspective. *Philosophia Mathematica*, 4(3):209–237, 1996.
17. Chris Mortensen. *Inconsistent mathematics*, volume 312. Springer, 1995.
18. John L Bell. From absolute to local mathematics. *Synthese*, 69(3):409–426, 1986.

19. Luis Estrada-González. Complement-topoi and dual intuitionistic logic. *Australasian Journal of Logic*, 9:26–44, 2010.
20. Saunders MacLane and Ieke Moerdijk. *Sheaves in geometry and logic: A first introduction to topos theory*. Springer Science & Business Media, 2012.
21. Jean-Pierre Marquis and Gonzalo Reyes. The history of categorical logic: 1963–1977. In Dov Gabbay, Akihiro Kanamori, and John Woods, editors, *Handbook of the history of logic*. Elsevier, 2011.
22. Johan Van Benthem. Program constructions that are safe for bisimulation. *Studia logica*, 60(2):311–330, 1998.
23. Johan Van Benthem. Extensive games as process models. *Journal of Logic, Language and Information*, 11(3):289–313, 2002.
24. Johan Van Benthem. *Logical dynamics of information and interaction*. Cambridge University Press, 2011.
25. Johan Van Benthem and Jan Bergstra. Logic of transition systems. *Journal of Logic, Language and Information*, 3(4):247–283, 1994.

Chapter 7
Paradox, Paraconsistency and Logical Revision

This chapter suggests that the approach suggested to this point provides a novel approach to understanding logical paradoxes. In particular, I argue that what is typically thought to be captured by paraconsistency and paracompleteness is best understood as a form of epistemological instability that can be dealt with in the context of the dialogical dynamics of reasoning. This is used to motivate an approach called paracoherence, which can deal with strengthened versions of paradox, and is compatible with a bivalent approach to local truth and falsity. This will be clarified by first considering logic and paradox in relation to standard arguments regarding logical revision. Arguments for and against logical revisionism share assumptions regarding logical rules that have been argued against in previous chapters. As such, I suggest that the interactive approach provides not a revision of logic, but an attempt to rethink logic from the ground up.

1 A (Standard) Inferentialist Approach to Paradox

In [1], Restall expands on the bilateralist approach to inferentialism that was discussed in Chap. 2 to deal more specifically with non-classical logics. In this setting, he argues that we should consider (logical) theories to be pairs of sets of sentences called BITHEORIES, rather than just sets of sentences. Such bitheories provide a way of "keeping track" of what holds and what fails to hold according to the theory. In keeping with the approach to logic by means of the coherence of assertions and denials, Restall defines bitheories as follows. First, he defines a logical theory in the usual way (see also above in Chap. 1):

Definition 76 A theory is a collection T of sentences that is closed under logical consequence. So T is a theory iff $(\forall\alpha)T \vdash \alpha$ then $\alpha \in T$.

J. Trafford, *Meaning in Dialogue*, Studies in Applied Philosophy, Epistemology and Rational Ethics 33, DOI 10.1007/978-3-319-47205-8_7

In keeping with the (standard) inferentialist spirit, Restall interprets this slightly differently to semantical approaches to closure under consequence, as:

> If some statement is a consequence of the theory, it is also a part of the theory. So, if you endorse the theory, commitment to this theory means that you are making a mistake if you deny any statement in the theory.

Restall goes on to define a COTHEORY in terms of what is to be rejected:

Definition 77 A cotheory is a collection U of sentences that is closed "over" logical consequence. So U is a theory iff $(\forall\alpha)\alpha \vdash U$ then $\alpha \in U$.

The idea here is that:

> [I]f some statement has the cotheory as a consequence, it is also a part of the cotheory. So, if you reject the cotheory, this rejection means that you are making a mistake if you assert any statement in the cotheory. The statements which have U as a consequence are unacceptable, granted commitment to deny U.

Taken together, Restall defines a BITHEORY, which consists of both a theory and cotheory:

Definition 78 $\langle T, U\rangle$ is a bitheory iff $(\forall\alpha)T \dashv\vdash \alpha$ then $\alpha \in T$ and $(\forall\alpha)\alpha \dashv\vdash U$ then $\alpha \in U$.[1]

Now, $\langle T, U\rangle$ provides us with directions on what should be asserted and denied, or, as discussed in Chap. 2, more accurately, what is undeniable (T), and what is unassertible (U). In this setting, we have a more flexible way of dealing with non-classical logics that provide an account of negation that does not coincide with the relationship of coherence between assertions and denials. For example, classically, it is the case that the assertibility of α is equivalent with the deniability of $\neg\alpha$, and the deniability of α is equivalent with the assertibility of $\neg\alpha$. So, in this sense, classical negation is symmetric, obeying the following rules:

$$\frac{\Gamma \dashv\vdash \alpha, \Delta}{\Gamma, \neg\alpha \dashv\vdash \Delta}\ (\neg\text{-L}) \qquad\qquad \frac{\Gamma, \alpha \dashv\vdash \Delta}{\Gamma \vdash \neg\alpha, \Delta}\ (\neg\text{-R})$$

Furthermore, any theory T that is closed under classical consequence is provably consistent such that T can not contain both α and $\neg\alpha$, for any α. Restall's structure accounting for bitheories provides a way of understanding this, since in classical logic, logical *consistency* is equivalent with a coherence relation between theories and co-theories:

Definition 79 (*Coherence*) For any set of formulas Σ, and bitheory $\langle\, T, U\,\rangle$ (where $T \cup U = \Sigma$), we say that the bitheory is COHERENT whenever it is the case that $T \cap U = \emptyset$.[2]

---

[1] Adopting the $\dashv\vdash$ version of the turnstile for the reasons discussed in Chap. 2.

[2] Bear in mind that this form of coherence is at odds with the interactive coherence described in preceding chapters, which we return to in Sect. 3.

With this, it is simple to see the equivalence of consistency and coherence for (symmetrical presentations of) classical logic.

Proof We refer here to the fact that theory and co-theory construction simulates a fairly standard Lindenbaum-Asser construction, together with the fact that any theory closed under classical logic is consistent as defined above. In the symmetrical calculus, as described above (in Chap. 2), it is possible to extend any theory, Γ, and cotheory, Δ, where *not* ($\Gamma \vdash\!\dashv \Delta$) to a maximal theory, $\Gamma' \supseteq \Gamma$, and cotheory, $\Delta' \supseteq \Delta$, with $\Gamma' \cup \Delta' = S$ and $\Gamma' \cap \Delta' = \emptyset$.[3] □

This shows us that, classically (as long as we are working within a symmetric sequent calculus), the relationship of coherence between assertions and denials coincides with the symmetry of negation.[4] It is, nonetheless, precisely this last theorem that the liar paradox threatens. The presence of a liar sentence renders classical logic both inconsistent (leading to triviality) and incoherent as follows.

First, introduce a transparent truth predicate into the logic, following Beall [6], which is "see through", such that $T(\ulcorner\alpha\urcorner)$ and α are intersubstitutable in all transparent contexts, and for all $\alpha \in S$ (where S is just a denumerable set of formulas).[5] Then, by reflexivity, this gives us familiar introduction and elimination rules:

- $T_I = T(\ulcorner\alpha\urcorner) \vdash \alpha$
- $T_E = \alpha \vdash T(\ulcorner\alpha\urcorner)$

Then, it is simple to construct a liar statement l, of the form $l : \neg T(\ulcorner l\urcorner)$. Now, when we attempt to construct a theory, Γ, and cotheory, Δ, we have a problem since l with either end up in both or neither (depending upon how we construe the process of adding formulas to these sets).[6] So, by these lights, classical logic expanded with T is not coherent. That is, it is simple to show, for example, that the formula l ends up on both sides of the turnstile, forcing it either to be added to both a theory and a

---

[3]Further details of this sort of construction can be found in [2], pp. 188–189; 201; and [3, 4].

[4]This result echoes a technical operation shown in a paper by Bendall [5], that it is possible to translate classical propositional logic to a logically equivalent system that contains a non-embeddable external negation whilst subtracting in the internal negation operator (and vice-versa). Perhaps more interestingly, in this context, is that he foregrounds the kind of "contradiction" discussed under the notion of coherence, below: "contradiction can then be construed in terms of conflicting judgment or belief rather than in terms of inconsistent propositional content and such notions as negative facts are thereby decisively banished." (p. 68). This seems to overstate matters in the specific case of classical logic that Bendall is working with, since rather than showing that contradictions in that system can be wholly understood in terms of what we are calling coherence, Bendall's result and our own show rather that contradictions in terms of coherence are equivalent with contradictions in terms of inconsistent propositional content.

[5]Corner brackets indicate a name-forming device.

[6]There are a multitude of options available in the literature attempting to prevent incoherence by, for example, restricting truth-predicate in some way. I will ignore these here since I am primarily interested in paraconsistent and paracomplete constructions. For example, the standard Tarskian response is to prevent a language from formulating the paradoxical sentence through the role of (infinite) metalanguages.

cotheory, or, by application of Cut to the final formulas, to neither (either way, the entire language is rendered trivial in the sense that any formula is derivable)[7]:

$$\dfrac{\dfrac{\dfrac{\dfrac{\dfrac{\dfrac{T(\ulcorner l \urcorner) \vdash T(\ulcorner l \urcorner)}{\neg T(\ulcorner l \urcorner) \vdash \neg T(\ulcorner l \urcorner)}}{l \vdash l}}{T(\ulcorner l \urcorner) \vdash l}}{\vdash \neg T(\ulcorner l \urcorner), l}}{\vdash l} \qquad \dfrac{\dfrac{\dfrac{\dfrac{\dfrac{T(\ulcorner l \urcorner) \vdash T(\ulcorner l \urcorner)}{\neg T(\ulcorner l \urcorner) \vdash \neg T(\ulcorner l \urcorner)}}{l \vdash l}}{l \vdash T(\ulcorner l \urcorner)}}{\neg T(\ulcorner l \urcorner), l \vdash}}{l \vdash}}{\vdash}$$

In the above terms, the typical non-classical response to the liar is to refuse the equivalence of coherence and negation-symmetry from the outset. So, for example, in a paracomplete logic such as intuitionistic logic L_I, $\alpha \vee \neg_I \alpha$ is not a theorem. In a paraconsistent logic such as co-intuitionistic logic L_C [8–10], it is possible that $\alpha \wedge \neg_C \alpha$ holds without trivializing the logic. As discussed above, intuitionistically, it is typical to define a negation as $\neg_I \alpha =_{df} \alpha \Rightarrow 0$. This negation satisfies $\alpha \wedge \neg_I \alpha = 0$, but $\alpha \vee \neg_I \alpha = 1$ need not hold, nor does $\neg_I \neg_I \alpha = \alpha$. So, negation is no longer symmetric. Co-intuitionistically, negation is usually defined as $\neg_C \alpha =_{df} 1 \leftarrow \alpha$, where $\leftarrow$ is a co-conditional connective. This negation satisfies $\alpha \vee \neg_C \alpha = 1$, but $\alpha \wedge \neg_C \alpha = 0$ need not hold, nor does $\neg_C \neg_C \alpha = \alpha$. Again, this negation is not symmetric.

As such, it is trickier to consider what is assertible and what is deniable in terms of the construction of a theory and cotheory for these logics. For example, it is typical, intuitionistically, to say that a formula α is refuted whenever there exists a proof of $\neg\alpha$. But, this is inequivalent to the above "refutation" interpretation of a sequent since $\vdash \neg\alpha \not\equiv \alpha \dashv$. It is, therefore, impossible to construct the class of negated formulas as dual to the class of assertible formulas by means of a cotheory. One way of thinking about this is, to follow Restall, in saying that such a *paracomplete* logic can not accept the classical rule ($\neg$-R). As such, we can construct a theory, and cotheory as above. But, now, the class of refuted formulas (i.e. those formulas for which their exists a proof of their negation), is distinct from the construction of a co-theory. For example, it may be the case that, for some α, neither α nor $\neg\alpha$ are proved; or, that $\neg\neg\alpha$ is proved, without α being proved also. So, there is an asymmetry between the two classes of formulas. Exactly the reverse is the case for co-intuitionistic logic, where such a *paraconsistent* logic can not accept the classical rule ($\neg$-L). As such, forming a bitheory as above, we would then have the case where, for some α, it is

[7] See [7, p. 144–5] for a similar construction.

not the case that whenever α is in a theory T that $\neg\alpha$ is in the cotheory U.[8] To sum up, where Γ is a theory and Δ a co-theory, we have the following situation:

- Classical: If $\alpha \in \Gamma$, then $\alpha \notin \Delta$, and, if $\alpha \in \Gamma$, then $\neg\alpha \in \Delta$ (and vice-versa)
- Intuitionistic: If $\alpha \in \Gamma$, then $\alpha \notin \Delta$ (and vice-versa), but, it may be the case that $\alpha \notin \Gamma$ and $\neg\alpha \in \Delta$.
- Co-intuitionistic: If $\alpha \in \Gamma$, then $\alpha \notin \Delta$ (and vice-versa), but, it may be the case that $\alpha \in \Gamma$ and $\neg\alpha \notin \Delta$.

It is this asymmetry of negation, which allows these logics to deal with statements such as l. I should say that I am glossing over a number of issues that arise for these logics, particularly with regard to the construction of strengthened liar sentences. Nonetheless, the general picture is adequate for the purposes of this discussion.

2 Logical Revisions Down the Rabbit Hole

There is, however, an obvious issue here, since liar-like statements are not (typically) understood to be simply formal mathematical problems. Rather, these have to do with ordinary language, and, as such, the negation-connective involved is also supposed to map, in some sense, to the use of the term "not" in natural language. For example, in Restall's presentation, the difference between the three positions consists in simply dropping one of the negation rules (or not), so, it looks like an advocate of paraconsistent and paracomplete logics will have to say that whatever is characterised by adopting both of the negation rules, it is not the correct account of negation. Moreover, since, for the (standard) inferentialist, the meaning of negation is determined by its logical rules, this is a little tricky. Does it entail, for example, that there is no such meaning determined by adopting both the left and right rules as Priest [11] argues? Or that there are three distinct meanings determined by the distinct adoption of the rules? As such, we run, quickly into the web of issues surrounding arguments for and against logical revision involving *meaning variance*.

Whilst the exegetical position is unclear regarding Quine's comment to wit "change of logic, change of subject", it has, without doubt, had a long-lasting legacy within the philosophy of logic. As Quine puts this in relation to a deviant system that would deny the law of non-contradiction (such as a paraconsistent logic):

> My view of the dialogue is that neither party knows what he is talking about. They think that they are talking about negation, ["$\neg$"], "not"; but surely the notion ceased to be recognisable as negation when they took to regarding some conjunctions of the form "$\alpha \wedge \neg\alpha$" as true, and stopped regarding such sentences as implying all others. Here, evidently, is the deviant logician's predicament: when he tries to deny the doctrine he only changes the subject [12, p. 81].

[8]It should not go unnoticed that the situation is similar to the distinction between designated and non-designated formulas in many-value semantics for paracomplete and paraconsistent semantics.

And in relation to a system that would drop the law of excluded middle (such as a paracomplete logic):

> ...what are we to say of revisions, such as imagined in the case of the law of excluded middle? Do we thereby change our theory or just change the subject, change the meaning of our words? This has been a recurrent challenge, and my answer is that in elementary logic a change of theory is a change of meaning. Repudiation of the law of excluded middle would be a change of meaning, and no less a change of theory for that [13, p. 270].

And, in contemporary form, as McGee states:

> An observation of Quine is worth repeating: we cannot hope to determine the correct rules of inference by a semantic investigation, trying to determine which rules are truth-preserving, because the same question 'What are the legitimate rules of inference?' is going to recur as we try to develop the metatheory. Instead, what determines what the connectives mean are the inferences in which we employ them; the rules of inference implicitly define the connectives. People who employ different rules of inference mean different things by connectives; so they are inevitably talking at cross-purposes [14, p. 206].

The latter is more specifically in line with the standard inferentialist account that a set of inferential rules determine the meaning of logical connectives. It looks as though the meanings determined by competing accounts of inferential rules will be distinct, and so any two distinct logical systems will not be dealing with the same "thing": they will be talking past each other. For example, any paraconsistent logic will deny the validity (in some contexts) of *ex contradictione quodlibet*: $\alpha, \neg\alpha \vdash \beta$. As such, whatever the meaning determined by the rules for $\neg$ is, since those rules must differ from the classical inferential rules, the meaning also differs from the classical meaning. This is a simplification of the argument, but, we can reconstruct the intended argument from Quine [15] as follows:

- In order for there to be a genuine disagreement between the classical logician and the paraconsistent logician, there must be a correct account of the meaning of negation, but a disagreement about the set of laws that govern negation.
- So, for genuine disagreement to be possible, there must be a correct account of the meaning of negation that is, in some way independent of the laws governing negation.
- But, there is nothing more to the meaning of negation than the inferential rules defining its use: 'There is no residual essence ...in addition to the sounds and notations and the laws in conformity with which a man uses those sounds and notations.' (p. 81)
- So, there can be no genuine disagreement. Rather, that disagreement will involve a change of subject since it involves a change in the rules governing the use of negation.

The third step is the key claim here, and, whilst Quine's behaviourism looms large, the contemporary formulations express a similar thesis in the form of inferentialism about meaning. As such, arguments around meaning variance are inevitably tangled up with the issues raised regarding inferentialist approaches to meaning that were discussed in Chap. 1. As Hartry Field points out, claims of meaning change are difficult to uphold:

> The question [of meaning change] is clear only to the extent that we know how to divide up such firmly held principles into those that are "meaning constitutive" or "analytic" and those which aren't, and this is notoriously difficult [16, p. 17].

The arguments of Chaps. 1 and 2 uphold this sentiment, whilst also expanding its reach. In this context, we may take those arguments to also mean that there can be *no possible* foundation for deciding the meaning of a logic operator by means of a set of inferential rules. Let us briefly rehearse some of the standard responses to the Quinean argument.

One option is to take a fairly standard inferentialist approach in which logical rules determine the meaning of a logical operator, whilst allowing that distinct sets of rules determine distinct meanings. So, for example, both the classical rules and the paraconsistent rules should be understood to determine meanings for negation that are "genuine" and distinct from each other. In which case, the logical operators must also be distinct, and so may be denoted with different syntactic forms (e.g. $\neg$ and $\backsim$). There are several problems with this approach already noted by Priest [11, Sect. 12]. The first is the simple point that it is difficult to know what to make of the use of "not" in natural language. If it is ambiguous between meanings, then content ascription becomes problematic since ordinary thinkers are likely to have very little understanding of the details of the inferential rules that are supposed to determine the content of the expressions that they are using. Hence, it will not be clear which logical operator agents are *actually* employing in their utterances concerning negation. Given also that there are many logics assigning different inference-forms to negation, and, according to this position the content of negation will differ accordingly, it looks like we will end up with, not just ambiguity, but a difficulty for communication with agents simply talking past each other. Further still, the problem with disambiguating which meaning of negation *is* being employed, is compounded when we consider that we will also want to know when each meaning *should* be employed. For example, we might think that intuitionistic negation ought to be used in certain contexts, and classical negation in others. Of course, this would mean that different classes of arguments will be considered valid according to which operator is used. But, this may lead to problems where we are reasoning with conjunctions of assertions. If we consider, for example, an inference involving negation that is valid according to classical logic, but invalid according to intuitionistic logic (perhaps involving the law of excluded middle), then we require some sort of (fairly stringent) criteria for discerning whether or not we are justified in performing that inference. Then, for a conjunction of assertions involving both intuitionistic and classical negations, this very quickly becomes problematic. In this specific setting, such problems are compounded by the well-known proof that intuitionistic negation collapses into classical negation in a standard proof system. According to Williamson [17] the collapse demonstrates that there cannot be two different "stocks" of logical constants (i.e. one classical and one intuitionistic), with different extensions. This is because, for any language in which negation has differing meanings, it is possible to construct a third language containing both meanings, but, as Williamson [17, p. 111] puts it, collapse demonstrates that the suggestion that the two meanings can occur in that third language is "provably false". Hand [18] makes a similar point:

> If each disputant admits the intelligibility of the opponent's view, then in particular the intuitionist must admit that there is a coherent classical conception of truth. It follows from this admission that a mixed system containing both negations is possible. Thus the malignancy of classical negation shows that the intuitionist must deny the intelligibility of distinctively classical pronouncements on truth and logic. From the intuitionist's viewpoint, there is no such thing as the classical conception.

As such, the idea that meanings are determined by inferential rules begins to look less attractive. One possible option would be to say that, in addition to distinct meanings of negation, the classical and intuitionistic logician are also employing different definitions of provability and truth. But, whilst this is surely correct, (though see [19] for a discussion of problems facing intuitionistic conceptions of truth), if the two meanings are supposed to "live together", then, as Eklund [20] points out, this faces serious problems if logicians require that valid arguments have normative traction.[9]

Say that we accept that negation meanings are unambiguous, then, like Priest, it looks like we would require some justification for saying that one meaning is "correct". Priest [11] canvases a number of arguments in support of this view, primarily arguing that classical (Boolean) negation is not the correct construal of the meaning of "not" in natural language. The idea that we can "read-off" the meaning of a logical operator from natural language is dealt with in Chap. 1, but it is worth explicating this further in this context. The suggestion here would be that there are distinct attempts to characterise negation, but the paraconsistent negation rules are the *correct* rules for determining its meaning.[10] But then, we are in exactly the place that we began in the introductory chapter, with each successive attempt to provide the justification of a set of inference rules leading to problems of circularity. In fact, in an earlier paper [24], Priest also suggests that determining a correct set of rules from a practice (in this case natural language use) is far from transparent. He considers an analogy with determining grammatical rules:

> There is no straightforward sense in which the rules can be read off from the practice. Neither is it much use asking people whether a certain grammatical rule is correct ...What we have, rather, is a set of data, namely a set of strings of words concerning which there is general agreement whether or not they are grammatical. The grammarian's task is then to construct a theory of grammar which accounts for the data [24, pp. 294–5].

Granting, for argument, that there are grammatical rules of the sort Priest is pointing towards, which are a matter of our intent, there is yet no separate grammatical realm

---

[9]Possibly the most prominent contemporary form of logical pluralism is due to Beall and Restall [21], which is supposed to be immune from problems of meaning variance. In [22], however, Hjortland shows that, to the contrary, this form of pluralism is wedded to meaning-variance.

[10]Dummett's argument against classical negation by means of harmonious constraints on inferential rules is a different way of approaching a similar suggestion:

> [the intuitionist] acknowledges that he attaches meanings to mathematical terms different from those the classical mathematician ascribes to them; but he maintains that the classical meanings are incoherent and arise out of a misconception on the part of the classical mathematician about how mathematical language functions [23, p. 17].

to which our theories of grammar aim at capturing.[11] Similarly for logical validity, we need not presuppose that there is a pre-existing "correct" account of the meaning of logical operators to which our logical theories aim. Rather, Priest is suggesting that we may presuppose only that there are substantive, non-arbitrary, theoretical constraints on the behaviour of logical operators, and that this behaviour will be captured by the correct logical rules. This is to take a different tack, suggesting that the task of formal logic is to develop systems that capture data from natural language reasoning, attempt to establish their validity (or otherwise) within logical theories, and also to formulate rules of inference for the behaviour of the logical constants involved. This view is often known as "anti-exceptionalism" (e.g. [25]), following the suggestion made by Haack [26] that '...logic is a theory, a theory on a par, except for its extreme generality, with other, "scientific" theories' (p. 26). The logical theory is subject to the same kind of constraints as any scientific theory, such as explanatory coherence and power, tractability, adequacy of fit to the relevant data, where paradoxes such as the Liar provide us with data for testing that adequacy.[12] Perhaps, on this story, a logical theory, (classical, intuitionist, paraconsistency and so on) is an attempt to establish both the canons of correct reasoning and also extensional adequacy for validity. As such, competing theories account for extensionally non-equivalent accounts of the class of valid arguments. The conflict, nevertheless, concerns the same subject matter.

Different logical theories count as valid different classes of arguments in ordinary language. This entails that a theory making substantive commitments regarding the meanings of the logical expressions that figure in those arguments (e.g. "not"), the rules that theory *supposes* to determine the content of an expression may turn out not to do so. As such, anti-exceptionalism takes a non-deductive approach to the justification of deductive logic, and has a significant set of hurdles to overcome in relation to the relationship between the way in which the normative account of logical validity is intertwined with this naturalistic account of explanation. For example, we would need an account of how the meaning of logical terms is stable across natural language use, whilst differing theories "home-in" on the correct explanation of that use, and, futhermore, whether or not that meaning is altered by the process of explanation itself.

The limitations of this approach are borne out by the views of its two main contemporary advocates, Graham Priest [27] and Timothy Williamson [28].[13] Both Priest and Williamson endorse an approach to logic that is "anti-exceptionalist". Williamson argues that classical logic is the correct logic on these grounds, whilst Priest argues that paraconsistent logic is a better fit according to the explanatory structure. As Hjortland [25] points out, both take exactly the same explanatory criteria

[11] Though, see Chap. 3 for references arguing that this approach to grammar is constitutively incorrect, due to the dynamic nature of grammar in interaction.

[12] See [11] for discussion of this view, which arguably originates with Quine.

[13] This is discussed by Hjortland [25].

for theory choice, yet they use these to ground distinct (and incompatible) logical theories.[14]

Even as a "fix" to the liar paradox, paracomplete and paraconsistent logics do not necessarily do as well as we might expect. This is because "revenge" paradoxes may arise depending upon how we are supposed to characterise their treatments of liar-like sentences. For example, if a paracomplete logician characterises a liar-like sentence as "indeterminate", then it is possible to construct a "revenge" paradox involving a sentence that says of itself that it is indeterminate. Similarly, for a paraconsistent logician, there is an issue with a sentence that says of itself that it is false *only*, and so does not conform to a treatment of self-referential sentences characterised as both true and false. The issue, for Priest, is that, whilst the sentence l may be taken to be both true and false, it is not possible to characterise sentences that are "just true".[15] One issue that arises is that, if it were possible for the paraconsistent theorist to express the notion of "just true", then it would also be possible to form "revenge" sentences involving these categories. That other logical approaches to paradox fall foul of revenge paradox is a common charge. Take, for example, Kripke's [34] theory, in which the above rules are said to partially define the truth predicate, whilst some sentences, such as l containing the predicate "true" are truth-value gaps. Supposing some rationale can be given for thinking that l falls into the category of a "gappy" sentence, l turns out not to be paradoxical. But, consider the sentence "$l_s : l_s$ is untrue". Paradox, seemingly, returns, given that a gappy sentence is, presumably, untrue. If l_s is true, then it is also untrue, yet if it is untrue (either false or gappy), then it is true. In [33], Priest's formulation of such revenge paradoxes is remarkably clear. A general schema for revenge problems is this: the language for which the solution is being given allows there to be a distinction made between those sentences that are true, and the rest. The Liar forces a sentence that is in the truths to also be in the rest. The original paradox assumed the rest to be equivalent to the false and solutions such as Kripke's suggest that the false is only a subset of the rest. Revenge problems show that the paradox can be reinstated for any such solution given the general formula: "$l_r : l_r$ is false or in the rest". But now, if Priest's language is capable of expressing that a sentence is "just true", or "just false", then sentences such as "$l_j : l_j$ is not just true" could be formed. This sentence, if treated in the same was as standard liar sentences quickly leads to overlap between just truth and just falsity, in which case "just true" does not seem to capture what was intended. The standard paraconsistent response is to accept the additional paradox as just an example of the way in which paradoxical sentences work (e.g. [11]), but this does seem to be intuitively unsatisfactory.[16] As such, the paraconsistent approach may also not even adequately solve the liar paradox since, as Soames [36, p. 7] puts it (in a more

[14] In part, as Hjortland notes, this could be put down to the fact that there is very little detail about how these criteria are supposed to be weighted and applied.

[15] For discussion of this issue, see [6, 29–33].

[16] There is also the additional issue of Curry's paradox, which has to do with a self-referential sentence that does not involve negation. In brief, if we have a transitive consequence relationship, and a detachable conditional that is residual of conjunction, then we can reason from a true statement to any arbitrary statement. As a result, the paraconsistent logician often weakens the conditional such

general context), 'the very activity of solving it in a particular case provides material for recreating it in a new and strengthened form ...once speakers are in a position to describe it as such, and to explain how the original version of the Liar is thereby blocked, they will be in possession of conceptual material that can be used to restate the paradox in a form resistant to the original solution'.

3 From Paradox to Paracoherence

Whatever the merits of the non-classical approach, there is a lingering problem which concerns the break between negation symmetry and the coherence of assertions and denials as a method of dealing with paradox. This is most clear when we consider how we should marshal our assertions in the context of these logics. For example, in a paraconsistent logic, we can not express rejection of a statement by means of its negation.[17] So, rejection must come apart from assertion of negation. This is required of both paraconsistent and paracomplete logics: $\neg\alpha$ can not be utilized to rule out that α may hold (paraconsistently), and neither can characterising certain sentences as "just false", as shown above. So, one option is to advocate "rejection" (or denial) where "deny α" is not reducible to the assertion of $\neg\alpha$.[18] Advocates of paraconsistent approaches, therefore, usually take assertion and denial to be exclusive states, so that denying α rules out the simultaneous assertion of α. Nonetheless, an agent can be committed to asserting certain contradictions. This is because asserting $\neg\alpha$ is not taken to be equivalent to denying α. A similar story can be told for the advocate of paracomplete logics where it is required that it must be possible to reject α without asserting $\neg\alpha$. Again, we can appeal to denial of α without also being committed to the assertion of its negation.

So, for the advocate of a paraconsistent logic, negation is weaker than denial, whilst for the paracomplete logician, negation is stronger than denial. As pointed out by Ripley [38], one reason that denial is important is that it is required for expressing disagreement. For example, since a paraconsistent logician can not disagree with the assertion α by asserting $\neg\alpha$ (since asserting $\neg\alpha$ does not rule out asserting α), so they must instead deny α, where the assertion and denial of α are exclusive states. Dually, a paracomplete logician cannot express disagreement with α by asserting $\neg\alpha$, so they must instead deny α (they may also deny $\neg\alpha$), where the assertion and denial of α are exhaustive states. Then, in order for assertion and denial to perform

(Footnote 16 continued)
that, for example, deduction theorem does not hold (universally) (see (e.g. [33]) and particularly the related discussion in [35]).

[17] For discussion, see [32].

[18] This is sometimes linked to arguments that there is pragmatic ambiguity of ordinary language negation. See [37] for cases in which the use of "not" in a natural language context indicates the rejection of an assertion without also indicating the assertion of the negation of the relevant sentence; e.g. "Some men are not chauvinists. All of them are"; "John isn't wily or crazy. He's wily and crazy". Though see [38] for arguments against this kind of data supporting ambiguity.

the roles that are required of them by advocates of paraconsistent and paracomplete logicians, we should still have available some form of coherence on those terms. For example, Ripley [38] suggests that the following two principles are widely assumed to hold by advocates of non-classical logics:

- The assertion of α and the denial of α are incompatible speech-acts;
- Having a settled opinion about α requires that one is willing to either assert α or deny α.

These together are equivalent with the definition of coherence give above in the context of Restall's framework. In case of classical logic, paradoxes involving negation coincide with paradoxes involving coherence of assertions and denials. This is no longer the case for non-classical logics, but, as might be expected, a strengthened paradox is then simple to construct. For example, we might follow Ripley [38] and just construct a statement whose assertion is equivalent to its own denial. Similarly, Murzi and Carrara [39] also show that by using analogues to Ripley's principles, and minimal resources, it is simple to construct a "denier" paradox suggesting that the first principle must fail for paraconsistent logics, and the second for paracomplete. In both cases, coherence is broken for non-classical logic, just as for classical logic.[19] Relatedly, Restall [1] shows that, by accepting certain basic principles, such as the rules supposed to define truth discussed above, it is simple to construct paradox in this context for either paracomplete or paraconsistent logics. So, the attempt to isolate paradoxes to those that arise for symmetric negation, from those that may arise at the level of coherence, breaks down.

To deal with the above, Ripley [38] argues for a position that he calls PARACOHERENCE, where '...a paraconsistent approach amounts to a paracoherent approach, in which one can be locally incoherent without global incoherence ...Taking the paracoherentist option I'm suggesting here amounts to choosing to maintain a certain amount of reflective tension'. In fact, this is not completely dissimilar to a comment made by Priest [11, Sect. 6.5] in response to paradoxes arising in relation to acceptance and rejection. There, Priest discusses clauses for what agents should accept and reject on the basis of evidence. He notes that a paradox arises for acceptance and rejection that mirrors the strengthened paradox often put in terms of "untruth" as opposed to truth. Priest argues that where there arises a conflict between acceptance and rejection, this is just a rational dilemma: 'Suppose that we have some α for which there is good evidence that it is both true and untrue. Then ...one ought to accept α and one ought to reject α. We have a dilemma, since we cannot do both'

[19]The additional issue that arises for advocates of logics where coherence and consistency come apart is that it is difficult to account for rational disagreement, see (e.g. [38]) for discusion and references to the relevant literature, where it is convincingly argued that non-classical logicians can not simply outlaw statements whose assertion is equivalent to its own denial. It is worth noting that a further complication arises for at least some non-classical logics, since it is not just that their interpretation of negation differs (e.g. [11]), but also denial does not carry the same meaning. For example, for the intuitionist, denying α is non-constructive such that α may be proved later, whilst for the co-intuitionist, denying α is synonymous with the constructive refutation of α (the status of α is fixed), whilst asserting α is non-constructive.

[11, p. 110]. The distinction that Ripley is suggesting is that, in some marginal cases involving paradoxical statements, agents are put in a disequilibrium position where, '...someone who adopts it disagrees with themself'. No additional characterisation of such states is provided, but I am not sure what it would be to "disagree with oneself", unless, of course, we have a story about the pragmatics of agreement and disagreement *between* agents, and in which these pragmatics are internalised by a single agent. Of course, this is exactly the kind of story that we have provided in this book. In this context, these disequilibrium states are no longer confined to the outskirts of discourse, but rather, they are at the base of every interaction where any assertion or denial is met with a symmetric, dual, move. At least to the degree to which these forms of disequilibria are understood to arise across all forms of reasoning, this coheres with the idea that paraconsistency and paracompleteness are best understood as forms of instability, involving "provisional states", which is advocated in [40]. The kind of statements that paraconsistent and paracomplete logics are built to deal with are often confined to paradoxes, it is argued there that such reasoning is also required for ordinary reasoning, particularly in the context of areas of inquiry such as scientific, or even everyday, belief. Moreover, it is also argued in [40], that in keeping with the inferentialist approach, such reasoning is required to deal less with alethic notions, and more with epistemological notions such as rationality, dialogue and scientific theorizing.[20] Aside from paradox, it is common to think that a domain concerning rationality such as scientific reasoning or dialogue may be inconsistent and incomplete. In these contexts, it is not unusual to have "some" reason (or some evidence) for the statement $\neg\alpha$, without also ruling out α. This is a state of non-equilibrium, where we do not have a settled (coherent) state regarding α. As Carnielli and Rodrigues [40] put it, 'There is no commitment to the truth of the sentences that express the contradiction. They may well be taken as a provisional situation that can be decided later, at least in principle'. It does not seem correct, then, to say that this is equivalent to a "true" contradiction:

> The act of setting up a non-explosive logic that does not produce triviality in the presence of contradictions implies neither a philosophical sympathy for contradictions, nor the rejection of contradiction avoidance as a criterion of rationality. We endorse the view that a contradiction may be taken as a provisional state that should, at least in principle, be eliminated by means of further investigation. In other words, the contradictions that are accepted in paraconsistent logics may be understood from a purely epistemological viewpoint [42].

Dually, it is not unusual to have "some" reason (or evidence) against a statement α, without ruling in $\neg\alpha$. Again, this is a state of non-equilibrium that should not be read alethically as saying that both α and $\neg\alpha$ are false. This is a provisional, or non-equilibrium state, precisely because upon further investigation, it may well be the case that a proof (or refutation) of α is found, or sufficient evidence to determine whether or not α is true, or α is false.[21]

---

[20] See also (e.g. [41]).

[21] They go on to note an analogy with intuitionistic logic, where, if there does not exist a constructive proof for some mathematical statement, then excluded middle does not hold for formulas involving

In the context that we have developed in previous chapters, we have a precise explanation by which we can fruitfully employ the notion of paracoherence since interactions are inherently predicated upon disequilibrium states. Moreover, we may also conjecture that it is the paracoherence of the symmetric relationship between assertions and (counterpart) denials that gives rise to the impetus towards justification and agreement. Of course, this is a rather different view of coherence to Restall's, which may, rather be understood as much more like a static snapshot of the dynamics of interaction. It is also a rather different view of paracoherence to Ripley's, which may, rather be understood as much more as a monological internalisation of the dynamics of interaction.

4 Paracoherent Interactions

The view adopted in this book is to take rather a different tack to the adoption of non-classical logics in the face of paradox. This is not a revision of logic, rather it is an inquiry into the heart of logic, where we find not a monological structure in static platonic beauty, but, rather, a messy dynamics that is driven by internal tensions. Whatever logic is, it is not just a matter of capturing valid syntactic forms of reasoning by means of autocratic rules. As such, the programme in consideration here is one, not of logical *revision*, but of the *reconstruction* of logic.

The diagnosis of paradoxes such as the liar by advocates of non-classical logics is to say that there is a problem with the rules taken to determine the meaning of negation, such that negation and coherence come apart. However, the fact that paradoxes involving coherence are simple to construct from principles that it looks like non-classical logicians require, suggests that the diagnosis may be wrong. As Ripley and Carnielli suggest, perhaps the correct diagnosis requires looking at the internal *tensions* that coherence brings with it, that is to say, the dynamics of reasoning. Recall that Restall's definition of coherence requires a "cut" rule:

$$\frac{\Gamma, \alpha \dashv\vdash \Delta \qquad \Gamma \dashv\vdash \alpha, \Delta}{\Gamma \dashv\vdash \Delta}\ (\text{Cut})$$

This is captured, informally, by the first of Ripley's constraints over coherence, that the simultaneuos assertion of α, and denial of α, is incoherent. The rule captures this notion of incoherence by the fact that empty sequent is entailed whenever we are in such a situation. If the rule is supposed to be in place at all stages of reasoning, including those including the sort of tensions observed by Ripley and Carnielli, then any such tension will immediately lead to incoherence. In other work, Ripley [7] pursues exactly this line of thought, investigating logics in which cut is not universally

(Footnote 21 continued)
it. But, this is a provisional state since, if we are able to provide a constructive proof at a later date, then excluded middle may be applied.

valid. This form of non-transitivity gives rise to a theory of paradoxical sentences in the context of a classical logic extended with a truth-predicate defined by the rules given above. In fact, this is precisely where we would expect Ripley's investigation to end up, insofar as he identifies the conditions under which paracoherence occur in an inferentialist framework that is *monological*. However, just as disagreeing with onself is practically nonsensical, the kind of tension that is identified by the notion of paracoherence is better explained by means of the *dialogical*. In this setting, the diagnosis of paradox as indicating dynamic tension also shows us that the notion of coherence at work in Restall's account is one that *eradicates the possibility of dynamic tension*. That is, the definition of coherence given in Restall (and Ripley) does not work in the context of dynamic interaction, since it would require a positive move involving some α, and a negative move involving α to simply rule each other out, as if nothing had occured. The obvious issue with this is that ruling things out does not come for free in the framework suggested here, rather, justification is required to rule things in or out. As such, and in order to bring to the surface this process by which moves are made as tests of each other, a form of coherence which is *predicated* upon the pitting of counterpart formulas against each other is required.

The definition of terminal-cut given in Chap. 4 mirror's Restall's, but in this setting it formalises the idea that an interaction shares focus such that a positive move in support of α^+ should be opposed to a negative move in support of α^-.[22] Terminal-cut, moreover, is admissible *subsequent* to the termination of an interaction, at which point the characterisation of an interaction may be observed more directly. So, for example, the set of (focal) formulas built up over the course of an interaction (A^+, A^-) should be coherent such that terminal-cut is admissible, and the interaction terminates with either:

$$\frac{A^-; \vdash\!\dashv; \qquad ; \vdash\!\dashv; A^+}{\alpha_m^-; \vdash\!\dashv;} \text{ (Terminal-cut)}$$

or;

$$\frac{A^-; \vdash\!\dashv; \qquad ; \vdash\!\dashv; A^+}{; \vdash\!\dashv; \alpha_m^+} \text{ (Terminal-cut)}$$

As we said in Chap. 4, the convergent formula (α_m^+, or α_m^-) of such an interaction that the formula is JUSTIFIED by the interaction, since it is this on which which the interaction terminates with an agreement between players. The above definition of a terminating coherent interaction ensures that, for every justified formula α, it has been tested by an attempt to justify the counterpart formula α°. Thus, terminal-cut operates as a means by which interactions can be said to *converge* towards justification. As such, if we think of this process of justification in terms of proofs and refutations, then the system effectively combines the motivations of *both* paraconsistent logics *and* paracomplete logics, whilst generalising the two. For example, the paraconsistent suggestion is generalised to the idea that (coherent) reasoning *always* involves tension

---

[22] Recall, for example, the discussion of the problem of cut in symmetric classical logic in Chap. 2.

between moves that are dual to each other, whilst the paracomplete suggestion is generalised to the idea that no formula is justified unless it is the result of a process of justification.

4.1 Local Truth and Falsity

Let us further consider the constructive nature of the interactive account under discussion here. Consider that in a typical constructive logic such as L_I, we say that a formula α is decidable iff $\alpha \vee \neg\alpha$ holds for α; we say that a formula α is stable iff $\neg\neg\alpha$ implies α. Neither of these is a theorem of L_I, though they hold in certain contexts. By analogy with this, in the current structure let us say that a formula α is *determined* whenever there exists a justification of that formula by means of a terminating coherent interaction (TCI) (bearing in mind that the justification may be positive, or negative). Let us call such a formula LOCALLY TRUE or LOCALLY FALSE, since assessments of the formula are always radically context dependent.

What is important about the set of formulae which are determined is that we are then in a position to *know*: the proof of α^+ *rules out* any further refutation attempts of α^-; or the refutation of α^- *rules out* any further proof attempts of α^+.[23] At termination, we have a conclusive result to the process of dialogue, which puts an end to doubts regarding α. Take, for example, an interaction that terminates with a positive justification of α. At that point, we have a set of formulas A^+ that provide the justification of α, together with a set of formulas A^- that are *tests* of A^+, whilst the formulas of A^+ are, by symmetry, also tests of A^-. The interaction (A^+, A^-) is said to both interpret *and* justify α such that it is locally true. So, in an abstract sense, this is just an extension of the constructive notion that a proposition is identified with its set of proofs. By definition, local truth and local falsity rule each other out, so, in this case we must be able to apply terminal-cut in this form (taking liberty with formalisation, as in Chap. 4):

$$\frac{A^-; \vdash\!\dashv; \qquad ; \vdash\!\dashv; A^+}{; \vdash\!\dashv; \alpha^+}\ \text{(Terminal-cut)}$$

Furthermore, at exactly this stage where we are no longer interested in the dynamics of interaction, terminal-cut operates much like Restall's cut. That is to say, at the place where reasoning may be considered as an object (subsequent to interaction), coherence in Restall's sense holds for it. In the terminology of Restall and Ripley, this suggests that paracoherence characterises interactions *prior to determination*, at which point coherence holds.

[23]The notion of a "conclusive defense" is also discussed in [43].

So, at the stage for which determination holds, local truth rules out local falsity, and vice-versa. So, we have a situation in which bivalence holds, which is, itself, *emergent* from a definitionally non-bivalent structure. As such, over these, and only these, formulas, we can examine their relationship in terms of a more objective relationship between propositions. In fact, we can express the relationship between the local truth and falsity of some proposition α by means of a kind of metalinguistic negation operation, (as suggested in the context of topos-theoretic structures in the previous chapter). Spelling out this relationship in this context: if, for some $\alpha^{+,-}$, α^+ is locally true, then α^- is *not* locally false; if α^- is locally false, then α^+ is *not* locally true. As such, we can define the following *external* negation operation, $\sim$:

Definition 80 For all and only those formulas of S which are locally true, or locally false, a metalinguistic negation operation, $\sim$ may be defined as follows:

- $\sim \alpha^+ := \alpha^-$
- $\sim \alpha^- := \alpha^+$
- $\sim\sim \alpha^{+,-} = \alpha^{-,+}$

Since $\sim$ is an external negation, it negates a formula as a whole, rather than some component of that formula. So, here, it carries the simple meaning that, for example, $\sim (\alpha^+ \wedge \beta^+)$ is equivalent to $(\alpha^- \wedge \beta^-)$. In English, this just says that, whenever "α and β" is locally true, we also know that "α and β" can not be locally false.

Whilst this form of bivalence holds for propositions, this approach radically alters the structure of paradox. Thinking of paracoherence as just what characterises interactions *in process* also makes transparent the way in which the supposed stability of monological approaches to logic will inevitably lead to paradoxical statements. Consider, for example, that, for any *determined* statement, not only is its determination always local, but also it must be justified by means of a process of TCI with a set of dual tests. But, in order that a statement pass all of the tests so that it can be the result of such an interaction, there is no way to construct paradoxical statements for propositions. The suggestion here is that what are commonly *called* paradoxical statements arise whenever the dynamics of interaction is rendered static and monological. Propositions, seen as the result of dynamic interactions, can not be involved in paradoxical statements by definition. It is also possible to see why negation is thought to be involved in paradox, since, negation consistency has typically been seen as commensurate with coherence between assertions and denials. Here, the duality between positive and negative moves gives rise to the emergent property of metalinguistic negation over propositions.

To show that this is a very different approach to paradox, consider a comparison with Priest's [44] generalised diagnosis of paradox by means of the "inclosure schema". Priest argues that there is a common structure to self-referential paradoxes called the inclosure schema. The inclosure schema takes two predicates P and Q, and a function f such that:

- $\Omega = \{y|P(y)\}$ exists and $Q(\Omega)$ (Existence)
- If $x \subset \Omega$ such that $Q(x)$ holds, then:

- (a) $f(x) \notin x$ (Transcendence)
- (b) $f(x) \in \Omega$ (Closure)

In general, the schema provides a function to Ω that takes us both inside, and outside of Ω. Whenever the conditions are satisfied, contradiction follows since Ω is trivially a subset of itself, and $Q(\Omega)$ holds, then we have both $f(x) \notin x$, and $f(x) \in x$. So, for example, Russell's paradox is generated by setting: $P(y)$ to be the predicate $y \notin y$; $Q(\Omega)$ the universal predicate true of any object; f the identity function. The liar paradox is generated by setting: $P(y)$ to be the predicate "y is true"; $Q(\Omega)$ is the predicate "x is definable"; $f(x)$ is the sentence "this sentence does not belong to the set x". So, $\Omega = \{y|P(y)\}$ is the set of true sentences, and $f(\Omega)$ is a form of the liar sentence that says "this sentence does not belong to the set of true sentences". According to Priest, this schema provides a structure by which to account of all forms of self-referential paradox.[24]

It is important to recognise that the notion of limit plays a central role in this approach, which Priest characterises as going *beyond* the limits of thought. For example, with the version of the schema instantiated by the liar, we must first accept a totality of true sentences in order to even define notions of transcendence and closure. The fact that certain things may be defined to live "on the border" of this totality is just an artefact of the decision to consider a domain to be predetermined prior to any kind of dynamic process of reasoning. This is important, since it means that Priest's dialetheism is much more conservative than it may seem to be at first sight. The idea that paradoxes occur only at the very limits of thought is one that also runs through far more conservative reponses to paradox, such as, for example, the blocking of paradoxes by fiat in ZF-set theory and Russell's theory of types. The idea that paradoxes can be found only at the "limits", brings with it the idea that they can also be sealed-off from the rest of our thinking.[25] This is how Wittgenstein [46] considered paradoxes, such that they are marginal, having little or nothing to do with ordinary language use: a contradiction is not to be understood as a 'local symptom of a sickness of the whole body' (p.104). In distinction, the approach to *paracoherence* put forward in this chapter might be thought of as arguing that antinomies are at the very basis of reasoning, and across the board. The pitting of counterpart formulas against each other is exactly the internal tension that propels the dynamics of reasoning forward. Paradoxes are simply the creatures that appear when we attempt to look at this structure as if it were both monological and static. This is to take an approach that is much closer to Hegel, who says that: '...thought, in its very nature is dialectical, and that, as understanding, it must fall into contradiction' [47, p. 15]. In fact, without the disequilibrium state that is constructed by means of the duality between positive and negative moves, there would be no impetus towards justification through reasoning together.

[24]See (e.g. [45]) for critical discussion of Priest's argument.

[25]See, for example, the extensive discussion on "classical recapture" (e.g. [6, 33]).

4.2 Logic and Truth

To make this distinction (and order of determination) a little more intuitive, it is helpful to return to the consideration of non-classical logics in the face of paradox. I will not attempt to provide a full account of truth (or paradox) here, but will mention a few issues. Since we are thinking of truth and falsity as radically local, and immanent to an operation between agents, it may be better to drop the terminology altogether insofar as it brings with it notions of universalism, objectivism, and representation. Moreover, as Raatikainen [19] points out, there is no simple way to identify truth with proof in constructive logics, since, both actualist and potentialist approaches bring different problems with them. Perhaps the closest approach to the ideas on offer here is in the pragmatic approach to truth argued for by Peirce (e.g [48]). For example, Peirce argues that:

> Truth is that concordance of an abstract statement with the ideal limit towards which endless investigation would tend to bring scientific belief, which concordance the abstract statement may possess by virtue of the confession of its inaccuracy and one-sidedness, and this confession is an essential ingredient of truth [48, Sect. 5.565].

In Peirce's approach, truth is defined as that which is believed at an ideal limit of inquiry:

> Different minds may set out with the most antagonistic views, but the progress of investigation carries them by a force outside of themselves to one and the same conclusion. This activity of thought by which we are carried, not where we wish, but to a foreordained goal, is like the operation of destiny. No modification of the point of view taken, no selection of other facts for study, no natural bent of mind even, can enable a man to escape the predestinate opinion. This great law is embodied in the conception of truth and reality. The opinion which is fated to be ultimately agreed to by all who investigate, is what we mean by the truth, and the object represented in this opinion is the real. That is the way I would explain reality [48, Sect. 5.407].

As Legg [49] points out, this is a very different approach to truth from the instrumentalism offered in many pragmatist approaches. Peirce's general idea can be fleshed out in terms of the suggestion that, ultimately, interactions converge such that local truths may become embedded in a wider structure that converges upon a more global account of truth. Importantly, this also allows us to make sense of an approach to truth that is somewhere between the ontological loading of correspondence theories of truth and deflationary theories of truth, since 'our holding p to be true means that we expect that future inquiry will converge on p' [49, Sect. 2]. To put this in the language that we have been using throughout this chapter; that α is locally true means that we expect future interactions to converge upon α. This approach to a form of ideal resolution over all interactions is, effectively, a way of ensuring that reasoning does not come to an end, much more than it is a 'perfect epistemic resting place' as Legg [49, Sect. 3] puts it in relation to detractors of Peirce's view. This also provides a way of ensuring that, whilst radically local, truth may yet be understood to be to some extent independent of agents' beliefs. As Peirce puts it: 'reality is independent, not necessarily of thought in general, but only of what you or I or any finite number of (persons) may think about it ...' [48, Sect. 5.408]. That is to say, it is the *infinity* of the community of inquiry, as Legg [49] describes it, which ensures that truth is never defined as static or independent of interaction between agents, nor is truth defined by the belief of a finite set of agents.

Let us briefly return to the way in which the liar paradox was constructed by means of rules defining a transparent truth-predicate. It should be clear, from the discussion so far, that we can not introduce a transparent truth-predicate across the entire structure, even if that truth is considered to be local. That is, simply because we have defined local truth by means of what is produced by a TCI. For the sake of argument, let us introduce local truth and falsity introduction rules as follows:

- $T_I = \overline{\alpha} \rightarrow T(\ulcorner \alpha \urcorner)$
- $F_I = \overline{\alpha^{\perp}} \rightarrow F(\ulcorner \alpha \urcorner)$

where $\overline{(-)}$ denotes a formula that is the result of a TCI, i.e. a proposition. To some extent, this follows the constructivist approach to truth (and, for us, falsity), such that, for a proposition α to be "true" (or "false") consists in there existing a terminating coherent interaction ending with a positive (or negative) focal formula α. These "rules" are admissible by the definition above, though they are effectively meaningless, telling us no more than we know already. Even so, it is certainly not obvious that truth and falsity elimination rules could be admissible. For example, let us consider elimination rules as follows:

- $T_E = (\ulcorner \alpha \urcorner) \rightarrow \overline{\alpha}$
- $F_E = (\ulcorner \alpha \urcorner) \rightarrow \overline{\alpha^{\perp}}$

The reason that these rules are inadmissible, even for local truth and falsity, and even only for propositions defined in terms of TCI, is that, according to the Peircean account outlined above, we would surely want to allow for the existence of statements that are (locally) true or false, but for which we do not have a TCI. In fact, absent this restriction, we would have no way of accounting for the idea of local truth *at all*, since we would be left with a situation in which truth and falsity may only be defined by means of whatever TCI we are currently interested in. Thus, the radical context dependence of interactions would then give rise to the radical context dependence of truth itself. It is only by allowing for there to be a wedge between local truth and falsity, and truth and falsity defined by means of convergence towards an ideal resolution, that it is possible to ensure that truth and falsity are not reduced to finite agential interactions. In this sense, again, it is a requirement of the interactive structure that antinomies are part and parcel of the way in which agents *approach* the truth.

References

1. Greg Restall. Assertion, denial and non-classical theories. In Francesco Berto, Edwin Mares, Koji Tanaka, and Francesco Paoli, editors, *Paraconsistency: Logic and Applications*, pages 81–99. Springer, 2013.

2. J Michael Dunn and Gary Hardegree. *Algebraic methods in philosophical logic*. OUP Oxford, 2001.
3. Jean-Yves Béziau. Sequents and bivaluations. *Logique Et Analyse*, 44(176):373–394, 2001.
4. Greg Restall. Truth values and proof theory. *Studia Logica*, 92(2):241–264, 2009.
5. Kent Bendall. Negation as a Sign of Negative Judgment. *Notre Dame Journal of Formal Logic*, 20(1):68–76, 1979.
6. J. C. Beall. *Spandrels of Truth*. Oxford University Press, 2009.
7. David Ripley. Paradoxes and failures of cut. *Australasian Journal of Philosophy*, 91(1):139–164, 2013.
8. Igor Urbas. Dual-intuitionistic logic. *Notre Dame Journal of Formal Logic*, 37(3):440–451, 1996.
9. Nicholas Goodman. The logic of contradictions. *Zeitschrift fur Mathematische Logic und Grundlagen der Arithmetik*, 27:119–126, 1981.
10. Cecylia Rauszer. Applications of Kripke models to Heyting-Brouwer logic. *Studia Logica*, 36(1-2):61–71, 1977.
11. Graham Priest. *Doubt Truth to Be a Liar*. Oxford University Press, 2006.
12. Willard Van Orman Quine. *Philosophy of logic*. Harvard University Press, 1986.
13. W. V. Quine. Two dogmas in retrospect. *Canadian Journal of Philosophy*, 21(3):265–274, 1991.
14. Vann McGee and Brian McLaughlin. Distinctions without a difference. *Southern Journal of Philosophy*, 33(S1):203–251, 1995.
15. W. V. O. Quine. *Philosophy of Logic*. Harvard University Press, Cambridge, MA, 1970. Second edition, 1986.
16. Hartry Field. *Saving Truth from Paradox*. Oxford University Press, Oxford, 2008.
17. Timothy Williamson. Equivocation and existence. *Proceedings of the Aristotelian Society*, 88(n/a):109–127, 1988.
18. Michael Hand. Classical and intuitionistic negation. *Southwest Philosophy Review*, 8(1):157–164, 1992.
19. Panu Raatikainen. Conceptions of truth in intuitionism. *History and Philosophy of Logic*, 25(2):131–45, 2004.
20. Matti Eklund. The multitude view on logic. In *New Waves in Philosophical Logic*, pages 217–240. Springer, 2012.
21. J. C. Beall and Greg Restall. *Logical Pluralism*. Oxford University Press, 2006.
22. Ole Thomassen Hjortland. Logical pluralism, meaning-variance, and verbaldisputes. *Australasian Journal of Philosophy*, 91:355–373, 2013.
23. Michael A. E. Dummett. *The Logical Basis of Metaphysics*. Harvard University Press, 1991.
24. Graham Priest. Two dogmas of quineanism. *Philosophical Quarterly*, 29(117):289–301, 1979.
25. Ole Thomassen Hjortland. Anti-exceptionalism about logic. In *St. Andrews*, 2015.
26. Susan Haack. *Deviant Logic, Fuzzy Logic: Beyond the Formalism*. University of Chicago Press, 1996.
27. Graham Priest. Revising logic. In P. Rush, editor, *The Metaphysics of Logic*, chapter 12, pages 211–223. Cambridge University Press, 2014.
28. Timothy Williamson. Logic, metalogic and neutrality. *Erkenntnis*, pages 1–21, forthcoming.
29. J. C. Beall. Shrieking against gluts: The solution to the 'just true' problem. *Analysis*, 73(3):438–445, 2013.
30. Francesco Berto. Absolute contradiction, dialetheism, and revenge. *Review of Symbolic Logic*, 7(2):193–207, 2014.
31. Anthony Everett. A dilemma for priest's dialethism? *Australasian Journal of Philosophy*, 74(4):657–668, 1996.
32. Terence Parsons. Assertion, denial, and the liar paradox. *Journal of Philosophical Logic*, 13(2):137–152, 1984.
33. Graham Priest. *In Contradiction: A Study of the Transconsistent*. Oxford University Press, 2006.
34. Saul A. Kripke. Outline of a theory of truth. *Journal of Philosophy*, 72(19):690–716, 1975.

35. J. C. Beall. Free of detachment: Logic, rationality, and gluts. *Noûs*, 49(1), 2013.
36. Scott Soames. *Understanding Truth*. Oxford University Press, 1999.
37. Jamie Tappenden. Negation, denial and language change in philosophical logic. In D Gabbay and H Wansing, editors, *What is Negation?*, pages 261–298. Springer, 1999.
38. David Ripley. Embedding denial. In Colin Caret and Ole Thomassen Hjortland, editors, *Foundations of Logical Consequence*, pages 289–309. Oxford University Press Uk, 2015.
39. Julien Murzi and Massimiliano Carrara. Denial and Disagreement. *Topoi*, (1):1–11, 2014.
40. Walter Carnielli and Abilio Rodrigues. On philosophical motivations for paraconsistency: An ontology-free interpretation of the logics of formal inconsistency. 2015.
41. Newton Da Costa and Steven French. *Inconsistency in science: A partial perspective*. Springer, 2002.
42. Walter Carnielli and Abilio Rodrigues. Towards a philosophical understanding of the logics of formal inconsistency. *Manuscrito*, 38(2):155–184, 2015.
43. Ralph H. Johnson. The ambiguous relationship between pragma-dialectics and logic. In F. H. van Eemeren, Peter Houtlosser, Haft van Rees, and A. M., editors, *Considering Pragma-Dialectics: A Festschrift for Frans H*, page 127. L. Erlbaum Associates, 2006.
44. Graham Priest. *Beyond the Limits of Thought*. Oxford University Press, 2002.
45. Haixia Zhong. Definability and the structure of logical paradoxes. *Australasian Journal of Philosophy*, 90(4):779–788, 2012.
46. Ludwig Wittgenstein. *Remarks on the Foundations of Mathematics*, volume 7. B. Blackwell, 1978.
47. Georg Wilhelm Friedrich Hegel. *Hegel's Logic, trans. William Wallace*. Oxford: Clarendon Press, 1975.
48. Charles Sanders Peirce. *Collected papers of Charles Sanders Peirce*, volume 5. Harvard University Press, 1974.
49. Catherine Legg. Charles peirce's limit concept of truth. *Philosophy Compass*, 9(3):204–213, 2014.

Chapter 8
Naturalization, Normativity, and Ethics

In this, concluding, chapter, I suggest a number of ways in which the view of logic suggested here can be broadly considered in relation to other theories of reasoning. To do so, the interactive approach discussed in this book will be situated in relation to the project of the naturalisation of logic argued for, primarily, by John Woods and Lorenzo Magnani. On their view, reasoning is typically non-deductive, and there may be no appeal to "reason" as a normative domain (typically supposed to be governed by logic) as holding over the way in which human agents *actually* reason. I argue that, whilst the interactive view shares some of the critical gesture of their work, it does, in fact, offer a means by which normative authority can, and indeed *should* hold so that reason and reasoning can not be collapsed into one another. In discussing these issues, I garner support from enactive and interactive theories of sub-personal activities, as well as the ethical and political dimensions of dialogue and social reasoning.

1 Naturalising Logic

A programme that bears many philosophical similarities to the one argued for in the previous chapters, yet with important differences, is the NATURALISATION of logic pursued primarily by John Woods (e.g. [1–3]) and Lorenzo Magnani (e.g. [4–6]). As such, it is useful to situate the interactive account of reasoning given here with their project. There is significant overlap between the critical approach that is taken to motivate the naturalisation of logic, and the criticisms of orthodox logic with which this book began. For example, Woods advocates an approach to logic that answers the challenge set out by Harman (discussed in the introductory chapter), which problematises the relationship between logic and reasoning. But, in distinction to the current volume, this is used as ammunition to argue for an account of reasoning and logic that goes hand-in-hand with the project of naturalised epistemology, which began in the work of Quine. This, according to Woods, brings with it a "third-

J. Trafford, *Meaning in Dialogue*, Studies in Applied Philosophy, Epistemology and Rational Ethics 33, DOI 10.1007/978-3-319-47205-8_8

way" approach to reasoning, which is often *non-deductive* and *non-inductive*. In this setting, 'consequence is the wrong paradigm' [3, p. 242], as Woods puts it, because the modelling of "premise-conclusion" reasoning does not usually operate according to the canons of deductive consequence. This latter point is practically the starting point for this monograph, so is something with which I wholeheartedly agree. Where I depart from Woods is in the collapsing of the normative role of reason into the natural processes of reasoning as carried out by *individual* agents. Nonetheless, the impetus to pursue the naturalized project is similar, in broad strokes to the arguments made at the beginning of this book:

> I want to know how the theorist knows that closure under consequence is normatively binding on the belief updates of real people [...] I have argued for many years that this question has yet to meet with anything close to an adequate answer, whether in heavy-equipment logic or informal logic. So let me simply state without much further ado:
> DOUBTFUL NORMATIVE LEGITIMACY: The normative legitimacy of heavy-equipment models of the intelligent behaviour of human individuals presently lacks a secure foundation. Their normative bona fides are in doubt. [2, p. 16]

This criticism of the normative legitimacy of logic clearly coheres with the general line of thought pursued here, particularly with the idea that 'orthodox logic has no people in it' [2, p. 12]. The response that Woods takes is *not* to attempt to repair the disparity between ordinary reasoning and the supposed normative authority of orthodox logic by making minor alterations to orthodox logic as has been pursued in non-monotonic and non-classical logics such as paraconsistency. For example, as Woods points out, even in the act of making such repairs, 'they will evade capture in the absence of a thorough and disciplined examination of the ground-data; and this alone means that logic will have to lighten up in its haughtiness towards the empirical' [2, p. 29]. I consider this latter point important, insofar as empirical data regarding the way in which agents reason is of obvious import for any project that takes reasoning seriously. The key difference between this, and the approach suggested in this book, is that I have argued that reasoning is irreducible to the individual, and, as such, is also irreducible to the empirical processes by which individual reasoning is considered to occur. Let us consider this in detail.

1.1 Collapsing the Normative into the Natural

The central claim in Woods' argument is that reasoning (and logic) should be based on empirical science to the point where there is a 'convergence of the normative on the normal' [6, p. 15]:

> As a first pass, and when there aren't particular reasons to the contrary, how we do reason from premises to conclusions is typically how we should reason. In other words, in matters of consequence-drawing there is a trending convergence between the normative and the normal, between what is usually done and what is rightly done [3, p. 52].

This convergence is supposed to operate on the kind of reasoning that Woods characterises as "premise-conclusion" reasoning, which involves an agent taking an input and producing an output.[1] As such, the revision of orthodox logic proposed is *grounded* in a renewed relationship with cognitive science allowing for the inferential behaviour of actual human agents to be characterised in "natural" terms. This requires, according to Woods, that logic is "empirically sensitive": 'At a minimum, the decision to naturalize the logic of reasoning is a decision to take into account well established lawlike results of the cognitive sciences' [3, p. 62].[2] Further still, logicians should, under pressure of empirical theories, attempt to explain, revise, and eliminate aspects of logical structures, so: 'a logic's empirical disconformities with the data and findings of the partner sciences be accounted for, under pain of having to give them up' [3, p. 6].

The normative-normal convergence, then, is distinct from similar approaches such as those involving a kind of reflective equilibrium.[3] On the latter approach, logical theories are still understood to exert normative pressure over agents insofar as agents' behaviour *implicitly* follows the correct logical processes according to the theory in question. This is not altogether dissimilar to the dispositionalist approach discussed in Chap. 1, wherein logic is supposed to provide norms over reasoning that are *external* to the process of reasoning, whilst these norms are also supposed to be implicit within those processes. As argued in Chap. 1, the fissure between the actions of human agents (and this alone is an idealisation in the form of a generalisation), and the supposed norms of logic, is so great as to always bring with it arguments from circularity. That is, in order to "write-away" all of those occasions in which we do not live up to the norms that are supposedly implicit in our behaviour, we end up simply "writing-in" a set of external norms that have no justification in that behaviour.[4]

Woods' response, unlike our own, is to jettison any appeal to the normative whatsoever, so that "reason" is identifiable with "reasoning". As such, the normative-normal convergence entails that 'in the absence of particular reasons to the contrary that humans reason well when they reason in the ways that humans normally reason in the conditions of real life' [3, p. 423]. Woods, rightly to my mind, suggests that it is important to consider the way in which cognitive science illuminates the tacit, and heuristic, processes of agents' activities so that we can understand the interplay between computational processes that are often automatic and non-conscious, with linguistic behaviour. Magnani [6], for example, argues that reasoning is often "dark"

[1] As Magnani points out, in this, and in many senses, the project of naturalisation is co-extensive with recent approaches to artificial intelligence that take a "bottom-up" approach in contrast to traditional symbol-processing approaches (e.g. [7]). This is discussed further in note 24.

[2] I will not take issue with this characterisation of the "lawlike" nature of cognitive science, though it is worth flagging up the difficulties in characterising cognitive science in these terms, and science more generally. For a recent discussion of related issues, see [8].

[3] See the discussion in [6].

[4] One response to this that we have not really mentioned so far is to say that logic is, to some extent, sui generis. This is perhaps most clear in the "universal logic" project begun by Jean-Yves Beziau, amongst others. This project is of clear mathematical interest, but it often does not have any clear articulation of the relationship between this technical study of the machinery of logic and the *use* of logic. Somewhat polemically, Woods [2] argues that the: 'factor of good-for recedes into the background, and technological self-study becomes sui generis, and withal the route to the upper elechons of academic achievement and repute.'.

(p. 124), in the sense that unconscious inferences are central to human reasoning, so we need to take into account the: 'unconscious, sublinguistic, inattentive, involuntary, automatic, effortless, non-semantic, computationally luxuriant, parallel, and deep' [3, p. 23].[5] With this in mind, there seems to be a fairly simple dilemma with respect to orthodox logic.[6] For, if there are supposed to be *rules* of logic that have to do with the way in which agents actually reason, then these rules look overly complex and draconian to actually be instantiated by agents. If, instead, we look to the unconscious and automatic ways in which agents reason *as already* reasonable, then this sort of reasoning does not appear to follow any sort of rule structure at all.

An immediate issue, however, with Woods' suggestion that we should collapse *reason* onto *reasoning*, is that there are many areas of reasoning in which it looks as though we would want to be able to *appeal* to some form of the normative that goes beyond that which is simply instantiated in the activities of agents. In fact, Magnani (e.g. [6], p. 15n.2), disagreeing with Woods, suggests that data on biases of judgment provide reason to think that there *some* areas of reasoning where it looks as though the normative and the natural do not coincide. There is a web of interrelated issues here, many of which are connected with the discussions of previous chapters. For example, as I argued in Chap. 1, it is not so simple to say that there is a normative-natural disparity in the context of data on biases of judgment, since this raises the spectre of rule-circularity. Equally, giving up any wedge between reasoning activities and some kind of norms does not also seem viable, nor necessarily motivated by the criticisms that Woods and Magnani marshall against "orthodoxy". For example, Woods [3] spends substantial time attempting to "explain away" traditional fallacies of reasoning, such as affirming the antecendent. Now, since reasoning is supposed, on their view, to include so called "fast and frugal heuristics", with limitations on processing time and incomplete knowledge, hasty generalization and abduction look like central aspects of how human agents actually reason. As Woods (and Magnani) note, these are not deductively valid inferences (though as I showed in the introductory chapter, they are also not at any disadvantage from the point of view of justification). But, on their approach, since such inferences are used by actual agents, they should not be considered to be fallacious: 'In a sense, there is nothing to correct, even when we are aware of the error in reasoning we are performing. Analogously, there is nothing to complain about ourselves, even if in some sense we are aware of the possible deceptive character of the reasoning and hypothetical cognition we are performing' [6, p. 24].[7] The dialectic of the argument moves from criticism of the supposed normativity of orthodox approaches to logic to the argument that there is

---

[5]Magnani suggests that this also gains support from Gabbay and Wood's [9] discussion of abductive reasoning where: 'they indeed stress the function of consciousness and indicate both its narrow bandwidth and its slow processing of information, an extraordinary quantity of information processed by the human system cannot be accessed by consciousness' (p. 18). It is often suggested, for example, that it is inherent in the structure of the construction of consciousness that vast swathes of information are suppressed.

[6]See [6, p. 19].

[7]Also: 'Traditionally, fallacies are considered mistakes that appear to be errors, attractive and seductive, but also universal, because humans are prone to committing them' [6, p. 21].

no normative role to be played by logic since *reason* is reducible to individual *reasoning*. From there, we are practically *forced* to accept that there are no normative criteria to assess the "deceptive" character of our reasoning. This is, to an extent, a question of authority. I, together with Woods and Magnani, have given up on the autocratic authority of deductive logic over the way in which we reason. But, where Woods and Magnani give up on *any* authority requiring a wedge between reason and reasoning, I have suggested that the norms *internal* to interactions may give rise to normative authority as holding over those agents involved in an interaction *insofar* as the interaction itself lives up to its own norms.[8]

For example, let us accept, as Woods and Magnani do, that the role of "fast and frugal" heuristics is more central to reasoning than is typically thought by standard accounts of logic. Thus, Woods and Magnani may also be assumed to be committed to some sort of complexification of the way in which such heuristics are often compared with slow and deliberative forms of reasoning. This follows, for obvious reasons, since the dual-process theory (mentioned in Chaps. 1 and 3) is *predicated upon* there being a way of explaining biases of judgement *as biases*. That is, there would be much less evidence supporting the positing of a dual-process theory, in which biases are measured up against the (typically classical) norms supposed to hold over Type-2 reasoning processes. As I mentioned in Chap. 3, and commensurate with the argument against dispositionalism in Chap. 1, in previous work with Alex Tillas [13], we identified four significant issues with dual process theories, which might be *thought* to add fuel to the fire for naturalized logic:

- Dual process approaches suggest a traditionally rationalistic approach to theory of mind in which biased intuitions are "controlled" by serialized, normative processes. This suggests that we have a standard normative theory (usually classical logic / probability theory) that is justified in some other way, but, as argued throughout, this is not the case.[9]
- The relationship between the processes of reasoning seems vastly more multifarious and mutually supportive than the dual systems approach allows. For example, it looks like systematic reasoning processes may also distort and bias heuristic reasoning processes, and, also post-rationalise heuristic-like processes. In addi-

[8]An additional issue for the collapse of the normative into the natural is, as argued for at length in the work of Wilfrid Sellars ([10], e.g.), it is far more difficult to extricate one from the other than it may seem. For example, Sellars' arguments regarding the theory-ladeness of experience clarify the way in which even basic empirical data is bound up with normative theoretical articulations. Pursuing this line of thought, Sellars' arguments may also be used to motivate an objection against the naturalisation project that follows a similar structure to arguments from circularity discussed in the introductory chapter and Chap. 1. Even identifying "patterns" of reasoning from empirical data presupposes some form of theory-laden, and typically normative, framework. This kind of argument in relation to the naturalisation of epistemology is explored in detail in [11]. In addition, as explored in [12], James O'Shea discusses Sellars' arguments (sympathetically) that: 'the normative dimensions of human conceptual cognition are not only consistent with, but in fact stand in intelligible systematic interrelations with an underlying naturalistic dimension of language / world or mind / world representational relations' (p. 152).

[9]Keith Stanovich [14] recognizes that there may be non-classical norms, but there is little work in the literature on non-standard normative theories.

tion, we have no device for telling which expressed beliefs represent an override by engaging reasoning, and which are mere rationalizations of heuristic processes (e.g. [15]).

- The intuitive pull of heuristic reasoning processes that appears to exist long after reflection is not really dealt with by dual systems accounts (i.e. they do not get the phenomenology right) (e.g. [15]).
- The dual-systems account is overly internalistic, not giving credecence to the external scaffolding of systematic reasoning processes.[10] In this respect, the role of learning, formalisation, and technologies (e.g. writing, maths) are not adequately understood within the framework. There are also reasoning processes that appear to begin under conscious, slow, control, but later become automated, such as learning logic and mathematics.

All of the above suggest that the distinction between non-biased processes as consciously controlled, as opposed to biased processes as unconscious and automatic, is too simplistic. Whilst these issues suggest that the idea of heuristics being measured up against some normative theory of reasoning (supposed to be implicit in more deliberative reasoning) is problematic, it does not follow that no appeal to norms is possible. For example, much of the literature on biases of judgement use the term "belief bias" (e.g. [16]) to characterise these discrepancies in reasoning. But, in the work of Dutilh-Novaes (e.g. [17]), it is argued, persuasively, that a better phrase describing the data is "doxastic conservativeness", which characterises the tendency of agents to bring to bear prior beliefs on the assessment of arguments. To illustrate this, she considers a well-known experiment, discussed in [16], where subjects presented with a deductively invalid syllogism are asked to assess its validity. Consider the following:

- 1. All living things need water
- 2. Roses need water
- C. Thus, roses are living things.

According to the obtained results, just 32 % of subjects said that this was not a valid argument, while 68 % responded that it is valid. The mood of the syllogism is AAA-2 (where the universal quantifiers are omitted in the second premise and the conclusion). A syllogism of the same mood was then presented to the subjects involving "alien" terms (involving an imaginary species, "Wampets", and an imaginary class, "Hudon"), and subsequently, they were asked to evaluate the following:

- 1. All animals of the hudon class are ferocious
- 2. Wampets are ferocious
- C. Thus, wampets are animals of the hudon class.

Here, 78 % of the subjects responded that the syllogism is invalid. Now, according to proponents of dual-process accounts, subjects in the first case would return the deductively invalid reply because of an intuitive pull, which overrides rational

[10] I discuss this further below.

processes. In contrast, on Dutilh-Novaes' account, on confrontation with the syllogism in question, subjects are attending to certain aspects of the premises, which are loaded with meaningful associations, including associations between roses and living things. Any "intuitive pull" is, rather, to be explained by the fact that these meaningful associations "pull" us in certain ways, and that this typically leads agents to make responses that are in line with a prior belief set.[11] As such, we can identify a *doxastic conservativeness* over agents' responses, and judgements, since pre-existing associations, doxastic states, emotions, and so on, play a significant role in "pulling" us to respond in ways that are conservative over what we already believe, feel, and so on.

If we accept this explanation (even partially), then this has two immediate consequences. First, that we may still explain what are typically called "biases", without any appeal to the external arbitrage of any normative structure. Second, we may yet require there to be some way of assessing beliefs, judgments, and so on, that goes beyond the way in which we currently do reason, since individual agents reason in a doxastically conservative manner. This latter point also provides a way of fleshing out the concern that sometimes agents seem to *require* appeals to the normativity of reason. For example, in Chap. 3, I briefly discussed Deborah Heikes' [18] argument that, without a viable account of reason, feminism lacks the means to argue for equality and emancipation. The relationship between such doxastic conservativeness and Heike's argument is clear: if we have no story about reason beyond the way in which we actually do reason, then the conservativeness of many social structures may well take on an appearance of implicit *correctness*, simply because these structures are entrenched, both cognitively, and socially.[12]

2 Norms, Interactions, and Power

2.1 Resuscitating Norms: The Universal in the Local

Both the interactive position argued for here, and Woods' project of naturalisation argue for a substantial, and foundational, shift in approach to logic and reasoning. The key distinction between the interactive approach argued for here and the naturalization project may be clarified by the Woods' proposition:

> Having and Drawing: Consequence-having occurs in logical space. Consequence-drawing occurs in a reasoner's mind. [3, p. 42]

[11] In [13], we argued that this is due to the automatic sub-activation of the association between tokens of the terms "roses" and "living things".

[12] On the notion of "generative entrenchment", see the work of William Wimsatt [19], and on the relationship between entrenchment, heuristics, and hegemonic power, see [20].

This proposition is part of a series of manifesto-like statements in which Woods draws out the distinction between a naturalized approach and "orthodoxy". He characterises orthodoxy as concerned with the *having* of consequences, whilst the naturalized project is concerned with the actual processes of drawing consequences by human agents. For Woods, in distinction to the project discussed in this book, such reasoning "occurs in a reasonser's mind". That is, the project of the reduction of reason to reasoning relies upon the simultaneous project of the reduction of reasoning to the *internal* activities of single agents.

From the outset, I have provided arguments that show that such, monological, approaches to reasoning are wrong-headed, and to an extent, the naturalisation project may even provide grist to the mill for this critical line of thought. For example, let us accept, as Woods and Magnani (and myself) argue, that any approach to logic should take into consideration the way in which human agents reason. Now, if we also take the view that the way in which human agents reason is primarily a matter of internalised judgments and activities, then it looks as though we may be forced to take a position where the normative is collapsed into the natural, and the ability for reasons to have any sort of authority is given up. Of course, what we do not want is to resurrect any external and autocratic authority for logic (whether representational or inferential), so we can not institute any normative criteria for reasons from an external point of view. The suggestion, instead, is to think about norms that are internal to *interaction amongst agents*, which provides agents with a means by which logic can be reconstructed to have normative traction in radically local ways.

As an example of norms that are *internal* to interactions, I have suggested that, for some α, any interaction can not terminate in agreement that α is assertible, and that α is deniable (at the same time). That is, in the terminology of the previous chapter, that α can not be justified locally true *and* locally false within a single terminating coherent interaction. As I mentioned in that chapter, this is distinct from, and prior to, anything that could be termed logical contradiction. Moreover, this is a norm that holds locally, between the agents involved in the interaction only, and only subsequent to the disequilibria of a symmetry in which assertions and denials are tested equally (in the subset of interactions that are coherent).

As I have argued, from the inside of an interaction, there are minimal norms of engagement between agents. For example, the norm of symmetry between agents suggests that all agents should be considered entitled both to put forward statements, and also to call them into question. As such, all statements may be thought of as open to testing, and that the utterance of a statement brings with it a normative commitment to provide reasons for that statement, or to provide reasons for its challenge (depending on the role being played).[13] This is part of the process by which to uphold symmetry, or equality between agents involved, such that control over the

[13] A similar norm is discussed in detail in [21, p. 136ff], though the norms for dialogue discussed there go far beyond those suggested here.

interaction is not imbalanced on behalf of any single agent.[14] It is notable, in this setting, that, in Laden's [22] work, "engaged conversation", which is just *reasoning* according to Laden, involves a commitment to the *mutual intelligibility* of statements. This latter commitment is also predicated upon a prior, underlying, commitment to construct a space of reasoning that is public, and "inviting" of others to share such a space. In this setting (which is not dissimilar to the interactions discussed here), the norm of symmetry may be further explicable in terms of the way in which any such public space of reasoning requires of us that we allow others to affect our participation in that interaction. Similarly, in what Laden calls "engaged reasoning", agents are further committed to "issue invitations" to others, which allow both themselves and others, to find a *common ground* relating to a set of shared concerns. As such, engaged reasoning requires of us that we are responsive to the reasons given by other agents as providing grounds for coming to agreement regarding the justification and testing of utterances. Importantly, these norms are both minimal *and* internal to the dynamics of interaction. Furthermore, just as we discussed in Chap. 2, these are norms which are *downstream*, in the sense that they come into play only subsequent to an utterance: something counts as a reason only in terms of the way in which it is treated subsequent to the process of interactive reasoning coming into play.

These internal norms may not capture what is usually intended by appeals to the normative, particularly in the domain of logic. This, of course, is deliberate. Nevertheless, it is central to the project offered here that these minimal, internal, norms also give rise to a reconstruction of logic subsequent to the dynamics of interaction. That is to say, local truth and local falsity, *insofar* as these are understood to be the result of terminating coherent interactions, may also be thought to *retroactively* hold as norms over the agents involved in that interaction. This is important from the point of view of any kind of normative *authority* that may be attributed to reason. For Laden, the status of authority, too, is dependent upon the process by which it is constructed: 'Authority ...is only eventually constructed when all come to agreement' [22, p. 67]. As Laden [22, p. 72] understands, the suggestion that the status of authority is determined *after* it is invoked, and *by* those over whom it holds is somewhat puzzling:

> How, it will be asked, could such a form of authority be normative at all and thus how could it be a genuine notion of authority? [...we need to] appreciate the kind of concept authority is. It is not a purely normative concept, but rather a descriptive term that refers to the normative order [...] to say that the authority of connection has forward-looking credentials and a distributed location does not imply that its invocation is not norm-governed. That is, even if you can undermine my authority through your response or lack of it, not everything you do to ignore me will have that effect [...] the process of drawing out these norms and eliciting their authority is not different from the process of reasoning itself, and thus it cannot be done in a manner that is final, monological, or decisive. [22, p. 72-3]

[14] That this is a norm, rather than an empirical description, is clear from the fact that many discourses are ideologically loaded, and weighted in favour of a single agent, or group of agents, as discussed in Chap. 3.

The suggestion is that, just like reasons, authority itself *if it is to have any normative traction at all*, is to be understood as operating *downstream*, not just from utterances, but from coherent interactions themselves. The fact that logical rules may be reconstructed to hold over agents involved in coherent interactions that have terminated allows us to consider the internal norms of the dynamics of interactions from the point of view of normative and valid local truths and falsities, which have retroactive authority. This form of authority, in distinction to more standard approaches to the authority of reasons, is elicited only through, and subsequent to, the structure of interaction.[15]

This is in distinction with the way in which authority has often been characterised in relation to reasoned argument. For example, in the account of argument proffered by Stephen Toulmin [24], reasoned authority requires formally-valid argument that links the belief in question to an unchallengeable starting point (e.g. p. 89). As such, the soundness of arguments is inextricable from arbitration by external authorities. This, as pointed out in [21] also requires arguments to bottom out in appeals to authority that can definitively legitimise a certain standpoint (e.g. p. 131), and, as such, fall prey to objections of circularity. In distinction, the kind of interactive reasoning here gives rise to a form of authority in which there is no ultimate arbiter, and no ultimate authority, other than the struggle to determine what form authority might take.

Nonetheless, the destabilisation of any appeal to objective correctness, or absolute authority, might be unsettling insofar as it looks as though it brings with it a kind of perspectival approach to (local) truth.[16] It is precisely this sort of perspectivalism, however, which the naturalization project is liable to run into, by reifying the role of individual reasoning activities. In distinction, whilst I have also argued against appeal to external arbitration, or objectivism for justification of reasoning, positioning reasoning as central to *social interaction* also allows us to consider norms that are grounded in those social practices. In this setting, I am in agreement with Brandom's arguments against any kind of *a*-perspectival view:

> Sorting out who should be counted as correct, whose claims and applications of concepts should be treated as authoritative, is a messy retail business of assessing the comparative authority of competing evidential and inferential claims [...] That issue is adjudicated differently from different points of view, and although these are not all of equal worth there is no bird's-eye view above the fray of competing claims from which those that deserve to prevail can be identified nor from which even necessary and sufficient conditions for such desserts can be formulated. [25, p. 601]

In our terms, it is through the ongoing, dynamic, process of reasoning that agents are capable of overcoming the individual perspective, such that it is, through a continual testing relationship between agents' utterances. As we have argued, whatever is taken to be true and meaningful is not determinable in advance of these processes, and,

---

[15]This bears some resemblance to the Spinozan conception of freedom in relation to constraints, discussed in (e.g. [23]).

[16]This is a view that has an affinity with Brandom's argument that 'there is never any final answer to what is correct' [25, p. 647].

as such, is also not possible to identify this with a single perspective. But, unlike Brandom, this process of determination is also not identifiable with the arbitration of a mythically homogenous community.[17] Rather, the view argued for in this book suggests that there is to be *no* privileged perspective since any perspective may be privileged from inside the context of an interaction, and the determination of truth and meaning is a function of the *shared* perspective that is constructed from the inside of those interactions. This, in Laden's terms, gives rise to a "plural subject" (e.g. [22], p. 177ff), where involvement in the space of engaged reasoning requires us to offer reasons to others that radically affect any agent involved, such that they are engaged towards a common focus. The role of this plural subject is to indicate the way in which agents shape the "we" of any interaction, such that reasons allow the other to also find common ground in productive interactional moves. Moreover, this plural subject may be one that is as yet actualised: 'we sometimes engage in deliberation with someone in the hopes of forming a relationship that does not yet exist' [27, p. 95].

Perhaps the obvious charge that may be brought against this view, particularly since we are currently foregrounding the role of authority, is that truth becomes dependent upon a struggle for power. I do think that the power and truth are interwoven insofar as any interaction inevitably requires of us to disbar appeals to an ultimate end of reasoning privileging a single perspective. It is, however, the ongoing nature of interactions that is constrained, not from without, nor from agents' intentionality, but from the *dynamics of the interactions themselves*. These *internal* norms provide the possibility for overcoming individual power relationships (and the struggles that ensue from them), and, furthermore, these may give rise to local truths that are rational insofar as they exert rational constraints over subsequent reasons and actions. The construction of authoritative norms *subsequent* to interaction also clarifies the fact that it is the *activity* by which such norms are constructed that is important, particularly since local truths may also be considered to be amenable to the dynamics of future interaction. Hence, far from invoking a mythically homogenous community, this dynamics *relies* upon the productive tensions of differing perspective. As Fosser puts it, describing Brandom's view, '[o]nly as engaged in activity with others does the possibility of being mistaken oneself become intelligible, because only then does one's perspective appear as a perspective' [26]. Whilst reasoning is inevitably intertwined with power insofar as any interaction is, it is also through interactive reasoning that such power relations may be overcome, questioned, and made transparent. As Fossen states: '[t]o engage discursively is also to open oneself up to criticism, and thereby to self-transformation' [26].

[17] As pointed out in [26], there is an odd tension between Brandom's account of intersubjectivity as "I-thou" rather than "I-we" (e.g. [25], p. 39), and the argument that the assessment of a community should be taken as correct: '[the] community's all-inclusive practical assessment cannot be mistaken' [25, p. 54].

2.2 Interactions Prior to Interaction

In the approach argued for in this book, meaning is never context-independent, rather the meaning of what agents say is dependent upon the way what is said is employed over the course of interactions.[18] Building upon the suggestions made regarding sub-personal processes in Chap. 3, we can clarify the relationship between empirically tractable individual processes, and the construction of meaning in a setting over which normative constraints hold. In Chap. 3, I suggested that there are good reasons to think that the norms emerging internally to interactions are not *necessarily* a matter of subject-level intentions, contra Grice.[19] In [29], for example, it is argued that coordination between agents is often sub-personal, involving mechanisms by which agents "synchronise" together, including emotional mechanisms (as discussed in Chap. 3). In [29], this was developed into a view of language as a *tool* for interaction, understood as a structure-building activity where 'this process of building up structure is what is taken to constitute natural-language syntax: syntactic mechanisms are procedures that define how parts of representations of content can be incrementally introduced and updated' [29, p. 572]. Most importantly, for our purposes, is that this 'activity intrinsic to language use' [29, p. 580] is not dependent upon the cognitively taxing processes of "mind-reading", and other attributions of intentionality to subjects. Rather, agents, in making utterances in interation may 'start off without fixed intentions, contribute without completing any fixed propositional content, and rely on others to complete the initiated structure, and so on' [29, p. 80].[20]

So, here, we have a clear link between the subpersonal, and pre-propositional processes and the constructive activity of language use between agents. These, underlying, processes might also be related to a family of views of cognitive activities often called "interactivism", "enactivism", and "interaction theory", in which cognitive activity (including the construction of meaning) is understood to be inextricably linked to agents' social and physical environment.[21] As such, the approach to meaning, on these views, is one of being "in-the-making", as Seibt [31] puts it: 'they are processes that begin in attunements and might, but need not, end in the formation of that which we consciously experience as propositional contents (beliefs, intentions etc.)' (p. 96–97). Often, on these approaches, as pointed out by Woods and Magnani (and discussed above), individual mental processes such as weightings in neural networks are taken to determine patterns of activation, which culminate in what we typ-

[18]In [22, pp. 89–90], Laden points to Stanley Cavell's work on reason, in which utterances are taken to be act-like, and their meaning is dependent upon the aims of these acts, though these aims may be unobvious even as the acts are made.

[19]Perhaps the most well-known account in which meaning is not dependent upon intentions is found in the work of Ruth Millikan, who argues that ordinary understanding also involves causal processes in the context of evolutionary functioning (e.g. [28]).

[20]Pezzulo [30] also discusses (often subpersonal) "coordination tools" which, he argues, help to explain the ease by which interactions occur.

[21]See the excellent discussion in [31] and [32].

ically call heuristics (seemingly "automatic" responses to stimuli).[22] However, these approaches do, also, emphasise interactive processes, which go beyond the merely causal-responsive structure of associations *even at the level of the sub-personal*. For example, according to the concept of "strong interaction" (e.g. [32, 33] meaning is always co-constituted in social interactions.[23] This theory draws upon evidence from developmental psychology and cognitive science to argue that interactions that take place in social contexts are grounded on unconscious coordination of movements, gestures, and speech acts.[24] As such, interactions are argued to 'have a transformative effect on individuals who engage in them' [32].

Whilst views such as "strong interaction" are not, often, connected with the forms of reasoning in interaction that we have been interested in, they are clearly consistent with it. In fact, it seems that interaction theory, and the interactive theory of grammar, provide us with the means to connect basic, subpersonal, processes of cause and response, with social processes of reasoning. Accepting that "thinking" is always constituted through agents' interactions with their social and physical environments, since agents are always already "in the world", also entails that these are 'extended in pragmatic and communicative practices that may further supervene on the tools, technologies and institutional practices through which we engage with the world' [32]. This connection is also borne out in Seibt's [31] discussion of interactive theories in relation to dialogue. There, she suggests that, over the course of dialogue, what may *begin* as causal-responsive processes of reaction to stimuli are brought to the surface and rendered transparent such that they are open to criticism (e.g. p. 99). Seibt goes on to argue that understanding in dialogical contexts is not to be understood as articulable in the form of propositional judgments, but rather 'it is imbued with the searching character of the dialogical experience, the felt tentativity of the effort of making sense that does not yet allow for settling down into judgment' (p. 94).[25] This, in Seibt's view, situates thinking in dialogue as akin to *spatial orientation*. This is in accord with Kant's account, is generalised to an abstract, and ubiquitous phenomenon experienced, for example 'when we search for the solution of complex problems or try to systematize a complex domain' [36, Sect. A 308-310]:

[22]Note, however, that Magnani does make some steps in this direction. For example: '[M]inds are extended and artificial in themselves. In this perspective logical and mathematical systems are the creative fruit of the interplay between internal and external representations and, once available, they represent an external materiality that plays a specific role in the interplay due to the fact that it (external materiality) exhibits (and operates through) its own cognitive constraints' [6, p. 18].

[23]Gallagher and Miyahara [32] define strong interation as: 'a mutually engaged co-regulated coupling between at least two autonomous agents where the co-regulation and the coupling mutually affect each other, constituting a self-sustaining organization in the domain of relational dynamics'.

[24]For details, and arguments for strong interaction, see [34, 35].

[25]Seibt suggests that this kind of understanding is a kind of "half-way house" between no meaning whatsoever, and full propositional content, and which is something that we are 'perfectly familiar with a kind of conscious cognitive processing that is neither a clear-cut instance of propositional knowledge nor a practical skill (knowing-how)' [31, p. 95].

> The insights that orientation provides are not classificatory judgments ("erkenntnis") based on criteria but a "felt need of reason" (*ibid.* 308), a knowledge of "how to go on", in which direction to proceed, both concretely and abstractly. [31, p. 95]

Throughout the preceeding chapters, I have emphasised this process of "going on", through the dynamics of reasoning together. It is in these *explicit* processes that meaning is constructed through continual engagement with other agents, where these processes may be thought to supervene upon the sub-personal process of interaction.[26] As such, we can make a clear connection between the domain that is supposed to call for the "naturalisation" of logic, whilst also highlighting the ways in which reason can *not* be reduced to reasoning, insofar as the latter requires an account of the normativity of thinking in interaction.

3 A (tentative Conclusion): From Logic to Freedom

It is, perhaps, quite odd to think that there might be any kind of link between logic and freedom. Logic, is often supposed to be free of the trappings of the world, merely a formal structure that has little to do with the materiality of agents' bodies, and social structures. However, for any account of logic, such as the one proposed here, that is grounded in *process*, and in *interaction*, it is inevitable that we should take into account that logic has consequences that are normative, and that this normativity is not grounded in justificatory structures or external authorities. Indeed, for any view (including Frege's), which takes logic to have normative traction over thinking, power and freedom are always already implicated within them. This dimension of logic is hidden by the reduction of the dialogical to the monological, and the viscious circularity, upon which they supposedly rest.

In the preceding chapters, I have argued that:

- Reasoning is inherently interactive, taking place within a socially embodied space;
- This space is not determined by logical rules;
- We can think of logical rules as emerging from the dynamics of interactions;
- These logical rules may then exert retroactive normativity over the agents involved in that interaction.

---

[26]This also has consequences for how we might approach reasoning in artificial processes. It seems to me that the connection that we are making here between sub-personal interaction, and linguistic interaction, may also find analogy with attempts to synthesise symbol-processing approaches to artificial intelligence with heuristic and dynamical approaches (on the latter, see [37, 38]). In this vein, Iizuka and Di Paolo [39] draw upon evolutionary robotics' approaches to argue that determining social relations can be generated from the dynamics of processes of interaction, such that artificial agents are capable of determining whether or not they are interacting with interactive or non-interactive (recorded) agents. They argue that this kind of capacity is not something that can simply be "built-into" a single, autonomous, agent, and also that successful interactions are dependent upon emergent properties of the coupling relationships *between* agents.

This is a very different approach to logic than orthodoxy would have it. Moreover, the way in which the normativity of logic operates here is not through an autocratic set of rules, but much more like an *ethical imperative*. Famously, Carnap stated that:

> In logic, there are no morals. Everyone is at liberty to build his own logic, i.e. his own form of language, as he wishes. All that is required of him is that, if he wishes to discuss it, he must state his methods clearly, and give syntactical rules instead of philosophical arguments. [40, Sect. 10]

Contra Carnap, attempts to justify rules purely syntactically were found wanting on many counts (in Chaps. 2 and 3). In these terms, the project under discussion here may be considered as one that attempts to account for logic *precisely in terms of morals*, or more specifically, the ethics of relationships between agents. For example, drawing upon Martin Buber's [41] work on the "I-thou" relationship, Seibt argues that, in dialogue 'the I is arrested in the experience of ongoing cognitive exploration of the Thou' [31, p. 92]. This is consistent with Laden's "plural subject", is constitutive of an engaged suspension of a single perspective amidst the ongoing engagement with the "other". It is in this setting that engaged reasoning occurs, such that logic may be reconstructed from terminating coherent interactions, and, as such, logic is fundamentally part-and-parcel with this rational, yet, non-individual, subjectivity.

In Laden's work, this line of thought is integrated with the suggestion that the plural subject is foundational in constructing a "common political will", whilst also allowing that this is often merely potential, rather than actual:

> [...] we should see it as inviting [...] one's deliberative partner to share a world-view - a space of reasons. Such an invitation can be extended both to someone with whom I already share a well-defined and mutually understood relationship and to one with whom I do not yet share anything but the possibility of forming such a relationship. [27, p. 95]

So, on Laden's view, this plural subject designates a relationship between agents in which the agents both sense that the relationship exists, and also gives rise to an understanding of what that relationship authorises (for those agents). In these terms, this approach may be understood as both corrective of, and indebted to, the discussion of subjectivity and freedom in the work of Michel Foucault (e.g. [42]). According to Foucault, accounting for human freedom requires us to consider the codes and disciplines that constrain and constitute human subjectivity, since there is no autonomous human subject that exists beyonds such constraints. This, post-structuralist, criticism of more traditional accounts of freedom is not incoherent with the approach to authority that we are interested in here. For example, we can not think of freedom, according to Foucault, as an inherent capacity or characteristic of a fully-formed subject.[27] As such, freedom is not to be thought of as the removal of external constraint, nor the ability of agents to act and reason despite such constraints, since these constraints are also considered to be the condition on the possibility of subjectivity itself (e.g. [43], p. 2).[28] So, the *feeling* of freedom, such as the freedom to

[27] See Oksala [43] for extended discussion of Foucault's account of freedom.

[28] This is also related to the classical distinction, traceable to Kant, between positive and negative freedom, discussed in detail in [44].

choose what we might consume as we are shuffled and buffetted around a shopping mall, may well be something like an illusion of freedom in which it is actually the case that a set of external constraints constitute our subjectivity: as consumers who have internalised external (material and ideological) power. In this sense, the theory of freedom as freedom *from all constraints* is something like the "ultimate" illusion. In the United Kingdom, there has been an advert recently in print media for e-cigarette technology, commonly known as vaping. This advert perfectly encapsulates the paradoxical nature of this appeal to an absolute freedom that is free from all constraints. The protagonist, presumably having laboured under years of nicotine addiction from cigarette smoking, is pictured "vaping" with a vista of a wide, and open, canyon appearing before him. The symbolism of the latter being an open and unconstrained future that he has been offered by …a new platform for nicotine addiction!

In contrast to the view of freedom as freedom from constraint, consider the following simplified piece of fiction. Take a hypothetical social group (including men, women, heterosexual, homosexual, transexual), over which hold the kind of structural constraints concerning gender and sexuality that are current in western, secular society, such as gendered pay gaps; heteronormative media; patriarchal ideologies, and so on. Now, consider that, through coherent interactions of "engaged reasoning" (to use Laden's terminology), this group comes to an agreement that the following proposition is justified: "all people are equal, and should be treated as such". Following this process of interactive reasoning, the proposition is treated as having authority over that group, and, as such, we may imagine a number of structural, material, and ideological changes being put in place. Treating the proposition as having authority by the social group has significant emancipatory impact upon those agents, and, as such, is a *self-imposed* constraint. This is a kind of freedom, then, which involves a group's self-imposed constraint, and it seems correct to say that this is a kind of freedom whether or not it *feels* as such to those over whom it holds.

The (tentative) suggestion, then, is that a form of subjectivity is constituted through interaction, just as is meaning, and normativity. Moreover, this is a new form of subjectivity, which is something that Laden's "plural subject" begins to capture. Thought of in terms of freedom, however, we can think of the construction of plural subjectivity as one that also reaches beyond the current constraints on the agents involved in interactive reasoning. For example, in accord with the discussion of authority above, we may say that a kind of freedom may be achieved via the binding of a group of agents to a norm that is constructed subsequent to agreement over coherent interactions between them. This, according to Foucault's understanding of subjectivity would then make for the internalisation of new codes and disciplines in the constitution of a new form of subjectivity. Moreover, this is a subjectivity that is capable of overcoming the constraints that are currently constitutive of what we now take subjectivity (for those agents) to be. Then, from a political point of view, this process of the constitution of new, *plural*, subjects according to the construction of authority as we have described it, is also a project of ongoing emancipation from that which currently constrains groups of agents. The project of interactive reasoning, then, would always be a project of the construction of new forms of agency which go beyond current constraints, such that agents' capacities for action are built upon and further augmented.[29] That this process is open-ended is clear, both from the fact that the social and physical environment is not static, and also because plural subjects' reasoning activities occur in the context of external, and internalised, constraints.[30]

Recall that, for Brandom, the space of reasons is supposed to be structured by a set of logical rules, and assertions made by agents are assumed to have propositional content, which may require reasons in support of that content. Laden, taking issue with this, takes an assertion to be much more like an "invitation" into the space of reasons, through which authority may be constructed via the exchange of social utterances.[31] The view adopted in this book both weakens the conditions upon engaging in this space of reasons, and strengthens the account of normative authority by means of the reconstruction of logic from terminating coherent interactions. This "space", is much more like a shared space of reason*ing*, which is, itself constituted by the interaction between agents. An assertion, or utterance, on this view has neither propositional content, nor is it quite like an invitation. Rather, it is more like a way of marking the interactive coordination to reason together, which is, itself, predicated on mechanisms of interaction that take place *prior* to these linguistic markers of interaction. The space, then, is not only not constituted by rules, but it is not even necessarily constituted by language, perhaps, instead, there is a topology of interaction that exists prior to the process of linguistic interchange altogether.[32] Nonetheless, the view adopted here is also stronger than Laden's with respect to the role of logic, since, for Laden, reasoning is never deductive. On the contrary, the normative structure of interactions, such that they may be coherent, gives rise to the reconstruction of logical rules, and even proofs and refutations with deductive validity. These, of course, differ in interpretation from orthodoxy, yet it is precisely through the symmetry of testing that we have argued that justification, and rational activity amongst agents is made possible. In this setting, the role of logic is manifest

[29] See [20] for discussion of related issues in the context of future-oriented forms of contemporary politics.

[30] Of course, not all interactions are, or perhaps *can be* coherent, due to these latter constraints, which often, we may assume ensure that there is a relationship of asymmetry between certain agents. But, equally, strategies by which it may become possible to engage in processes of interaction that are symmetric between agents are part-and-parcel with the process by which interactions may occur in the first place.

[31] See, for example, (e.g. [22], p. 60).

[32] Consider the following, fairly prosaic, example. One morning, after waking, I walked upstairs and started to watch a television programme called "Saturday Kitchen", which is a regular weekly cookery show in the United Kingdom, and whose long-time presenter had recently left the series. A while later, I heard my partner waking, and as she began to ascend the stairs, she uttered "he's not as good as the old one". This, radically contextual, utterance, was clear from *inside* the space of interaction between myself and my partner, and, as is clear, the utterance can not be what began, or constructed, that space. Rather, it seems much more viable to say that the space was already in place, through sub-personal, and conscious, perception and awareness of my partner, the environmental context, and so on. The utterance, was much more a marker that we might engage in reasoning regarding the appraisal of the presenter in relation to the one who had left the week prior.

as that which is generated by coherent interactive reasoning, whilst also allowing for the construction of local truths and falsities, which have normative authority over the agents involved.

I began this book by introducing inferentialism as an attractive alternative to representationalism, but which itself is beset by problems. These include that logic and reasoning are disconnected, that the justification of logical rules is circular, that an inferentialist epistemology is too demanding, and that an inferentialist approach to logic fails to fully determine the meaning of logical connectives. As I showed in Chaps. 1 and 2, I agree that these issues are unlikely to be overcome by standard approaches to inferentialism, based in monological, static, rule-governed approaches. Instead, and through argument regarding the *role* of proofs, refutations, assertions, and denials, I have argued that meaning is something that we construct *together*, not by objective proofs or representational fiat. Instead, it is built-up through a process of interaction between agents. So, on this view, we do not take inferential rules to govern reasoning, but rather, more minimal constraints on the dynamics of reasoning in interaction to justify rules subsequent to that interaction, where these interactions are cooperative engagements between agents. Where we end up, on this approach, is with a form of inferentialism that both deals with the issues that we began with, and also provides a novel account of the justification and normativity of logical rules that leads us to a new way of thinking about the relationship between logic and reasoning. By considering logic and reasoning to be fully embedded within the dynamics of interaction, whilst also not giving up on the possibility of the construction of norms that have authority over us, we also find, perhaps surprisingly, that logic has a role to play in the constitution of freedom and political will. There is far more to be said on all of these issues, and particularly this relationship between logic and freedom, which deserves much greater attention. As with any thought, written or otherwise, this book is itself part of a dialogue with varied agents, in varied fields, both in text and in person. In this, my hope is that these interactions continue to further investigate the machinery of reasoning; where reasoning is always reasoning *together*.

References

1. John Woods. *Paradox and Paraconsistency: Conflict Resolution in the Abstract Sciences*. Cambridge University Press, 2003.
2. John Woods. Advice on the logic of argument. *Revista de Humanidades de Valparaíso*, (1):7–34, 2013.
3. John Woods. *Errors of reasoning: Naturalizing the logic of inference*. College Publ., 2013.
4. Lorenzo Magnani. *Abductive cognition: The epistemological and eco-cognitive dimensions of hypothetical reasoning*, volume 3. Springer Science & Business Media, 2009.
5. Lorenzo Magnani. The eco-cognitive model of abduction. *Journal of Applied Logic*, 13(3):285–315, 2015.
6. Lorenzo Magnani. Naturalizing logic: Errors of reasoning vindicated: Logic reapproaches cognitive science. *Journal of Applied Logic*, 13(1):13–36, 2015.
7. Rodney A Brooks. Achieving artificial intelligence through building robots. Technical report, DTIC Document, 1986.

8. James Ladyman. *Every Thing Must Go: Metaphysics Naturalized.* Oxford University Press, 2007.
9. Dov Gabbay and John Woods. Advice on abductive logic. *Logic Journal of the Igpl*, 14(2):189–219, 2006.
10. Wilfrid Sellars et al. Empiricism and the philosophy of mind. *Minnesota studies in the philosophy of science*, 1(19):253–329, 1956.
11. James O'shea. *Wilfrid Sellars: Naturalism with a normative turn*, volume 6. Polity, 2007.
12. James O'Shea. Prospects for a stereoscopic vision of our thinking nature: On sellars, brandom, and millikan. *Humana Mente Journal of Philosophical Studies*, 21:149–172, 2012.
13. J. Trafford and A. Tillas. Intuition and reason: Re-assessing dual-process theories of cognition. *Teorema: Revista internacional de filosofía*, XXXIV(3), 2015.
14. Keith E Stanovich. *Who is rational?: Studies of individual differences in reasoning*. Psychology Press, 1999.
15. Jonathan Evans. *Bias in human reasoning: Causes and consequences.* Lawrence Erlbaum Associates, Inc, 1989.
16. Walter C Sá, Richard F West, and Keith E Stanovich. The domain specificity and generality of belief bias: Searching for a generalizable critical thinking skill. *Journal of educational psychology*, 91(3):497, 1999.
17. Catarina Dutilh Novaes. *Formal Languages in Logic: A Philosophical and Cognitive Analysis.* Cambridge University Press, 2012.
18. Deborah K. Heikes. *Rationality and Feminist Philosophy*. Continuum, 2010.
19. William C. Wimsatt. *Re-Engineering Philosophy for Limited Beings: Piecewise Approximations to Reality*. Harvard University Press, 2007.
20. Nick Srnicek and Alex Williams. *Inventing the Future: Postcapitalism and a World Without Work*. Verso Books, 2015.
21. Frans H. Van Eemeren and Rob Grootendorst. *A Systematic Theory of Argumentation: The Pragma-Dialectical Approach*. Cambridge University Press, 2006.
22. Anthony Simon Laden. *Reasoning: a social picture*. Oxford University Press, 2012.
23. Matthew J. Kisner. *Spinoza on Human Freedom: Reason, Autonomy and the Good Life*. Cambridge University Press, 2013.
24. Stephen Edelston Toulmin. *Knowing and Acting: An Invitation to Philosophy*. Macmillan, 1976.
25. Robert Brandom. *Making It Explicit: Reasoning, Representing, and Discursive Commitment.* Harvard University Press, 1994.
26. Thomas Fossen. Politicizing brandom's pragmatism: Normativity and the agonal character of social practice. *European Journal of Philosophy*, 20(4):371–395, 2012.
27. Anthony Simon Laden. *Reasonably radical: Deliberative liberalism and the politics of identity.* Cambridge Univ Press, 2001.
28. Ruth G. Millikan. *Language, Thought and Other Biological Categories*. MIT Press, 1984.
29. Eleni Gregoromichelaki and Ruth Kempson. Grammars as processes for interactive language use: incrementality and the emergence of joint intentionality. In *Perspectives on linguistic pragmatics*, pages 185–216. Springer, 2013.
30. Giovanni Pezzulo. Shared representations as coordination tools for interaction. *Review of Philosophy and Psychology*, 2(2):303–333, 2011.
31. Johanna Seibt. Chapter five. intercultural dialogue and the processing of significance: Cognition as orientation. In Jesper Garsdal and Johanna Seibt, editors, *How is Global Dialogue Possible?: Foundational Reseach on Value Conflicts and Perspectives for Global Policy*, pages 85–116. De Gruyter, 2014.
32. Shaun Gallagher and Katsunori Miyahara. Neo-pragmatism and enactive intentionality. In Jay Schulkin, editor, *New Directions in Philosophy and Cognitive Science: Adaptation and Cephalic Expression*. Palgrave Macmillan, 2012.
33. Hanne de Jaegher, Ezequiel di Paolo, and Shaun Gallagher. Can social interaction constitute social cognition? *Trends in Cognitive Sciences*, 14(10):441–447, 2010.

34. Shaun Gallagher. The practice of mind: Theory, simulation or primary interaction? *Journal of Consciousness Studies*, 8(5-7):83–108, 2001.
35. Matthew Ratcliffe. *Rethinking Commonsense Psychology: A Critique of Folk Psychology, Theory of Mind and Simulation*. Palgrave Macmillan, 2007.
36. Immanuel Kant. *Metaphysical Foundations of Natural Science*. Cambridge University Press, 2004.
37. Randall D. Beer. Computational and dynamical languages for autonomous agents. In Tim van Gelder and Robert Port, editors, *Mind as Motion: Explorations in the Dynamics of Cognition*, pages 121–147. MIT Press, 1995.
38. Randall D. Beer. Dynamical approaches to cognitive science. *Trends in Cognitive Sciences*, 4(3):91–99, 2000.
39. Hiroyuki Iizuka and Ezequiel Di Paolo. Minimal agency detection of embodied agents. In *Advances in Artificial Life*, pages 485–494. Springer, 2007.
40. Rudolf Carnap. *The Logical Syntax of Language*. London, K. Paul, Trench, Trubner & Co., Ltd., 1937.
41. Martin Buber. *I and Thou*. Scribner, New York, 1970.
42. Michel Foucault. *Discipline and Punish*. Vintage Books, 1977.
43. Johanna Oksala. *Foucault on Freedom*. Cambridge University Press, 2005.
44. Isaiah Berlin. *Two concepts of liberty: an inaugural lecture delivered before the university of Oxford on 31 october 1958*. Clarendon, 1959.

MIX
Papier aus verantwortungsvollen Quellen
Paper from responsible sources
FSC® C105338

If you have any concerns about our products,
you can contact us on
ProductSafety@springernature.com

In case Publisher is established outside the EU,
the EU authorized representative is:
Springer Nature Customer Service Center GmbH
Europaplatz 3, 69115 Heidelberg, Germany

Printed by Libri Plureos GmbH
in Hamburg, Germany